Radiopharmaceuticals and Other Compounds Labelled with Short-lived Radionuclides

Other Titles of Interest

JOURNALS

Applied Radiation & Isotopes

Computerized Tomography

Radiation Physics & Chemistry

Int. J. of Radiation Oncology–Biology–Physics

Int. J. of Nuclear Medicine & Biology

BOOKS

HEALY: Plutonium: Health Implications for Man

DEWHURST: An Introduction to Biomedical Instrumentation

HALL: Radiation and Life

KIEFER & MAUSHART: Radiation Protection Measurement

Radiopharmaceuticals and Other Compounds Labelled with Short-lived Radionuclides

Editor

MICHAEL J. WELCH

The Mallinckrodt Institute of Radiology,
Washington University School of Medicine,
St Louis, Missouri, U.S.A.

PERGAMON PRESS

OXFORD · NEW YORK · TORONTO · SYDNEY · PARIS · FRANKFURT

U.K.	Pergamon Press Ltd., Headington Hill Hall, Oxford OX3 0BW, England
U.S.A.	Pergamon Press Inc., Maxwell House, Fairview Park, Elmsford, New York 10523, U.S.A.
CANADA	Pergamon of Canada Ltd., 75 The East Mall, Toronto, Ontario, Canada
AUSTRALIA	Pergamon Press (Aust.) Pty. Ltd., 19a Boundary Street, Rushcutters Bay, N.S.W. 2011, Australia
FRANCE	Pergamon Press SARL, 24 rue des Ecoles, 75240 Paris, Cedex 05, France
WEST GERMANY	Pergamon Press GmbH, 6242 Kronberg-Taunus, Pferdstrasse 1, West Germany

First edition 1977

Library of Congress No. 76-26764

Published as a special issue of the International Journal of Applied Radiation and Isotopes, Volume 28, Numbers 1/2, and supplied to subscribers as part of their subscription.

Printed in Great Britain by A. Wheaton & Co. Ltd., Exeter

ISBN 0 08 021344 8

Contents

Editorial

NUCLEAR Medicine, which may be defined as the application of radionuclide techniques to the investigation of human disease, is a rapidly expanding speciality. All nuclear medicine procedures have two essential requirements, an appropriate radiation detection device and a suitable radionuclide preparation or *radiopharmaceutical.* In general, the characteristics of the detection device determine the precision and sensitivity of the procedure while the radiopharmaceutical determines the type of functional and/or anatomical information which is produced. In recent years the increased availability of radionuclides of short half-life from generator systems or from charged particle accelerators has led to a dramatic increase in both the types of radiopharmaceutical available and in the scope of the biomedical investigations available to the nuclear medicine physician. In this Special Issue, our Guest Editor, Professor MICHAEL WELCH has assembled a series of papers, by many distinguished authors, which review the current "state of the art" in the field of short-lived radiopharmaceuticals and illustrate the exciting potential for radiopharmaceutical development.

DAVID M. TAYLOR

International Journal of Applied Radiation and Isotopes, 1977, Vol. 28, p. 3. Pergamon Press. Printed in Northern Ireland

Introduction

MICHAEL J. WELCH
Mallinckrodt Institute of Radiology, Washington University School of Medicine, St. Louis, Missouri, U.S.A.

ALTHOUGH two major publications on radiopharmaceutical development[1,2] have been published in the last few years, the field of nuclear medicine in general and radiopharmaceutical development in particular have been stimulated by the innovations in the field of transmission computerized axial tomography. As pointed out by TER-POGOSSIAN,[3] "the strength of nuclear medicine can best be realized by the use of radiopharmaceuticals designed to identify function rather than the more conventional armamentarium." Although for function studies the ideal tracers are compounds labeled with the short-lived radionuclides oxygen-15, nitrogen-13 and carbon-11, not all institutions have or are likely to own accelerators capable of producing these radionuclides. To parallel developments utilizing the short-lived radionuclides, therefore, work is proceeding on methods to label biologically active compounds with more easily obtainable nuclides, such as iodine-123, indium-111, gallium-67, gallium-68, bromine-77 and technetium 99*m*. The work with gallium-68 shows the potential of the new positron-imaging devices[4,5] being used in institutions that do not have the capability of producing the short-lived positron radionuclides, oxygen-15, nitrogen-13, carbon-11 and fluorine-18. These imaging devices couple the advantages of tomography with the ability to label compounds designed to measure specific physiological functions.

Developments with other radionuclides appear directed toward the development of metabolic analogs. These include various technetium chelates, various agents designed to measure thrombolytic activity and compounds specifically designed to label blood cellular components.

The development of radiopharmaceuticals designed to measure specific functions needs a parallel development of techniques for the quality control of radiopharmaceuticals. Although pharmaceutical quality control has been discussed in detail in the past,[2] measuring and ensuring radiochemical purity is of particular importance for these new labeled radiopharmaceuticals.

In this special issue, through both review and contributed articles, potential applications and developments in labeling with short-lived radionuclides whose use is restricted to institutions with accelerators are discussed. The current and potential use of generator-produced radionuclides as well as other radionuclides of half-life up to four days, and the problems of quality control of such labeled compounds are examined. A short half-life was defined as four days in order to include indium-111 and gallium-67, but to exclude iodine-131, whose decay characteristics render the radionuclide less than ideal for *in vivo* use.

The contributions to this special issue illustrate the exciting potential of radiopharmaceutical development, despite the threat of transmission computerized axial tomography as an alternative to certain radionuclide studies. It is hoped that the compounds discussed will be utilized in spite of the concerns of MCAFEE[6] that the "regulatory crisis" is severely limiting the approval of radiopharmaceuticals in the United States.

REFERENCES

1. *Radiopharmaceuticals and Labeled Compounds*, Vol. 1 and 2. IAEA, Vienna (1973).
2. *Radiopharmaceuticals* (Edited by SUBRAMANIAN G., RHODES B. A., COOPER J. F. and SODD V. J.), Society of Nuclear Medicine, 1975.
3. TER-POGOSSIAN M. M. In *The Year Book of Nuclear Medicine 1976* (Edited by QUINN III J. L.), pp. 9–14. Year Book Publishers, Chicago (1976).
4. BURNHAM C. A. and BROWNELL G. L. IEEE Trans. Nucl. Sci. NS-19-3, 201 (1972).
5. TER-POGOSSIAN M. M., PHELPS M. E., HOFFMAN E. J. and MULLANI N. A. *Radiol.* **114**, 89–98 (1975).
6. MCAFEE J. G. *Am. J. Roentgenol.* **126,** 908–909 (1976).

International Journal of Applied Radiation and Isotopes, 1977, Vol. 28, pp. 5–11. Pergamon Press. Printed in Northern Ireland

Potentials and Problems of Short-lived Radionuclides in Medical Imaging Applications*

PAUL V. HARPER
Departments of Surgery and Radiology, The University of Chicago, and The Franklin McLean Memorial Research Institute,† Chicago, Illinois 60637, U.S.A.

(*Received* 14 *February* 1976)

Various considerations involved with the production, processing and synthesis and applications of short-lived nuclides are presented. In particular, the use of rapid synthesis and the unique uses of short-lived agents to present static equilibrium images from which dynamic information can be derived are examined. This approach appears in principle to be generally applicable in situations where the physiologic processes in question have durations comparable to the half lives of the nuclide used.

WHILE many of the obvious advantages of short-lived radionuclides for imaging studies have long been appreciated, only recently have applications been described in which the short physical half-life is a necessary property for making the measurement desired. Developments in the areas of production techniques, instrumentation, chemistry and clinical application have usually gone hand in hand and these various interdependent efforts are often difficult to separate.

Historically, during the past decade, the gradual replacement of ^{131}I by ^{123}I and $^{99m}TcO_4^-$ for thyroid work, and by ^{99m}Tc as the general purpose label, of ^{85}Sr by ^{87m}Sr, ^{18}F, and by the ^{99m}Tc phosphate compounds for bone imaging, reflects the effort to reduce radiation absorbed dose and to increase photon flux in order to improve image quality by making use of short-lived agents.

Other dosimetric properties of the nuclide, i.e. the presence or absence of particle radiations and the biological residence times of the radiopharmaceutical preparation, enter into the picture and may become the dominating factors in particular situations. It has been proposed that when an *in vivo* procedure is carried out using a radionuclide, other things being equal, the time of the measurement following the administration of the nuclide should approximate the average life of the nuclide in order to obtain the greatest information for the smallest radiation absorbed dose.[1] Unfortunately, the decay rates of available radionuclides often do not match the time scale of the desired measurement. For instance, the long-lived, slowly excreted ^{75}Se-L-selenomethionine, the only readily available amino acid analog, is used for short-term measurements. In addition, ^{133}Xe, ^{127}Xe and ^{203}Hg chlormerodrin are examples of long-lived agents which are excreted rapidly. This appears to provide substantial advantages, i.e. long shelf-life, low radiation dosage. Such applications however have substantial disadvantages, in that large amounts of long-lived material are introduced into the environment via the patient in a manner not always easy to control. In addition, in these agents, minor distribution components may have long residence times which may become the dominating factors in the dosimetry. This is particularly true in the case of ^{203}Hg-chlormerodrin. The use of short-lived nuclides avoids these difficulties and reduces the contamination problems sub-

* This work supported in part by Contract NI HV-52980, Radiologic Imaging Center Grant 18940-04 and SCOR-Ischemic Heart Disease Contract No. H1-17648.

† Operated by the University of Chicago for the Energy Research and Development Administration under Contract No. E(11-1)-69.

stantially. A major spill of a radionuclide with even an intermediate half-life is a serious laboratory disaster, while with shorter-lived agents this problem is greatly reduced.

A number of elements of biologic interest have only short-lived radioisotopes. These are oxygen, nitrogen, carbon and fluorine. The first three of these materials require on-site production facilities. Sufficiently efficient production methods for ^{18}F exist so that an order of magnitude of physical decay during shipment can be tolerated; however, the expense involved for such "over production" is not trivial. The longest lived isotopes of the other substances are the 2.0 min ^{15}L, 10 min ^{13}N and 20.4 min ^{11}C. While a compact cyclotron ($\$5 \times 10^5 <$ cost $< \$1 \times 10^6$), is the most effective production device for these agents in a medical environment, physics-oriented accelerators (tandem Van de Graaff generators) have been used effectively when such equipment is available in reasonable proximity to the user's location. Modest energy and beam current requirements (10 μA 6–8 MeV deuterons) are capable of producing usable amounts of these agents, although much more flexible and efficient production is possible with multiple particle cyclotrons which are capable of producing curie quantities of ^{13}N and ^{11}C by the (p, α) reaction on ^{16}O and ^{14}N respectively and of ^{18}F by (d, α) on ^{20}Ne.

The need for efficient use of such materials has stimulated a growing technology for the rapid synthesis of radiopharmaceuticals. It is usually necessary to administer several millicuries of an agent to achieve the desired result; thus preparation times of more than a few minutes result in significant decay losses which in turn require starting with a much higher initial activity with its attendant problems. While some organic syntheses are rapid, the Grignard reactions[2] for instance, organic methods are in general slow and alternative approaches have been sought. Enzymatic methods appear encouraging as they permit the rapid production of fairly complex biological compounds in acceptable pharmaceutical form. This has been done with nitrogen-13 compounds producing glutamate and asparagine from ^{13}N ammonia and alanine by transmethylation of pyruvate using ^{13}N glutamate as the ^{13}N donor.[3–5]

The use of enzymatic methods for the fixation of CO_2 as aspartate has been attempted,[6] and it seems likely that the use of enzymes may prove to be a generally useful approach either alone or in connection with organic methods where it may, for instance be useful in introducing an amino group to produce an optical isomer directly. The fixation of oxygen-15 by an enzymatic method has become the routine approach for making labeled ^{15}O water by exchange in the presence of carbonic anhydrase.[7] This method also works *in vivo* since inhaled $C^{15}O_2$ is almost instantly converted to $H_2{}^{15}O$ when absorbed in the lungs. In general, the enzymatic methods permit the production of very high specific activity products by the introduction of a high specific activity label. Additional chemical substances in the final preparation consist only of the precursors, which, under these circumstances, are usually non-toxic metabolites. The removal of the enzyme by column chromatography or precipitation is rapid and complete. If an immobilized enzyme is used, this becomes unnecessary.

A more complex biological synthesis using an intact organism[8] has also proved effective in the production of ^{11}C glucose.

Efforts to speed up conventional organic methods such as the Strecker synthesis for amino acids by using high temperature and pressure are likewise proving fruitful[9] and this approach, in connection with an enzymatic method to separate the resulting optical isomers, is an additional promising option.

Sometimes it has been found possible to make use of the reactions of radiation chemistry to produce a desired compound. A good example is the preparation of ^{13}N-ammonia by the deuteron bombardment of methane.[10] The systematic exploration of the effects of pressure and composition in various gas targets is another approach which is being carried out. It seems likely that this method will be useful only for rather simple molecular species.

Much of this effort is covered in recent extensive reviews and conferences.[11,12]

The production of short-lived radionuclides from generators has a long history beginning with ^{222}Rn. The use of this approach has been largely responsible for the present popularity of short-lived agents. The ^{132}Te–^{132}I generator produced a 2.3 hr agent very useful for thyroid uptake measurements, but it was not until ^{68}Ge–^{68}Ga, ^{99}Mo–^{99m}Tc and ^{113}Sn–^{113m}Im generators became available that nuclear medicine turned substantially in this direction.

The availability of high activities which could be administered with reasonable radiation doses led to developments in instrumentation to make use of these capabilities. The autofluoroscope was developed to make use of the 2-min ^{137}Ba activity from the 30-year ^{137}Cs parent. The thick crystals and parallel data acquisition system were chosen to give high detector efficiency and rapid counting capability for dynamic studies, design capabilities which are only recently being approached by single crystal camera systems. The development of dynamic heart imaging from radiocardiology using a single crystal detector to the multiple chamber dynamic analysis has put substantial pressure on instrument manufacturers to increase data acquisition and storage rates, which have become the limiting factors in such studies. The entry of a number of new manufacturers into the camera field has produced a remarkable and healthy competition and substantial technical improvement in these areas in the past few years.

The introduction of the ultra short-lived, generator-produced agents such as ^{81m}Kr[13] and ^{82}Rb,[14] has introduced a further step in the direction of rapid imaging which is currently being explored.

Of the substances mentioned above, ^{15}O, ^{11}C, ^{13}N, ^{18}F, and the two generator produced short-lived nuclides, ^{68}Ga and ^{82}Sr, decay by positron emission, so that comments on the potentials and problems involved with positrons are appropriate in the present discussion.

Positrons, being particle radiation, result in high deposition of energy in tissue where the source activity is located, so that the absorbed dose is intrinsically fairly high even with a short half-life material. The positrons travel a short distance in tissue before slowing down to thermal energy where the annihilation process, for the most part, occurs. The path length results in a small apparent broadening of the effective object distribution of the nuclide.[15] Detection of the annihilation radiation with ordinary gamma imaging equipment may be carried out; however, special collimation is necessary because of the high energy 511 keV. This approach has been extensively explored in connection with bone imaging with ^{18}F. Ordinarily a scanner is the most efficient device to use because its thick crystal is more efficient than the $\frac{1}{2}$-in. thick camera crystal. Collimation of the high energy radiation for a camera poses several problems. Available lead collimators are either very thick with low efficiency, or have thick septa with a reduced number of holes which reduces the efficiency and introduces a substantial hole pattern into the final image because of the high intrinsic resolution of the scintillation camera at this energy. Tungsten collimation is effective and reduces these problems somewhat, but is expensive.[16] Removal of the hole pattern from the image has been possible by movement of the collimator in a manner analogous to the Bucky grid.[17] Pin hole collimation has been used effectively.

The improvement in energy resolution with increasing energy permits effective removal of the scattered photons from the positron annihilation radiation arising in the source by pulse height analysis, so that the effective source contrast is improved over that obtained with lower energy gamma rays.

The detection of positron annihilation radiation in coincidence provides a form of electronic collimation which has additional advantages as well as some limitations. The photon pair originate at the annihilation event between the positron and an ambient electron and leave in essentially opposite directions. The deviation of the angle between the two photons from 180° is approximately 0.6° due largely to the thermal motion of the particles and constitutes the principal intrinsic uncertainty in this mode of imaging.

A variety of configurations exists for coincidence imaging. The sensitivity is always intrinsically low in spite of the absence of mechanical collimation because of the requirement that both photons of an annihilation pair must avoid scatter or absorption in the object, and both must be detected in the instrument to produce a proper detection event, and this must be accomplished against a background of unwanted primary and scattered radiation, which give rise to random coincidence events due to the non-zero width of the timing windows. As a result, serious count rate limitations exist in such a system, which only recently have given way to some extent to instrumental development. This has been accomplished by using multiple detectors to increase the effective solid angle of acceptance so that the count rate limitation in any one detector is small;[18,19] or, in the case of the single crystal camera system[20,21] by limiting input singles counts by excluding scattered radiation using graded absorbers and by using fast–slow coincidence to eliminate most of the remaining unwanted singles events prior to the slower pulse positioning operation in the camera analogue network. These various approaches have succeeded in making positron coincidence imaging reasonably competitive from the count rate point of view.

The principal intrinsic advantage of positron coincidence detection is that since it deals only with unattenuated radiation from the source distribution, the results obtained are depth independent. Moreover, the best resolution of such a system is, from geometric considerations, at a point midway between the detectors rather than at the camera face as with the ordinary gamma camera. As a result, reconstruction of a radionuclide distribution by any of the various available methods becomes much more readily possible, free from the effects of attenuation, and when normalized against a transmission image is capable of giving quantitative results when compared to a standard source.

The use of quantitative reconstructive imaging appears at first to be incompatible with the use of short-lived or ultra short-lived agents because high count density records are in general required for such applications. This incompatibility vanishes however, when equilibrium images are used.[22,24]

Furthermore, the administration of a short-lived nuclide by whatever route until a steady state is achieved results in a situation where dynamic biological parameters may be determined from the steady state parameters of the equilibrium. Since the steady state may in principle be maintained indefinitely, no time limitation is introduced and conventional stationary imaging approaches may be used including quantitative reconstructive methods. This approach is unique to short-lived emitters. The necessary conditions to be met under these circumstances is that the physical decay time of the agent should be of the same approximate magnitude as that of the physiologic function under investigation.[25–30]

For example, consider in a simplistic way the blood flow into a tissue volume of interest at equilibrium. The quantity of activity in the volume V may be determined from quantitative reconstructive imaging methods to be Q. Since Q at equilibrium is constant, the amount of activity entering V and leaving V must be equal. The amount entering via the blood stream is the arterial concentration C_a times the flow F. This must equal the loss by physical decay plus the amount leaving by washout and failure of extraction.

This relation may be expressed as:

$$C_aF = Q\lambda_D + C_aF(1-E) + Q\lambda_w,$$

where E = extraction ratio, λ_D is the physical decay constant and λ_w is the (mono)exponential washout constant.

Implementing this for a diffusible tracer, say ^{82}Sr, in the myocardium, the washout will be very slow compared to the physical decay so that λ_w becomes negligible. Assume an extraction ratio of ≈ 0.7:

$$\lambda_D = 0.55 \text{ min}^{-1} \text{ for } {}^{82}\text{Rb}.$$

Then

$$0.7\ C_aF = 0.55\ Q$$

$$F = \frac{0.55}{0.7}\frac{Q}{C_A}.$$

Perfusion, $F/V \simeq 0.8\,(C_t/C_a)$ where $C_t = Q/V$ the concentration in the tissue of the myocardium. The ratio C_t/C_a should be measurable in the image with fine enough 3D resolution. If F/V is assumed to be normal, i.e. $\sim$100 ml/min/100 g, then $C_t/C_a \simeq 1.2$; at 50 ml/100 g, $C_t/C_a = 0.62$, assuming the extraction is unchanged. Thus the ratio C_t/C_a appears to be a sensitive measure of perfusion.

In the case where a rapidly diffusible tracer is used such as $H_2{}^{15}O$ where the washout is rapid and E may be assumed equal to 1, the balance equation is:

$$C_aF = Q(\lambda_D + \lambda_w)$$

since $\lambda_w = F/V$ assuming equal partition of the

tracer between blood and tissue, $F = Q\lambda_D/[C_a - (Q/V)]$ and since $Q/V = C_t$, the tissue concentration of the tracer, perfusion $= F/V = \lambda_0/[(C_a/C_t) - 1]$.

For ^{15}O, $T_{1/2} = 2.03$ min $C_a/C_t \approx 1.34$ in the normal myocardium for $F/V = 100$ ml/m/100 g. For $F/V = 50$, $C_t/C_a = 1.68$. If the ratio C_t/C_a can be measured in the image with fine enough 3D resolution, it should also provide a measure of regional perfusion. This might be made more sensitive by using ^{14}O, $T_{1/2}$ 1.18 min. The corresponding values of C_t/C_a are then 1.58 for $F/V = 100$ and 2.17 for $F/V = 50$. Combining the results from ^{82}Rb and $H_2{}^{15}O$ in principle permits determination of the distribution of rubidium extraction, although it does not seem likely that at the present state of the art sufficiently precise measurements are possible.

An implementation has been described in the lung (without the reconstruction) by FAZIO and JONES[31] using ^{81m}Kr administered continuously by inhalation to define the average regional turnover (ventilation) in different regions in the projected lung image, and a similar approach has been used to measure cerebral blood flow in children.[32]

A somewhat different implementation is described by JONES and MATTHEWS[13] in which the introduction of the short-lived agent ^{81m}Kr is independent of flow, resulting from the decay of the parent ^{81}Rb lodged in the tissue. Here the rate of appearance of the ^{81m}Kr is proportional to the amount of ^{81}Rb present which may be monitored independently. The observed parameter, the concentration of ^{81m}Kr in tissue in equilibrium with its parent, is dependent on flow. In the absence of flow, the ^{81m}Kr disappears only by physical decay while in the presence of flow it is also washed out. Thus $P = \lambda_D Q + \lambda_w D$ where P is the production rate proportional to $[^{81}Rb]$. In the absence of flow $\lambda_w = 0$, so that

$$\frac{Q_{\text{no flow}}}{Q_{\text{flow}}} = 1 + \frac{\lambda_w}{\lambda_O} = 1 + \frac{1}{\lambda_D}\frac{F}{V}.$$

Under these conditions, a normal myocardial flow will reduce the ^{81m}Kr level in the equilibrium image about 25% below the no-flow level.

The use of the 24 min $^{79}Rb \rightarrow 55$ sec ^{79m}Kr would make this approach much more sensitive to flow because of the longer half-life of the Krypton-79*m*, although the low abundance of the krypton radiation would require the use of solid state detectors in order to resolve the 127 keV photopeak from the background continuum.

In general, as illustrated in the above examples, when the introduction of the label is independent of flow the longer half-life will produce greater object contrast with respect to flow. When the introduction of the tracer is dependent on flow, the shorter half-life agent will give a result more sensitive with respect to flow.

The generalization of this type of approach to any form of physiological turnover appears in principle to be possible.

The simultaneous administration of the two oxygen isotopes ^{14}O and ^{15}O has been used to estimate the transport of oxygen to different parts of the body by measuring the ratio of the two isotopes as they decay at different rates.[33]

For routine application of a short-lived agent, one of the principal requirements is a reliable source of supply. In a practical context the use of short-lived agents in critically ill patients imposes substantial logistical problems, although the capability of making sequential measurements to follow the course of a disease process or to observe the effect of stress on thereapeutic efforts would seem to be one of the main clinical advantages of short-lived agents.

Measurement of regional cerebral blood flow, in the presence of serious head injury, for instance to determine progression of change due to swelling, while not at present a routine procedure, certainly appears to be a desirable goal. Dedicated or mobile equipment and an available supply of radionuclide for such a process would be necessary on short notice. A possibly viable option in this situation would be the availability of a ^{81}Rb-^{81m}Kr generator to be used routinely during the day for lung studies and available otherwise for cerebral flow studies. Intra-arterial injection would be necessary in the left ventrical or root of the aorta. Ordinary static images in the usual projections should give a reasonable estimate, although not quantitation of the relative regional cerebral blood flow.

To use the more sophisticated option of $H_2{}^{15}O$ as a diffusible tracer administered as CO_2 to produce an equilibrium state which could be displayed by reconstructive techniques would require the ready availability of both special imaging equipment, a computer, and a suitable accelerator.

In the study of cardiac patients, particularly the evaluation of patients for bypass surgery, the documentation of regional myocardial perfusion and perfusion reserve is a principal need for which there is no technique available at present which is generally accepted as giving good data. The alternative approach of xenon washout has well known imaging, detection and physiologic problems. The uptake of potassium ion[34] and NH_4^+ ion[35] has been shown to be proportional to perfusion in the absence of hyperemia when compared to measurements made with microspheres in the normal myocardium and in the tissues adjacent to a transmural infarct. The uptake of such agents under conditions where uptake could be quantitated should give a reasonably good measure of relative tissue perfusion.

To distinguish between ischemic tissue and infarct or scar, comparison of localization at rest and under stress should show alteration in ischemic tissue. The use of a label and procedure which would permit such an evaluation at a single setting would be most desirable, especially if it were non-invasive.

For studies conducted in connection with cardiac catheterization procedures, both the agents ^{13}N and ^{11}C have half-lives too long to be used for a resting plus a sequential stress study as 40–50 min must elapse between administrations even of ^{13}N to avoid interference. This would make the whole procedure last 2–4 hr which is not a compatible procedure for a busy catheterization laboratory. Here a possible option exists of intracoronary injection of ^{81}Rb[36] which should localize largely in the myocardium. The level of the daughter radioactivity, ^{81m}Kr, should then faithfully reflect changes in perfusion with only the delay necessitated by the 13-sec half-life which reached 95% of final value in one minute. The direct intracoronary injection should avoid the difficulty encountered in the intravenous injection where extraneous activity essentially destroys the capability of making quantitative uptake estimates.[37]

In this presentation we have attempted to outline some of the problems, capabilities, advantages, and future possibilities of short-lived radionuclides used in imaging procedures and quantitative physiological measurements in nuclear medicine in the context of clinical practice and imaging capabilities as they exist today and as they appear to be headed in future developments.

Although we have emphasized the possible uses of reconstructive quantitative imaging in connection with equilibrium imaging, it is clear that useful information may be also obtained by ordinary static imaging techniques although at a lower level of accuracy.

REFERENCES

1. WAGNER H. N., JR. and EMMONS H. *Radioactive Pharmaceuticals* (Edited by ANDREWS G. A., KNISELEY R. M. and WAGNER H. N., JR.), pp. 1–32. USAEC Division of Technical Information, Report CONF 651111 (1966).
2. WINCHELL H. S. and WINSTEAD M. B. UCRL 19420 (1969).
3. HARPER P. V., LATHROP K. A., KRIZEK H., LEMBARES N. and DINWOODIE R. *Radiopharmaceuticals* (Edited by SUBRAMANIAN G., RHODES B. A., COOPER J. F. and SODD, V. J.), pp. 180–183. Soc. Nucl. Med. Publishers, New York (1975).
4. COHEN M. B. *et al. Proc. 1st World Cong. Nucl. Med.*, p. 907 (1974); *Radiopharmaceuticals* (Edited by SUBRAMANIAN G., RHODES B. A., COOPER J. F. and SODD V. J.), pp. 184–188. Soc. Nucl. Med. Publishers, New York (1975).
5. GELBARD A. S., CLARKE L. P. and LAUGHLIN J. S. *J. nucl. Med.* **15,** 1223 (1974).
6. HARA T., TAYLOR C., LEMBARES N., LATHROP K. A. and HARPER P. V. *J. nucl. Med.* **12,** 361 (1971).
7. WEST J. B. and DOLLERY C. T. *J. appl. Physiol.* **17,** 9 (1962).
8. RAICHLE M. F., LARSON K. B., PHELPS M. E., GRUBB R. L., JR., WELCH M. J. and TER-POGOSSIAN M. M. *Am. J. Physiol.* **228,** 1936 (1975).
9. WASHBURN L. C., SUN T. T., WIELAND B. W. and HAYES R. L. *J. nucl. Med.* **16,** 579 (1975).
10. TILBURY R. S., DAHL S. R., MONAHAN W. G. *et al. Radiochem. Radioanalyt. Lett.* **8,** 317 (1971).
11. WOLF A. P., CHRISTMAN D. R., FOWLER J. S. *et al. Radiopharmaceuticals and Labelled Compounds* (pp. 345–381), Vol. 1. IAEA, Vienna (1973).
12. SUBRAMANIAN G., RHODES B. K., COOPER J. F. and SODD V. J. (Editors). *Radiopharmaceuticals.* Soc. Nucl. Med. Publishers, New York (1975).
13. JONES T. and MATTHEWS C. M. E. *Nature*, Lond. **230,** 119 (1971).
14. YANO Y. and ANGER H. *J. nucl. Med.* **9,** 412 (1968).
15. PHELPS M. E., HOFFMAN E. J., HUANG S.-C. and TER-POGOSSIAN M. M. *J. nucl. Med.* **16,** 649 (1975).

16. HARPER P. V., SCHWARTZ J., BECK R. N. *et al. Radiology* **108,** 613 (1973).
17. BRUNSDEN B., HARPER P. V. and BECK R. N. *J. nucl. Med.* **16,** 517 (1975).
18. BURNHAM C. A. and BROWNELL G. L. *IEEE Trans. Nucl. Sci.* **19,** 201 (1972).
19. TER-POGOSSIAN M., PHELPS M., HOFFMAN E. and MULLANI N. *Radiology* **114,** 89 (1975).
20. MUEHLLEHNER G. *J. nucl. Med.* **16,** 653 (1975).
21. MUEHLLEHNER G., BUCHIN M. P. and DUDEK J. H. *IEEE Trans. Am. Nucl. Soc.* NS-23, **1,** 528 (1976).
22. JONES T., CLARK J. C., HUGHES J. M. and ROSENZWEIG D. Y. *J. nucl. Med.* **11,** 118 (1970).
23. HOOP B. *et al. IEEE Trans. Am. Nucl. Soc.* NS-23, **1,** 584 (1976).
24. RUSS G. A., BIGLER R. E., MCDONALD J. M., TILBURY R. S. and LAUGHLIN J. S. *J. nucl. Med.* **15,** 529 (1974).
25. JONES T., CHESLER D. A. and TER-POGOSSIAN M. M. *Br. J. Radiol.*, in press (1976).
26. JONES T., MCKENZIE C. G., MOSS S., BUCKINGHAM P. D. and CLARK J. C. Presented at the 12th Symposium on *Radioisotopes in Clinical Medicine and Research*, Bad Gastein, 1976. To be published in *Strahlentherapie.*
27. JONES T., JONES H. A., RHODES, C. G., BUCKINGHAM P. D. and HUGHES J. M. B. *J. clin. Invest.* **57,** 706 (1976).
28. FAZIO F., MACARTHUR C. G. C., RHODES C. G., JONES T. and HUGHES J. M. B. *Radioaktive Isotope in Klinik und Forschung* (Edited by HOFER R.) Vol. 12, p. 41. Egermann, Vienna (1976).
29. TURNER J. H., SELWYN A. P., JONES T., EVANS T. R., RAPHAEL M. J. and LAVENDER J. P. *Cardiovascular Res.*, in press (1976).
30. TURNER J. H., SELWYN A. P., JONES T., EVANS T. R., RAPHAEL M. J. and LAVENDER J. P. *Radioaktive Isotope in Klinik und Forschung* (Edited by HÖFER R.) Vol. 1, p. 427. Egermann, Vienna (1976).
31. FAZIO F. and JONES T. *Br. Med. J.* **265,** 673 (1975).
32. ARNOT R. N., GLASS H. I., CLARK J. C., DAVIS J. A., SCHIFF D. and PICTON-WARLOW C. G. *Radioaktive Isotope in Klinik und Forschung* (Edited by FELLINGER K. and HOFER R.), p. 60, Vol. 9. Urban & Schwartzenberg, München, Berlin and Vienna (1970).
33. NICKLES R. J. and AU Y. F. *Phys. Med. Biol.* **20,** 54 (1975).
34. PROKOP E. K., STRAUSS H. W., SHAW J. *et al. Circulation* **50,** 978 (1974).
35. WALSH W. F., HARPER P. V., RESNEKOV L. and FILL H. R. *Circulation* **54,** 266 (1976).
36. HOLMAN L. Personal communication.
37. RICH B., LEMBARES N., HARPER P. V., LATHROP K. A. and ATKINS F. *Radiopharmaceuticals* (Edited by SUBRAMANIAN G., RHODES B. A., COOPER J. F. and SODD V. J.), pp. 174–179. Soc. Nucl. Med. Publishers, New York (1975).

International Journal of Applied Radiation and Isotopes, 1977, Vol. 28, pp. 13–20. Pergamon Press. Printed in Northern Ireland

A Look at ^{13}N and ^{15}O in Radiopharmaceuticals

MARIA G. STRAATMANN
Mallinckrodt Institute of Radiology, 510 South Kingshighway, St. Louis, Missouri 63110, U.S.A.

(*Received* 20 *May* 1976)

Radiopharmaceuticals and labeled compounds incorporating the short-lived nuclides nitrogen-13 and oxygen-15 are reviewed. The necessity for ingenuity in synthetic procedures using rapid, efficient techniques is emphasized as well as requirements for isotope production. Current biological, physiological and clinical applications of compounds labeled with ^{13}N and ^{15}O are included in light of a resurgence of interest in positron-emitting radionuclides.

THOUGH the majority of biological materials are made up of carbon, nitrogen and oxygen, and the ability to study metabolic processes *in vivo* by external monitoring would be greatest for compounds incorporating radioactive forms of naturally occurring elements, the combination of requirements for the nuclide in a biological study and the availability of radioactive isotopes of these three limits their use on any widespread basis. Besides carbon-11 (the production and utility of which is discussed elsewhere in this volume) the only radioisotopes that can be considered are the accelerator produced, short-lived oxygen-15 (half-life, 2.04 min) and nitrogen-13 (half-life, 9.96 min). Both are positron-emitters, (β^+_{max} ^{15}O, 1.7 MeV; β^+_{max} ^{13}N, 1.19 MeV) which may be detected *in vivo* by means of the 0.511 MeV annihilation radiation. The efficient use of these isotopes was limited for many years due to lack of optimal detector systems for this high energy emission. In recent years, however, new developments in instrumentation[1,2] have caused a renewed interest in positron-emitters due to the high inherent resolution that may be achieved by utilizing coincidence detection. Accordingly, determinations of positron ranges and their effect on spatial resolution in scintigraphic images have been recently reported for these nuclides.[3,4]

The short half-lives of ^{13}N and ^{15}O indicate that these isotopes must be produced in a facility near to that in which they will be used. This has been accomplished principally by an inhouse cyclotron as discussed in several reviews on medical cyclotrons.[5–7] Due to the greater availability of Van de Graaff accelerators, a number of authors have recently discussed the use of these machines in the production of ^{13}N and ^{15}O.[8–12] The employment of an electron accelerator designed for radiation therapy in photonuclear reactions to produce medically useful nuclides has been examined by WELCH[13]. Though yields are low, such an approach provides another potential source for short-lived isotopes in the absence of a cyclotron.

The short half-lives of these isotopes are most responsible for difficulties associated with their use. Obviously, complicated organic syntheses incorporating ^{13}N and ^{15}O are impossible and the work done in the field of radiochemical labeling thus far has led to simple, inventive and above all, rapid techniques for the application of these nuclides. A number of reviews in the recent literature concern themselves with the practical problems involved.[14–19]

Nitrogen-13

^{13}N-*gases*. Nitrogen-13 is available through several nuclear reactions, but most commonly used in the production of ^{13}N-labeled gases is the $^{12}C(d, n)^{13}N$ reaction. BUCKINGHAM and CLARK[20] describe a method for providing saline solutions of nitrogen-13 for clinical use in which the target material is a graphite matrix bombarded in the presence of carbon dioxide as a sweep gas. In the heat generated by the cyclotron beam, the CO_2 erodes the target by the reaction

$$C + CO_2 \rightarrow 2CO,$$

allowing ^{13}N trapped in the carbon lattice to escape. (Previous low yields of ^{13}NN from a graphite target were due to the ^{13}N atoms being confined to the molecular lattice, leading to combination with carbon to form ^{13}N-cyanide. This same effect was noted by WELCH and LIFTON[21] in a study of ^{13}N generated in inorganic carbides. They demonstrated that when recoil nitrogen is induced into the carbide lattice, carbon–nitrogen bonds are formed and that the yield of $C^{13}N$ is related to crystal structure.)

The target effluent was passed through a copper oxide furnace to convert any CO formed to CO_2. The CO_2 was absorbed by sodium hydroxide solution and the ^{13}NN dissolved in saline. Yields of 300 μCi/ml were obtained using a 14 MeV deuteron beam current of 60 μA.

WELCH[22] reported on the production of ^{13}NN from the deuteron bombardment of carbon dioxide in which the ^{13}N atoms formed were believed to react with traces of N_2 impurity according to the scheme:

$$^{13}N + N_2 \rightarrow N^{13}N^* + N$$
$$N^{13}N^* + CO_2 \rightarrow N^{13}N$$
$$N^{13}N^* + CO_2 \rightarrow N^{13}NO.$$

An intermediate excited species of molecular nitrogen ($N^{13}N^*$) was thought to be involved. The gas was prepared for clinical use by passage over hot cupric oxide followed by absorption of CO_2 on soda lime and dissolution of the $N^{13}N$ in saline.

CROUZEL and COMAR[23] irradiated high purity carbon dioxide with deuterons, condensed the CO_2 and ^{13}N in liquid nitrogen, eliminated the CO_2 by pumping with a Toepler pump and obtained 30 mCi of ^{13}NN from an irradiation of 40 min by a 20 μA, 9 MeV beam. The dissolved ^{13}NN in physiological saline had a final activity of 1 mCi/ml.

The various ways of producing ^{13}N nitrogen solutions using target materials CO_2, graphite, and activated charcoal as conducted at Hammersmith Hospital have been thoroughly examined.[18] Production parameters, advantages and disadvantages of each technique are discussed.

Nitrogen gas labeled with ^{13}N has been utilized in a variety of physiological investigations, primarily for the study of pulmonary function due to its low tissue solubility. ROSENWEIG *et al.*[24] demonstrated uneven ventilation within and between regions of the lung. Airway closure and regional ventilation were illustrated by GREENE *et al.*[25] by ^{13}N gas studies, as was lung dysfunction in workers exposed to *Bacillus subtilis* enzyme.[26] More recently, ^{13}NN has been used to evaluate small air-ways function in myocardial infarction[27] and the strength of pulmonary vascular response to regional alveolar hypoxia.[27]

Biological studies requiring much higher levels of gaseous ^{13}N activity have led to new means of production. AUSTIN *et al.*[29] have reported on a batch process in which ^{13}NN with specific activity of $\sim$20 mCi/ml has been generated by the $^{13}C(p, n)^{13}N$ nuclear reaction using an amorphous carbon target enriched to about 97% ^{13}C and a proton bombarding energy of 11 MeV. After irradiation the powder in the target (value $\simeq$ \$3.30) is mixed with CuO powder and KNO_3 in a Dumas combustion technique and the resultant gas pressed into 1 ml vials. Processing takes from 12 to 15 min from the end of bombardment. This high activity product has been employed in investigations of biological nitrogen fixation.[30,31]

PARKS *et al.*[32] synthesized ^{13}N-labeled atmospheric gases via proton irradiation of a high pressure oxygen target by the $^{16}O(p, \alpha)^{13}N$ nuclear reaction. This method produces ^{13}N atoms in a chemical environment favoring the formation of carrier-free nitrogen oxides. The primary products, $^{13}N_2$, $^{13}N_2O$ and $^{13}NO_2$, were found in a ratio of 8:1:3.5 although only 10% of the induced activity was recovered. After irradiation until an equilibrium production rate was established, 5 mCi of $^{13}NO_2$ could be collected within 10–15 min, a yield of about 1 mCi/μA. The gases formed were intended for application to air pollution studies.

^{13}N-*ammonia*. One of the most heavily exploited forms of nitrogen-13 for medical studies is that of ^{13}N-ammonia which has been produced by a variety of methods. WELCH and LIFTON[21] induced ^{13}N activity in inorganic carbides by the $^{12}C(d, n)^{13}N$ nuclear reaction. Deuteron bombardment of Al_4C_3 in particular led to large amounts of $^{13}NH_3$ and some methylamine when the irradiated powder was dissolved in HCl and distilled from a basic solution. The resulting mixture was shown by radio-gas-chromatography to contain only 7% CH_3NH_2. Subsequent analysis by liquid chromatography showed this impurity to be closer to ~20% of the total activity. It was concluded that amines were converted to ammonia by heat decomposition on the injection block and that purities determined by gas chromatography were suspect.(33) The carbide technique suffers the disadvantage that activated aluminum products (^{28}Al) are also formed in the process, and though the preparation itself is not affected, an unnecessary radiation dose is received by the chemist carrying out the processing.

An alternative procedure introduced by TILBURY *et al.*(34) involves the deuteron bombardment of methane. The methane in an open circuit flowed through the pyrex glass lined tube of the target chamber into water or isotonic saline to dissolve the ammonia. Analysis by radio-gas-chromatography indicated 95% $^{13}NH_3$. STRAATMANN and WELCH(33) adapted the procedure to include a distillation step. Using a glass target and a circulating system with the activity being trapped in acid after the target, the ^{13}N activity was found to contain labeled impurities as detected by a liquid chromatographic analysis that were converted to $^{13}NH_3$ in the distillation process. Typically 30 mCi/ml can be prepared from a 15 min bombardment by 30 μA of 7 MeV deuterons.

All of the above techniques for producing $^{13}NH_3$ result in solutions containing carrier ammonia. Aluminum carbide contains nitrogen as an impurity in levels as high as 0.2% depending upon the batch of carbide used. During dissolution, ammonia is formed in concentrations approaching 0.02 M in 5 ml. Presumably the ammonia found in methane preparations arises from radiolysis of trace nitrogen impurities in the methane followed by consecutive reactions with hydrogen atoms also formed radiolytically from the hydrocarbon. However, LATHROP *et al.*(35) suggested using the $^{16}O(p, \alpha)$ ^{13}N nuclear reaction on an oxygen gas target (or on water); in this case no carrier problem exists. The bombarded gas was bubbled through NaOH to convert oxides of nitrogen to nitrates or nitrites which were then reduced with titanous chloride and the $^{13}NH_3$ was isolated by steam distillation. GELBARD *et al.*(36) examined the methane/^{13}N reaction for producing ammonia, comparing the preparation with that using the (p, α) reaction on water. The reduction of nitrogen oxides was accomplished using zinc dust in sodium hydroxide solution and the $^{13}NH_3$ distilled over in a stream of nitrogen. VAALBURG *et al.*(37) have recently described an improved method for ^{13}N-labeled ammonia production via the $^{16}O(p, \alpha)^{13}N$ on water at a bombarding energy of about 19 MeV. In addition, reduction by Devarda's alloy proved to be more efficient and reproducible than other techniques. They indicate values 2.5 times higher than previously reported activity yields at the higher proton energy.

The ^{13}N-ammonia produced by these various methods has been utilized in a number of clinical investigations. MONAHAN, TILBURY and LAUGHLIN(38) demonstrated the uptake of $^{13}NH_3$ in dogs by the myocardium, liver and kidney. This was followed by an examination of the clinical feasibility of myocardial imaging by HARPER *et al.*(39,40) using a Nuclear Chicago HP Anger Camera, illustrating the utility of $^{13}NH_3$ for detecting myocardial infarction. These authors linked ammonia uptake by the myocardium to the conversion of ammonia to glutamine by demonstrating a reduced incorporation of $^{13}NH_3$ when methionine sulfoximine (a glutamine synthetase inhibitor) had been preinjected. Although they suggested that heavy smokers demonstrated a higher pulmonary uptake than non-smoking subjects, there is probably a more complex reason for this effect. Studies performed at the Mallinckrodt Institute of Radiology have produced the anomalous pulmonary uptake in non-smokers as well, when the $^{13}NH_3$ was analyzed at >95% purity by liquid chromatography.(41) HOOP *et al.*(42) refined myocardial imaging by using $^{13}NH_3$ and a multi-crystal positron camera.

Effects of blood pH on brain, blood, and liver concentrations of injected ^{13}N-ammonia in dogs has been evaluated,(43) and applications of ^{13}N to central nervous system studies discussed.(44) In animals and patients attempts have been made to use $^{13}NH_3$ to image metabolic processes using emission tomography.[45,46] These authors interpret the distribution of ^{13}N in the myocardium and brain as being indicative of blood perfusion in these tissues. Another physiological investigation proposes the diagnostic use of $^{13}NH_3$ in normals and patients with cirrhosis of the liver by rectal administration of the ammonia.[47]

Ammonia has become so well-accepted as a radiopharmaceutical for myocardial imaging that it has been used as a standard of comparison for other potential positron-emitting imaging agents.[48]

$^{13}NH_3$ *as a synthetic precursor*. In addition to its direct utilization, $^{13}NH_3$ has been applied as a synthetic precursor in a number of labeling procedures, most often in the production of ^{13}N-amino acids. The amino acids have been labeled by enzymatic syntheses, characterized by their speed and selectivity. Whereas conventional synthesis can lead to racemic mixtures, an enzyme reaction results in only the optical isomer of interest, specifically labeled under mild conditions. The first such amino acids were ^{13}N-glutamine and ^{13}N-glutamate.[16,33,49,50] ^{13}N-glutamine was produced in this laboratory in an incubation mixture (37°C) containing 3–4 units glutamine synthetase (EC 6.3.1.2), 10 μmoles ATP, 50 μmoles Mg^{2+}, 24 μmoles cysteine, 50 μmoles L-glutamic acid, $^{13}NH_3$ in pH 7.2, imidazole buffer to a total volume of 1 ml. A reaction time of 15 min was deemed optimal. ^{13}N-glutamine was separated and purified by passage at pH 5 through 2 cc Amberlite CG-4B resin. The ^{13}N-glutamic acid preparation was similar: incubation at 25°C for 10 min, 10 μl 50% glycerol solution of glutamic acid dehydrogenase (EC 1.4.1.3), 5×10^{-3} M α-ketoglutarate, $^{13}NH_3$, 1.2×10^{-4} M NADH, 10^{-5} EDTA, in pH 8.0 *tris* buffer to a total volume of 1 ml.[33]

The first reported yields for these two amino acids using $^{13}NH_3$ from the bombardment of methane were from 5–25% incorporation of the label.[16,49] This was shown to be a carrier ammonia problem that could be overcome by pre-treatment of the target chamber followed by distillation of the once formed $^{13}NH_3$.[33] To circumvent this problem, most subsequent labeling attempts have made use of $^{13}NH_3$ produced by the proton bombardment of water.[16,50] Either of these procedures results in yields of >95% labeling efficiency.

Lathrop *et al.*[16] observed ^{13}N-glutamate localization in myocardium, liver, kidneys and pancreas. However Cohen *et al.*[49] demonstrated that pancreatic uptake was time dependent and suggested that glutamine and glutamate labeled with ^{13}N would not suffice for static imaging as their uptake by myocardium was significantly less than that exhibited by $^{13}NH_3$. Gelbard *et al.*[50] investigated the tissue distribution of these ^{13}N-amino acids synthesized with high levels of activity to demonstrate a different metabolic pattern among $^{13}NH_3$, ^{13}N-glutamine and ^{13}N-glutamic acid in dogs.

Cohen *et al.*[51] produced ^{13}N-alanine from a sequential enzymatic synthesis in which ^{13}N-glutamate was added to 0.1 ml of a preincubated at 38°C pyruvic acid (0.1 M) mixture in 0.2 ml 0.01 M tricine buffer at pH 8.0 containing 8 units of glutamic-pyruvic transaminase. Further incubation continued at 37°C for 7 min. Separation was accomplished using an AG50W-X8 ion exchange resin. The ^{13}N-alanine was evaluated as a pancreatic imaging agent. To improve the yield by decreasing the time for synthesis and to reduce the risk of pyrogenic or antigenic response to the enzyme, the enzymes used were immobilized on a solid support to produce pharmaceutical quality ^{13}N-alanine.[52]

Gelbard *et al.*[53] synthesized ^{13}N-asparagine from $^{13}NH_3$ in an enzyme reaction mixture containing: 20 μmoles aspartic acid, 15 μmoles ATP, 15 μmoles $MgCl_2$, 1 μmole ammonium acetate and 5 mU of asparagine synthetase. The solution was incubated for 10 min at 37°C and the amino acid separated by an ion exchanger. Tissue distribution studies indicated that asparagine concentrated in heart and liver to a greater extent than NH_3 or alkaline metals.

Production of other ^{13}N-amino acids is possible via glutamate dehydrogenase reactions with $^{13}NH_3$. Straatmann and Welch[33] observed ^{13}N-valine from α-ketoisovalerate, ^{13}N-alanine from pyruvate and ^{13}N-leucine from α-ketoisocaproate in yields varying from 6.8 to 26%. Although conditions were not optimized for these reactions, the feasibility of the labeling technique was demonstrated. In addition,

GELBARD *et al.*[50] speculated that ^{13}N-glutamine could be used to synthesize amino sugars, purines, pyrimidines, nicotinamide nucleotides or other amino acids via amidotransferases.

HARPER[19] suggested that $^{13}NH_3$ could serve as a route to ^{13}NN if oxidized by sodium hypochlorite; this of course would form ^{13}NN solution at a very high concentration.

In a unique synthesis, PETTIT[54] *et al.* passed $^{13}NH_3$ over gallium and cobalt oxides to produce $HO^{13}NO$, nitrous acid. This was bubbled into 1-(2-chloroethyl)-3-cyclohexylurea in formate buffer to form 3-cyclohexyl-1-nitrosourea-^{13}N in a yield of 0.2 μCi/mg 1 hour after end of bombardment.

GERSBERG *et al.* used $^{13}NO_3$ produced by 14.5 MeV proton bombardment of water as a tracer for direct quantitative measurements of denitrification rates of water-logged rice soils. This technique has the advantage of measuring denitrification rates of natural systems over short time intervals, without changing the concentration of nitrate in the system.

Recoil synthesis with ^{13}N. Attempts to directly label organic molecules with recoil ^{13}N atoms have been largely unsuccessful. WELCH and STRAATMANN[55] undertook a study of the reactions of ^{13}N with acetic acid, acetaldehyde and ethanol in which these compounds were bombarded with 7 MeV deuterons to produce ^{13}N by the $^{12}C(d, n)^{13}N$ nuclear reaction. It was shown that in the liquid phase, recoil ^{13}N atoms generally react with organic compounds to form fragmented products ($^{13}NH_3$, $^{13}NO_3^-$ and $C^{13}N^-$). No products were observed that could arise from the occurrence of an NH substitution reaction. Unlike methyne and methylene, which in recoil studies have been shown to insert, the nitrene radical appears only to abstract hydrogen to eventually form ammonia. STENSTRÖM[51] bombarded inulin and insulin with 185 MeV protons and found fragmentation and severe radiolysis of the products. KASCHE[58] reported labeling α-chymotrypsin with ^{11}C and ^{13}N by a similar technique.

Oxygen-15

The variety of compounds that can be labeled with ^{15}O is even more limited than in the case of ^{13}N. In general, this isotope is only amenable to production methods involving direct atom combination in the accelerator target area with rapid conversion via heat or exchange in a flow system. Procedures for ^{15}O labeling have been adequately discussed in several recent reviews.[5,17,18,22,44]

The formation of $O^{15}O$ of high specific activity in nitrogen carrier gas can be accomplished by the deuteron bombardment of nitrogen containing a trace of oxygen. This labeled gas can then be converted to $C^{15}O$ and $CO^{15}O$ by passage over activated charcoal at 900 and 400°C respectively.[59] Another route to $CO^{15}O$ is via the irradiation of $N_2 + CO_2$ gas mixtures. An exhaustive discussion of the pertinent and necessary parameters to be considered in the production of ^{15}O gases can be found in a monograph by CLARK and BUCKINGHAM[18].

These ^{15}O-gases have had their most extensive application in the further formation of ^{15}O-labeled compounds. Oxyhemoglobin and carboxyhemoglobin are labeled by dissolving $O^{15}O$ and $C^{15}O$ respectively in blood.[18,22,60] These gases must be free of any $CO^{15}O$ as it readily dissolves in water or blood to be rapidly converted to $H_2{}^{15}O$[61,62] via the reaction sequence:

$$CO^{15}O + H_2O \rightleftarrows H_2CO_2{}^{15}O \rightleftarrows CO_2 + H_2{}^{15}O.$$

The ^{15}O-labeled compounds mentioned above have been utilized in widely divergent physiological and clinical studies. The short half-life has proven to be an advantage in *in vivo* investigations related to cerebral metabolism, allowing rapid sequential injections. Water labeled with ^{15}O for intra-arterial injection has been employed in the determination of regional cerebral blood flow.[63] This value in conjunction with experiments utilizing ^{15}O-oxyhemoglobin and ^{15}O-carboxyhemoglobin has in turn been applied to the estimation of regional cerebral oxygen metabolism[64] and cerebral blood volume.[65] Radiolabeled water has been used as a standard for the evaluation of blood brain barrier permeability due to the assumption that water is freely diffusible in brain tissue. A number of studies with $H_2{}^{15}O$ have shown this not to be the case.[66–68]

Recently, a method has been reported for the assessment of regional oxygen extraction in the brain by the continuous inhalation of oxygen-15.[69]

Pulmonary function has also been examined utilizing ^{15}O-labeled compounds, studies mostly done between 1960 and 1964.[70] At that time, ^{133}Xe became available in high concentration solutions and largely supplanted the use of the ^{15}O gases.[5] This trend is being reversed due to increased interest in positron-emitting isotopes, however, and new applications of ^{15}O are appearing in the literature: regional distribution of pulmonary blood flow and ventilation was determined after myocardial infarction using $O^{15}O$ and $CO^{15}O$;[71] wash-out studies of ^{15}O-labeled O_2, CO and CO_2 were conducted in relation to pulmonary blood flow;[72] and $H_2{}^{15}O$ was used to measure the distributions of extra-vascular fluid volumes in isolated perfused lungs.[73]

Cardiac function has been examined following the inhalation of $CO^{15}O$ as a non-invasive technique for introducing a tracer ($H_2{}^{15}O$) via the pulmonary venous blood. This technique has been investigated for its potential use in left heart radiocardiograms and myocardial blood flow,[74] as well as in the quantitative estimation of these cardiac parameters: (1) cardiac output, (2) mean central transit time, (3) left ventricular ejection fraction, and (4) regurgitant fraction.[75]

A method has been suggested for tagging ozone with ^{15}O for use in biomedical studies.[76] Pure oxygen was irradiated with bremsstrahlung from the Lawrence Livermore 100 MeV Linac in the reaction $^{16}O(\gamma, n)^{15}O$. The resultant gas is ^{15}OO, ^{15}OOO and unactivated O_3. These components are separated and analyzed by a series of traps and coincidence detector counting vessels in a flow system. Preliminary experiments with small dogs were carried out to determine lung function in the presence of ozone. The Anger positron camera was used as a detecting device.

Summary

Nitrogen-13 and oxygen-15, as radioactive forms of atoms naturally found in biological molecules, are uniquely suited to a tracer role in biomedical investigations. Despite their short half-lives (which in some applications have served to an advantage), a number of sophisticated molecules have been labeled due to ingenuity on the part of chemists in devising rapid, efficient syntheses. With the renewed interest in positron-emitters in medical studies, these isotopes may well increase in importance in years to come.

Acknowledgement—The author extends thanks to Dr. Michael J. Welch for helpful suggestions on the manuscript, Ms. Barbara Asberry for aid in assembling references and Ms. Diane Zaltsman for graciously typing the text.

The author's research was supported by NIH Grant No. HL13851.

REFERENCES

1. Burnham C. A. and Brownell G. L. IEEE Trans Nucl Sci, NS-19-3, 201 (1972); Brownell G. L., Burnham C. A., Hoop B., Jr. and Kazemi H. In *Medical Radioisotope Scintigraphy 1972*, Vol. I. IAEA, Vienna, 313 (1973).
2. Ter-Pogossian M. M., Phelps M. E., Hoffman E. J. and Mullani N. A. *Radiology* **114,** 89 (1975).
3. Phelps M. E., Hoffman E. J., Huang S. C. and Ter-Pogossian M. M. *J. nucl. Med.* **16,** 649 (1975).
4. Cho Z. H., Chan J. K., Ericksson L., Singh M., Graham S., MacDonald N. S. and Yano Y. *J. nucl. Med.* **16,** 1174 (1975).
5. Glass H. I. and Silvester D. J. *Br. J. Radiol.* **43,** 589 (1970).
6. Silvester D. J. In *Radiopharmaceuticals and Labelled Compounds*, Vol. I. IAEA, Vienna, 197 (1973).
7. Gelbard A. S., Hara T., Tilbury R. S. and Laughlin J. S. In *Radiopharmaceuticals and Labelled Compounds*, Vol. I. IAEA, Vienna, 239 (1973).
8. Cohen B. L. and Canada T. R. *Phys. Med. Biol.* **17,** 416 (1972).
9. Phelps M. E. and Wieland B. W. *Phys. Med. Biol.* **18,** 284 (1973).
10. Rydin R. A. and Engler R. A. *Phys. Med. Biol.* **19,** 884 (1974).
11. Reed M. F., McEllistrem M. T. and Pettit W. A. *Phys. Med. Biol.* **20,** 493 (1975).
12. Mantel J., Damico C. J. and Kurtzman R. S. *Am. J. Roent.* **112/4,** 707 (1971).
13. Welch M. J. Conf 730301, *Proc Int. Conf. Photonuclear Reactions and Applications* (Edited by Berman B. L.), 1179 (1973).
14. Glass H. I. In *Medical Radioisotope Scintigraphy 1972*, Vol. 2. IAEA, Vienna, 299 (1973).
15. Wolf A. P., Christman D. R., Fowler J. S. and Lambrecht R. M. In *Radiopharmaceuticals and Labelled Compounds*, Vol. 1. IAEA, 345 (1973).
16. Lathrop R. A., Harper P. V., Rich B. H.,

DINWOODIE R., KRIZEK H., LEMBARES N. and GLORIA I. In *Radiopharmaceuticals and Labelled Compounds*, Vol. 1. IAEA, 471 (1973).
17. HOOP B., JR., LAUGHLIN J. S. and TILBURY R. S. In *Instrumentation in Nuclear Medicine* (Edited by HINE G. and SORENSON J.), Vol. 2, p. 407. Academic, New York, (1973).
18. CLARK J. C. and BUCKINGHAM P. D. *Short-lived Radioactive Gases for Clinical Use*. Butterworth, Boston (1975).
19. HARPER P. V., LATHROP K. A., KRIZEK, H., LEMBARES N. and DINWOODIE R. In *Radiopharmaceuticals* (Edited by SUBRAMANIAN G., RHODES B. A., COOPER J. R. and SODD V. J.), p. 180. Society of Nuclear Medicine, New York (1975).
20. BUCKINGHAM P. D. and CLARK J. C. *Int. J. appl. Radiat. Isotopes* **23,** 5 (1972).
21. WELCH M. J., LIFTON J. F. *J. Am. Chem. Soc.* **93,** 3385 (1971).
22. WELCH M. J. In *Prog. Nucl. Med.* Vol. 3, (Edited by HOLMAN B. L.), p. 37. Karger, Basel and University Park Press, Baltimore (1973).
23. CROUZEL C. and COMAR D., *Radiochem Radioanal Letters* **20,** 273 (1975).
24. ROSENZWIEG D. Y., HUGHES J. M. ᾿ and JONES T., *Respir. Physiol.* **8,** 86 (1969/70).
25. GREENE R., HOOP B. and KAZEMI H. *J. nucl. Med.* **12,** 719 (1971).
26. SHORE N. S., GREENE R. and KAZEMI H. *Environ. Res.* **4,** 512 (1971).
27. HALES C. A. and KAZEMI H. *N. Engl. J. Med.* **290,** 761 (1974).
28. HALES C. A., AHLUWALIA B. and KAZEMI H. *J. appl. Physiol.* **38,** 1083 (1975).
29. AUSTIN S. M., GALONSKY A., BORTINS J. and WOLK C. P. *Nucl. Instr. Meth.* **126,** 373 (1975).
30. THOMAS J., WOLK C. P., SHAFFER P. W., AUSTIN S. M. and GALONSKY A. *Biochem. Biophys. Res. Comm.* **67,** 501 (1975).
31. WOLK C. P., AUSTIN S. M., BORTINS J. and GALONSKY A. *J. Cell. Biol.* **61,** 440 (1974).
32. PARKS N. J., PEEK N. F. and GOLDSTEIN E. *Int. J. appl. Radiat. Isotopes* **26,** 683 (1975).
33. STRAATMANN M. G. and WELCH M. J. *Radiat. Res.* **56,** 48 (1973).
34. TILBURY R. S., DAHL J. R., MONAHAN W. G. and LAUGHLIN J. S. *Radiochem. Radioanal. Lett.* **8,** 317 (1971).
35. LATHROP K. A., HARPER P. V., RICH B. H., DINWOODIE R., KRIZEK H., LEMBARES N. and GLORIA I. In *Radiopharmaceuticals and Labelled Compounds*, Vol. 1. IAEA, Vienna, 471 (1973).
36. GELBARD A. S., HARA T., TILBURY R. S. and LAUGHLIN J. S. In *Radiopharmaceuticals and Labelled Compounds*, Vol. 1. IAEA, Vienna, 239 (1973).
37. VAALBURG W., KAMPHUIS J. A. A., BEERLING-VAN DER MOLEN H. D., REIFFERS S., RIJSKAMP A. and WOLDRING M. G. *Int. J. appl. Radiat. Isotopes* **26,** 316 (1975).
38. MONAHAN W. G., TILBURY R. S. and LAUGHLIN J. S. *J. nucl. Med.* **13,** 274 (1972).
39. HARPER P. V., LATHROP K. A., KRIZEK H., LEMBARES N., STARK V. and HOFFER P. B. *J. nucl. Med.* **13,** 278 (1972).
40. HARPER P. V., SCHWARTZ J., BECK B. S., LATHROP K. A., LEMBARES N., KRIZEK H., GLORIA I., DINWOODIE R., MCLAUGHLIN A., STARK V. J., MAYORGA A. and BROOKS H. L. *Radiology* **108,** 613 (1973).
41. Unpublished data, Mallinckrodt Institute of Radiology, St. Louis, Missouri.
42. HOOP B., JR., SMITH T. W., BURNHAM C. A., CORRELL J. E., BROWNELL G. L. and SANDERS C. A. *J. nucl. Med.* **14,** 181 (1973).
43. CARTER C. C., LIFTON J. F. and WELCH M. J. *Neurology* **23,** 204 (1973).
44. WELCH M. J., EICHLING J. O., STRAATMANN M. G., RAICHLE M. E. and TER-POGOSSIAN M. M. In *Noninvasive Brain Imaging*, p. 25. Society of Nuclear Medicine, New York (1975).
45. TER-POGOSSIAN M. M., HOFFMAN E. J., WEISS E. S., COLEMAN R. E., PHELPS M. E., WELCH M. J. and SOBEL B. E. Proc. Conf. *Cardiovascular Imaging and Image Processing Theory and Practice—1975* **72,** 277 (1976).
46. PHELPS M. E., HOFFMAN E. J., COLEMAN R. E., WELCH M. J., RAICHLE M. E., WEISS E. S., SOBEL B. E. and TER-POGOSSIAN M. M. *J. nucl. Med.*, in press (1976).
47. HAZENBERG H. J. A., GIPS C. H., BEEKHUIS H., SCHUURMAN J. J. and VAALBURG W. In *Recent Progress in Ammonia Metabolism* (Edited by IMLER M., SZAM I.), Verlag Gerhard Witzstrock GmbH, Baden-Baden, Brussels, 31 (1974).
48. BUDINGER T. F., YANO Y. and HOOP B. *J. nucl. Med.* **16,** 429 (1975).
49. COHEN M. B., SPOLTER L., MACDONALD N., MASUOKA D. T., LAWS S., NEELY H. H. and TAKAHASHI J. In *Radiopharmaceuticals and Labelled Compounds*, Vol. 1. IAEA, 483 (1973).
50. GELBARD A. S., CLARKE L. P., MCDONALD J. M., MONAHAN W. G., TILBURY R. S., KUO T. Y. T. and LAUGHLIN J. S. *Radiology* **116,** 127 (1975).
51. COHEN M. B., SPOLTER L., MACDONALD N., CHANG C. C. and TAKAHASHI J. In *Radiopharmaceuticals* (Edited by SUBRAMANIAN G., RHODES B. A., COOPER J. F. and SODD V. J.), Society of Nuclear Medicine, New York, 184 (1975).
52. COHEN M. B., SPOLTER L., CHANG C. C., MACDONALD N. S., TAKAHASHI J. and BOBINET D. D. *J. nucl. Med.* **15,** 1192 (1974).

53. Gelbard A. S., Clarke L. P., Laughlin J. S. *J. nucl. Med.* **15,** 1223 (1974).
54. Pettit W. A., Mortara R. H., Digenis G. A. and Reed M. F. *J. Med. Chem.* **18,** 1029 (1975).
55. Gersberg R., Krohn K., Peek N. and Goldman C. R. *Science,* in press.
56. Welch M. J. and Straatmann M. G. *Radiochimica Acta* **20,** 124 (1973).
57. Stenstrom T. Ph.D. Thesis, Uppsala University, 1970.
58. Kasche V. *Radiochem. Radioanal. Lett.* **3,** 51 (1970).
59. Welch M. J. and Ter-Pogossian M. M. *Radiat. Res.* **36,** 580 (1968).
60. Ter-Pogossian M. M., Eichling J. O., Davis D. O. and Welch M. J. *J. Clin. Invest.* **49,** 381 (1970).
61. Welch M. J., Lifton J. F. and Ter-Pogossian M. M. *J. Lab. Compounds* **5,** 168 (1969).
62. Welch M. J., Lifton J. F. and Seck J. A. *J. Phys. Chem.* **73,** 3351 (1969).
63. Ter-Pogossian M. M., Eichling J. O., Davis D. O., Welch M. J. and Metzger J. M. *Radiology* **93,** 31 (1969).
64. Carter C. C., Eichling J. O., Davis D. O. and Ter-Pogossian M. M. *Neurology* **22,** 755 (1972).
65. Eichling J. O., Raichle M. E., Grubb R. L., Larson K. B. and Ter-Pogossian M. M. *Circ. Res.* **37,** 707 (1975).
66. Raichle M. E., Eichling J. O. and Grubb R. L. *Arch. Neurol.* **30.** 319 (1974).
67. Eichling J. O., Raichle M. E., Grubb R. L. and Ter-Pogossian M. M. *Circ. Res.* **35,** 358 (1974).
68. Raichle M. E., Eichling J. O., Straatmann M. G., Welch M. J., Larson K. B. and Ter-Pogossian M. M. *Am. J. Physiol.* **230,** 543 (1976).
69. Jones T., Chesler D. A. and Ter-Pogossian M. M. *Br. J. Radiol.* **49,** 339 (1976).
70. West J. B. *Prog. Atom. Med.* **2,** 39 (1968).
71. Kazemi H., Al Bazzaz F., Parsons E. F. and Hoop B., Jr. In *Dynamic Studies with Radioisotopes in Medicine,* IAEA, Vienna, 819 (1971).
72. Schmidt W. W., Hoop B., Jr. and Kazemi H. *J. appl. Physiol.* **35,** 655 (1973).
73. Jones T., Jones H. A., Rhodes C. G., Buckingham P. D. and Hughes J. M. B. *J. Clin. Invest.* **57,** 706 (1976).
74. Jones T., Levene D. L., Greene R. In *Dynamic Studies with Radioisotopes in Medicine.* IAEA, Vienna, 751 (1971).
75. Watson D. D., Gelband H., Tamer D. R., Janowitz W. R., Kenny P. J., Sankey R. R., Finn R. D., Hildner F. J., Greenberg J. J. and Gilson A. J., in press (1976).
76. Meyer P., Conf. 730301. Proc. Int. Conf. *Photonuclear Reactions and Applications* (Edited by Berman B. L.), 1065 (1973).

International Journal of Applied Radiation and Isotopes, 1977, Vol. 28, pp. 21–24. Pergamon Press. Printed in Northern Ireland

A System for Oxygen-15 Labeled Blood for Medical Applications

RAJESHWARI SUBRAMANYAM, WILLIAM M. BUCELEWICZ, BERNARD HOOP, Jr. and STEPHEN C. JONES

Physics Research Laboratory, Massachusetts General Hospital, Boston, MA 02114, U.S.A.

(*Received* 20 *May* 1976)

Oxygen-15 labeled compounds in blood have been used successfully for cerebral circulation and cerebral oxygen metabolism measurements. The present paper describes a system for the rapid sequential production of ^{15}O-HgB, $C^{15}O$-Hgb and $H_2{}^{15}O$ in blood under sterile and pyrogen-free conditions. A tonometer has been adopted for labeling blood without hemolysis and foam production.

INTRODUCTION

Radioactive oxygen-15 (^{15}O) is the longest lived positron-emitting radionuclide of oxygen ($T_{1/2} = 2$ min).[1,2] Oxygen-15 labeled gaseous molecular oxygen ($^{15}O_2$), carbon monoxide ($C^{15}O$) and carbon dioxide ($C^{15}O_2$) administered by inhalation have been used for clinical investigations of pulmonary and cardiac malfunctions.[3–5] Regional cerebral blood flow and regional oxygen utilization rates have been determined successfully by extracranial monitoring of radioactivity following the intravascular administration of labeled oxyhemoglobin (^{15}O-Hgb) and water ($H_2{}^{15}O$) in blood.[6,7] Such studies often require rapid sequential preparation of short-lived radiopharmaceuticals with high specific activities. This necessitates an efficient labeling system from which labeled compounds suitable for human use are made available directly.

Previously described methods for labeling involve bubbling the radioactive gases through blood.[6] It was found in our laboratory that this procedure causes intense frothing and hemolysis of thc blood. To overcome these disadvantages, a tonometer has been adopted for the purposes of labeling and the method has proven to be simple and efficient.

The present paper describes a labeling system for the production of ^{15}O-Hgb, $H_2{}^{15}O$ and $C^{15}O$-Hgb in whole blood under sterile and pyrogen-free conditions.

PROCEDURE AND RESULTS

Production of oxygen-15

Oxygen-15 is produced by deuteron irradiation of gaseous nitrogen containing 2–3% oxygen via the $^{14}N(d, n)^{15}O$ reaction.[8–10] The radiolabel scavenged by oxygen is extracted as $^{15}O_2$ under continuous flow of the target gas. Typical yields of gaseous $^{15}O_2$ activity are about 200 mCi/l of the target gas at a flow rate of 0.5 l/minute and a deuteron beam current of 50 μA.

The irradiated gas is directed into the labeling system (Fig. 1) which differs in several respects from that described elsewhere.[6] It has three loops, two of which contain charcoal furnaces heated to suitable temperatures for conversion of $^{15}O_2$ to $C^{15}O$ and $C^{15}O_2$. The use of independent loops eliminates contamination and permits easy switching from one labeled gas to another via solenoid valves. The labeled gas emerging from a loop can be returned to the cyclotron target box directly or through a tonometer for further irradiation. Alternately, it can be piped to the positron camera laboratory for studies involving inhalation of these gases.

In order to obtain high specific activity of labeled blood, higher levels of $^{15}O_2$ activity are required. This is accomplished by minimizing the total volume of the labeling system and recirculating a fixed volume of gas through the target box under continuous bombardment.

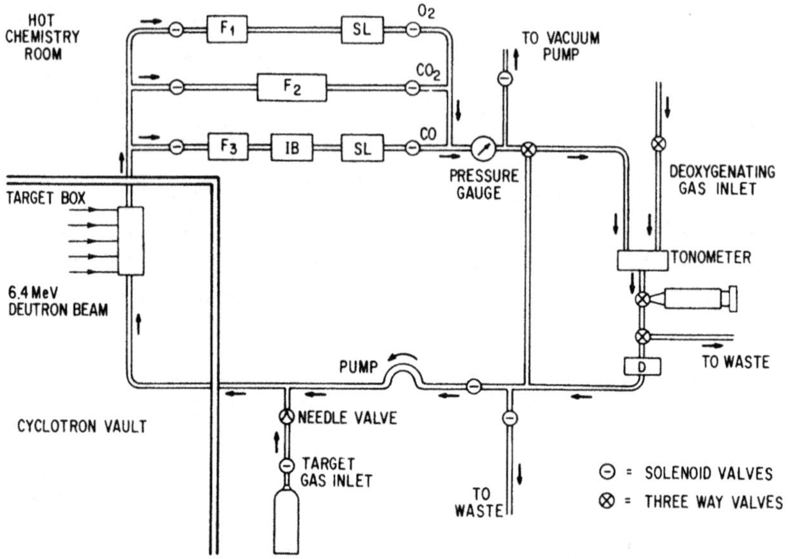

FIG. 1. A schematic diagram of the blood labeling system. IB is an ice bath and F_1, F_2 and F_3 are activated charcoal furnaces at room temperature, 600 and 1000°C respectively.

Total volume of $^{15}O_2$ and $C^{15}O$ loops are 670 and 970 ml respectively.

Radiochemical quality control of labeled $^{15}O_2$, $C^{15}O$ and $C^{15}O_2$ is performed routinely by radio-gas-chromatography. A frequent change of soda-lime traps and cleaning of the system has been found to be mandatory for high $C^{15}O$ yields.

^{15}O-*Hemoglobin*

The hemoglobin is easily labeled with ^{15}O by exposing heparinized blood to $^{15}O_2$; however careful handling of the blood is needed to prevent denaturation. Instead of bubbling the gas directly through blood which frequently causes hemolysis, it is preferable to pass the gas over a large surface area of blood. A IL-237 tonometer, (Instrumentation Laboratory, Lexington, MA 02173) which is usually used for deoxygenating blood has been adopted for this purpose. The tonometer (Fig. 2) consists of a cylindrical glass vessel placed over a distilled water pool. A thermostated water bath surrounding the vessel insures that the blood remains at the preselected temperature (between 37 and 40°C). Hemolysis of blood due to drying during deoxygenation is prevented as the deoxygenating gas is humidified by bubbling through the distilled water. The gas inlet used for deoxygenating gas cannot also be used for the radioactive gases because the activity will be lost in the distilled water pool due to their solubilities. Therefore, a special lucite lid with an inlet and an outlet has been built to fit the tonometer.

During gas flow the tonometer is operated on a 3 second cycle; the glass vessel is rotated for $1\frac{1}{2}$ sec forcing the blood outward from the center of the sample vessel to form a thin fresh film on the walls. After the rotation stage, the vessel stops for $1\frac{1}{2}$ sec during which time the

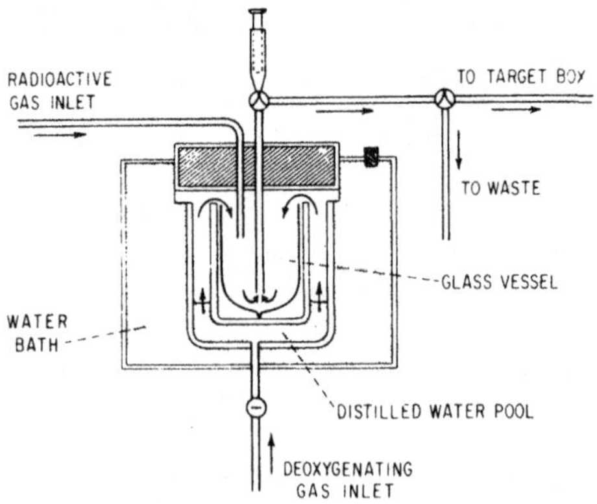

FIG. 2. A schematic diagram of the tonometer consisting of a sample vessel, water bath and gas inlets–outlets.

blood slides back from the walls into the center. This procedure causes the labeling of the blood to occur without hemolysis or foam production.

On passing the $^{15}O_2$ over blood, the hemoglobin reacts with oxygen to form labeled oxyhemoglobin. Specific activity of blood can be increased if all the non-radioactive oxygen atoms are removed first by deoxygenating blood with gaseous nitrogen containing 9% CO_2. The addition of 9% CO_2 to the deoxygenating gas is to adjust PCO_2 value to that in arterial blood. The increase in specific activity of blood is approximately 100% if a deoxygenating period of 20 min or more is used.

$C^{15}O$-*Hemoglobin*

Labeling of hemoglobin with $C^{15}O$ is similar to that already described for ^{15}O-Hgb. Labeled molecular oxygen is first converted to $C^{15}O$ by passing it over activated charcoal heated to 1000°C. An ice bath and a soda lime trap are also included in the loop. The former rapidly cools the target gas preventing conversion of $C^{15}O$ to $C^{15}O_2$ by further reaction with O_2 and the latter removes traces of contaminant CO_2. Deoxygenated blood is exposed to $C^{15}O$ for several minutes under continuous irradiation and recirculation.

$H_2{}^{15}O$ *in blood*

The production of $H_2{}^{15}O$ in blood is the simplest of the three labeling procedures. $^{15}O_2$ is first converted to $C^{15}O_2$ by passing it over activated charcoal heated to 600°C, and then over blood in the tonometer. The fast exchange of ^{15}O between H_2O and CO_2 is assisted by the enzyme action of carbonic anhydrase present in the blood and leads to the formation of labeled water.

A typical production sequence proceeds as follows—the recirculating system and the target box is first evacuated and then slowly filled with the target gas. A steady flow of gas via one of the three loops to waste is maintained while the cyclotron is turned on and tuned to obtain a steady deuteron beam. There upon the gas inlet and waste lines are closed to form a closed loop system. The gas is recirculated under continuous bombardment bypassing the tonometer while the blood is being deoxygenated. After 6 min, the active gas is recirculated through the tonometer for about 3 min. The labeled blood is withdrawn by a syringe. Typical activities obtained are 1 mCi/ml and 0.5 mCi/ml for ^{15}O-Hgb and C^{15}O-Hgb respectively. In case of $H_2{}^{15}O$ a 10 second exposure to active gas yields 1 mCi/ml of blood.

DISCUSSION

The labeling system described has been in use over a period of one year for the routine production of labeled compounds for pulmonary function and cerebral circulation studies in an animal model.[11,12]

The tonometer is used for the dual purpose of deoxygenating and labeling; it permits labeling of blood-volumes ranging from 0.5 to 8 ml without hemolysis. In addition, the system has been shown to produce pyrogen-free labeled compounds.

Acknowledgements—We wish to thank Drs. D. J. Hnatowich and G. L. Brownell for their helpful discussions and encouragement throughout the project. This work was supported by Public Health Project POI-NS10828.

REFERENCES

1. *Nuclear Data Tables*, Part 4. U.S. Atomic Energy Commission, Government Printing Office, Washington, D.C. (1960).
2. West J. B. In *Progress in Atomic Medicine* (Edited by Lawrence John H.), Vol. 1, p. 39. Grune & Stratton, New York (1965).
3. Dollery C. T. and West J. B. *Circulation Res.* **8**, 765 (1969).
4. Dyson N. A., Hugh-Jones P., Newberry G. R., Sinclair J. D. and West J. B. *Br. med. J.* **1**, 231 (1960).
5. Dyson N. A., Hugh-Jones P., Newberry G. R. and West J. B. Proceedings of the second international conference: United Nations Peaceful Uses of Atomic Energy, Geneva, Sep. 1958, Pergamon Press, New York (1959).
6. Ter-Pogossian M. M., Eichling J. O., Davis D. O., Welch M. J. and Metzer J. M. *Radiology* **93**, 31 (1969).
7. Ter-Pogossian M. M., Eichling J. O., Davis D. O. and Welch M. J. *J. Clin. Invest.* **49**, 381 (1970).
8. Welch M. J. and Ter-Pogossian M. M. *Radiat. Res.* **36**, 580 (1968).
9. Clark J. C., Matthews C. M. E., Sylvester D. J. and Vonberg D. D. *Nucleonics* **25**, 54 (1962).

10. Clark J. C. and Buckingham. In *Short-lived Radioactive Gases for Clinical Use.* Butterworth, London (1975).
11. Ahluwalia B., Kanarek D., Jones T., Hoop B., Brownell G. L. and Kazemi H. *J. nucl. Med.* **15,** 474 (1974).
12. Kanarek D., Ahluwalia B., Latty A., Erdman J., Kazemi H. and Shannon D. C. *Fed. Proc.* **35,** 792 (1976).

International Journal of Applied Radiation and Isotopes, 1977, Vol. 28, pp. 25–28. Pergamon Press. Printed in Northern Ireland

Production of ^{13}N-Molecular Nitrogen for Pulmonary Positron Scintigraphy

STEPHEN C. JONES, WILLIAM M. BUCELEWICZ, ROBERT A. BRISSETTE, RAJESHWARI SUBRAMANYAM and BERNARD HOOP, JR.

Physics Research Laboratory, Massachusetts General Hospital, Boston, MA 02114, U.S.A.

(*Received* 20 *May* 1976)

A system for the production of ^{13}N labeled molecular nitrogen (^{13}NN) dissolved in physiological saline is described. The $^{12}C(d, n)^{13}N$ reaction with 50 μA of 6 MeV deuterons from a small medical cyclotron is used to produce ^{13}NN as the sole product in the target chamber. Carbon dioxide is both the target and the sweep gas in a flow system that terminates with continuous absorption of CO_2 in an NaOH bubbler. The specific activity of ^{13}NN is raised in relation to its relative volume as impurity in the CO_2. After CO_2 absorption, the residual ^{13}NN is shaken with physiological saline to produce 20 ml of sterile, pyrogen-free injectate with a specific activity of 270 μCi/ml. Radio-gas-chromatography is used for the development and quality control of the technique. ^{13}NN in solution is used for measurements of pulmonary function with positron scintigraphy.

INTRODUCTION

MOLECULAR nitrogen labeled with ^{13}N ($T_{1/2}$= 9.96 min) is a useful radiopharmaceutical agent for the clinical study of lung air volume, blood flow, ventilation and airway closure. Nitrogen-13 labeled N_2 (^{13}NN) may be introduced intravascularly as dissolved gas in physiological solution or by inhalation. Due to its low solubility in blood and tissue, the distribution of ^{13}NN in the body is confined largely to the air volume of the lungs. Administered intravenously, dissolved ^{13}NN is transported to the lungs where it diffuses from the pulmonary capillary bed into the lung air spaces. The amount of ^{13}NN appearing in a given region of the lung during a breath hold is proportional to blood flow to that region. Similarly, the amount of ^{13}NN appearing in a given region of the lung during inhalation of gaseous ^{13}NN is related to ventilation of that region. Inhalation of gaseous ^{13}NN can also be used to measure airway closure.

Since its introduction for the measurement of regional ventilation at Hammersmith Hospital, London,[1] ^{13}NN has been used in the measurement of airway closure,[2–4] in the analysis of regional shifts in perfusion due to regional alveolar hypoxia,[5] and for the measurement of pulmonary perfusion volume and ventilation in patients with a variety of pulmonary diseases.[6,7]

In these applications, the principal advantage of ^{13}NN is the lower solubility of N_2 in blood and tissue compared to that of ^{133}Xe, the most widely used gaseous agent for pulmonary studies.[8,9] A smaller amount of ^{13}NN is dissolved in plasma and tissue (especially adipose tissue).

A further advantage of ^{13}NN is its short physical half-life; ^{13}N poses a less severe personnel radiation hazard than does ^{133}Xe. In addition, ^{13}N decays almost entirely by positron emission (with an electron capture intensity = 0.2%) and may therefore be imaged with positron scintigraphic techniques.[10]

Although a variety of charged-particle induced nuclear reactions for the production of ^{13}N for medical applications have been explored, the $^{12}C\ (d, n)^{13}N$ production reaction ($Q = -0.3$ MeV) has proved particularly attractive.[11] The total reaction cross section is of the order of 0.1 barn for deuteron energies between 1 and 15 MeV.[12–15]

Early production methods for gaseous ^{13}NN

using the $^{12}C(d, n)^{13}N$ reaction have been described.[16,17] ^{13}NN in solution has been produced by BUCKINGHAM and CLARK using 15.4 MeV deuterons on graphite with CO_2 sweep gas[18] and 6.3 MeV deuteron irradiation of CO_2.[19] A syringe system employing a shaking mechanism and a buffer solution reservoir was used to dissolve the ^{13}NN in solution.[18,19]

This report describes an alternative method of preparing ^{13}NN in solution following 5.8 MeV deuteron irradiation of CO_2, with continuous absorption of CO_2 from the irradiated gas. The system described results in high absolute and specific activities of ^{13}NN in solution.

The system described herein is part of a radioactive gas processing system for the production of a variety of short-lived radioactive gases and simple labeled compounds. In the present case, the target chamber employed is also suitable for the production of ^{15}O labeled gases and consequently switching from one radioactive gas to another may be accomplished more conveniently and rapidly.

METHOD OF PRODUCTION

Nitrogen-13 labeled molecular nitrogen is produced in the Massachusetts General Hospital Cyclotron via the $^{12}C(d, n)^{13}N$ reaction by deuteron irradiation of gaseous carbon dioxide. A schematic diagram of the production system is shown in Fig. 1. A 6.4 MeV deuteron beam is directed through a 0.025 mm aluminum foil window into a 0.2 l reactangular gas target chamber of depth 16 cm. Gaseous CO_2 is passed through the target chamber at 0.5 psig, degrading the deuteron beam from 5.8–3.5 MeV.[20] Beam currents are maintained below 50 μA in order to minimize the possibility of foil rupture. Both high purity CO_2 (99.99% CO_2) and technical grade CO_2 (~99% CO_2) have been used. Recoil ^{13}N is scavenged principally by molecular nitrogen present as an impurity in the target gas and carried by the target gas from the chamber. The activated gas is conducted through 10 m of 3 mm i.d. polyethylene tubing to a tube furnace containing 4–8 mesh Cu metal beads in a 2 × 18 cm quartz tube at 900°C. The Cu metal beads are pretreated with O_2 at 0.5 l min for one hour to oxidize their surface. The CuO furnace serves to re-oxidize radiolytically produced CO to CO_2 and is useful for about five hours of production time. A modified 2 l gas washing bottle containing 6N NaOH is used to absorb the CO_2. During 10–15 min of continuous irradiation, CO_2 is absorbed in the NaOH. A residual gas bubble consisting mainly of air impurities and including ^{13}NN collects at the top of the gas washing bottle. The gas bubble is drawn into a sterile syringe containing 5–20 ml of sterile de-gassed physiological saline which has been cooled to approximately 0°C to increase dissolution of ^{13}NN.

A typical production sequence proceeds as follows: The target box and conduit lines are purged with CO_2 via valve V1 to remove previous target gases. The CO_2 flow rate is adjusted to about 35 ml/min and irradiation with 30–50 μA deuterons commences. The irradiation is

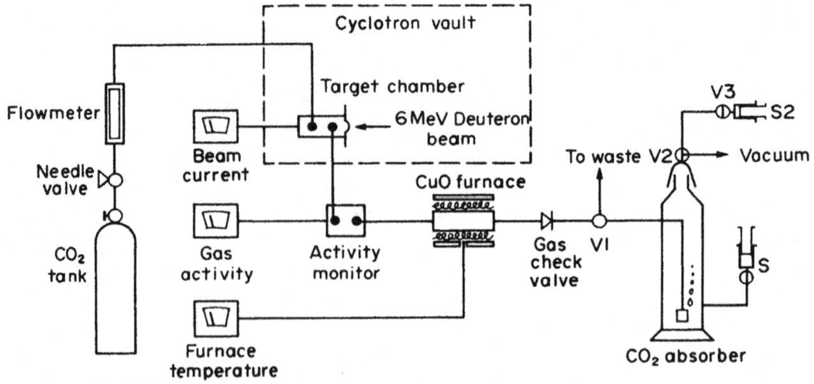

FIG. 1. A schematic diagram of the section of the gas handling system devoted to the production of ^{13}NN in solution.

continued for at least 20 min. When activity is needed, the washing bottle is completely filled with NaOH from the reservoir syringe S and the irradiated gas is then shunted from waste into the NaOH bubbler, via valve V1. Typically, a 20 min collection results in the accumulation of 30 mCi ^{13}NN in one ml of residual gas (employing technical grade CO_2). Occasionally, the gas flow is increased momentarily to flush the activity in the gas line from the cyclotron into the NaOH absorber and to increase the activity in the residual gas bubble. After the line between V2 and V3 has been evacuated, the collected gas is transferred via V2 and V3 to syringe S2 containing physiological saline. The syringe is shaken for a few seconds to dissolve the ^{13}NN and the remaining ^{13}NN is withdrawn into a storage syringe for decay or subsequent use. The pH of the saline solution is then checked for neutrality before clinical use. Approximately 4 mCi of ^{13}NN are dissolved in 15 ml of saline (270 μCi/ml).

RADIOCHEMICAL ANALYSIS

Analysis of the irradiated CO_2 and of the residual gas after CO_2 absorption was performed, using conventional radio-gas-chromatography. A 2 m, 30/80 mesh, 5A molecular sieves column (Perkin-Elmer Corp.) and a 3 m, 60/80 mesh Porapak-Q column (Waters Associates) were used in a Perkin-Elmer 154D gas chromatograph at room temperature with thermal conductivity detection for chemical analysis and with He carrier gas. Fractionated gases from the chromatograph were passed through a shielded loop viewed by a NaI detector for radiochemical analysis. In samples of the irradiated CO_2, essentially all of the activity within experimental error was ^{13}N labeled molecular nitrogen. No stable or labeled NO or NO_2 were detected before or after CO_2 absorption.

The irradiated CO_2 contained 1% CO as a result of equilibrium between radiolysis of CO_2 to CO and subsequent reconversion of CO to CO_2.[21,22] The CuO furnace converts the CO to CO_2 so it can be absorbed by the NaOH; otherwise the volume it represents in the residual gas bubble descreases the specific activity of the bubble and hence, the specific activity of dissolved ^{13}NN.

Recoil $^{13}N^*$ interacts rapidly with N_2 impurity in the target gas:[23–26] $^{13}N^* + N_2 \rightarrow {}^{13}NN + N$. At low target radiation doses, obtained with lower beam currents or faster gas flow rates, a maximum of 1% of ^{13}N activity was associated with an undetectable chemical amount of N_2O via a mechanism such as $^{13}NN + CO_2 \rightarrow {}^{13}NNO + CO$.[23] However, $^{13}N_2O$ is largely decomposed by the high average radiation dose (>20 eV/molecule) delivered by the deuteron beam under the irradiation conditions employed.[26]

RESULTS AND DISCUSSION

Using the excitation function of Brill and Sumin[13], and the d*E*/d*x* tables of Williamson *et al.*,[28] the saturation activity for our target system was calculated to be 4.46 mCi/μA, in excellent agreement with the measured value of 4.5 mCi/μA.

Although the production of ^{13}NN in solution is a batch process, the absorption of CO_2 and the simultaneous concentration of ^{13}NN are performed continuously. The specific activities obtained with this system are 35% higher than those reported by Clark and Buckingham[19]. Specific activities of 270 μCi/ml are obtained routinely in 20 ml of physiological saline. The minimum time necessary from cyclotron start-up to radiopharmaceutical delivery is 30 minutes.

Radiochemical quality control of the gaseous ^{13}NN is performed routinely with radio-gas-chromatography. In addition, the technique has been shown to produce a pyrogen-free and sterile injectate. A Notice of Claimed Investigational Exemption for a New Drug (IND) was filed with the United States Food and Drug Administration and IND number 7675 for human use of the ^{13}NN produced by this technique was assigned on 23 February, 1971. The technique has been used since September, 1972 for the routine delivery of ^{13}NN in gaseous form and in solution to approximately 210 patients.

The principal application of ^{13}NN at our laboratory has been the routine clinical measurement of airway closure[2,3,4] and ventilation and perfusion[6] with the MGH positron camera.[29]

Acknowledgements—The authors wish to thank Drs. R. GREENE, C. A. HALES and H. KAZEMI for their expression of clinical need and Drs. G. L. BROWNELL and D. J. HNATOWICH for their perceptive comments. The work was supported by USPHS grant HE-06664 and US ERDA contract AT(11-1) 3333. One of us (S. C. JONES) wishes to acknowledge the support of the Health Sciences Fund of M.I.T.

REFERENCES

1. MATTHEWS C. M. E., DOLLERY C. T., CLARK J. C. and WEST J. B. In *Radioactive Pharmaceuticals* (Edited by ANDREWS G. A., KNISELEY R. M. and WAGNER H. N., JR.), p. 567. U.S. Atomic Energy Commission, Oak Ridge, Tenn. (1966) (CONF-651111).
2. GREENE R., HOOP B. JR. and KAZEMI H. *J. nucl. Med.* **12,** 719 (1971).
3. SHORE N. S., GREENE R. and KAZEMI H. *Environ. Res.* **4,** 512 (1971).
4. HALES C. A. and KAZEMI H. *New Engl. J. Med.* **290,** 761 (1974).
5. HALES C. A., AHLUWALIA B. and KAZEMI H. *J. appl. Physiol.* **38,** 1083 (1975).
6. AHLUWALIA B., HALES C., KAZEMI H. and BROWNELL G. L. Positron camera imaging for the assessment of lung function. *1st World Cong. Nucl. Med.* Oct., Tokyo (1974).
7. ROSENZWEIG D. Y., HUGHES J. M. B. and JONES T. *Respir. Physiol.* **8,** 86 (1969).
8. KAZEMI H., PARSON E. F., VALENCA L. M. *et al. Circulation* **41,** 1025 (1970).
9. BALL W. C., STEWART P. B., NEWSHAM L. G. S. *et al. J. clin. Invest.* **41,** 519 (1962).
10. BROWNELL G. L. and BURNHAM C. A. In *Instrumentation in Nuclear Medicine* (Edited by HINE G. J. and SORENSON J. A.), Vol. 2. Academic Press, New York (1974).
11. HOOP B., JR., LAUGHLIN J. S. and TILBURY R. S. In *Instrumentation in Nuclear Medicine* (Edited by HINE G. and SORENSON J.), Vol. 2, p. 407. Academic Press, New York (1973).
12. WILKINSON D. H. *Phys. Rev.* **100,** 32 (1955).
13. BRILL O. D. and SUMIN L. V. *Atom. Energ.* **7,** 377 (1959).
14. WOHLLEBEN K. and SCHUSTER E. *Radiochim. Acta* **8,** 78 (1967).
15. JASZCZAK R. J., MACKLIN R. L. and GIBBONS J. H. *Phys. Rev.* **181,** 1428 (1969).
16. BUCKINGHAM P. D. and FORSE G. R. *Int. J. appl. Radiat. Isotopes* **14,** 439 (1963).
17. NICHOLAS D. J. D., SILVESTER D. J. and FOWLER J. F. *Nature, Lond.* **189,** 634 (1961).
18. BUCKINGHAM P. D. and CLARK J. C. *Int. J. appl. Radiat. Isotopes* **23,** 5 (1972).
19. CLARK J. C. and BUCKINGHAM P. D. *Short-lived Radioactive Gases for Clinical Use*, p. 171. Butterworths, London (1975).
20. EVANS R. D. *The Atomic Nucleus,* p. 653. McGraw Hill, New York (1955).
21. ANDERSON A. R. and BEST J. U. F. *Advances in Chemistry* **82,** 231 (1968).
22. DOMINEY D. A. and PALMER T. F. *Dis. Farad. Soc.* **36,** 35 (1963).
23. STATNICK R. M., KASHIHIRA N. and SCHMIDT-BLEEK F. *J. Inorg. Nucl. Chem.* **31,** 878 (1969).
24. WELCH M. J. *Chem. Commun.* **21,** 1354 (1968).
25. ANBAR M. *J. Phys. Chem.* **70,** 2052 (1966).
26. SCHMIED H. and KOSKI W. S. *Am. Chem. Soc. J.* **82,** 4766 (1960).
27. JONES F. T. and SWORSKI T. J. *J. Phys. Chem.* **70,** 1546 (1966).
28. WILLIAMSON C. F., BOUJOT J. P. and PICARD J. Rapport CEA-R3042 (1966).
29. BURNHAM C. A. and BROWNELL G. L. *IEEE Trans. Nucl. Sci.* **19,** 201 (1972).

International Journal of Applied Radiation and Isotopes, 1977, Vol. 28, pp. 29–48. Pergamon Press. Printed in Northern Ireland

Carbon-11 and Radiopharmaceuticals*

ALFRED P. WOLF and CAROL S. REDVANLY†
Chemistry Department, Brookhaven National Laboratory, Upton, NY 11973, U.S.A.

(*Received* 21 *May* 1976)

INTRODUCTION

AN UNIDENTIFIED light element radionuclide with a half-life of about 20 min was first observed in 1934 in a target of boron oxide which had been bombarded with 0.9 MeV "deutons" for 15 min at 5–10 μA.[1] Two of the authors of that report, LAURITSEN and CRANE[2a] subsequently identified this new nuclide as carbon-11, and in the same year[2b] prepared the same nuclide by proton bombardment of boron oxide. They stated that most of the carbon-11 was as ^{11}CO or $^{11}CO_2$ although their experimental evidence indicated that what they had collected was $^{11}CO_2$. The nuclear reactions suggested as being responsible were $^{10}B(d, n)^{11}C$ and $^{11}B(p, n)^{11}C$. Decay occurred by positron emission. It remained to YOST, RIDENOUR and SHINOHARA[3] to demonstrate that the result of the deuteron bombardment of B_2O_3 at 0.9 MeV gave carbon-11 via the $^{10}B(d, n)^{11}C$ reaction, that the half-life of carbon-11 was ~20.5 min and that the carbon-11 gases produced were in fact a mixture of ^{11}CO and $^{11}CO_2$. It is perhaps interesting to note in these days of multichannel analyzers, computerized positron tomography and the other wonders of technology that: "The presence, amount, and half-life of the radioactive material were determined by quartz-fiber electroscopes constructed and kindly loaned by Professor LAURITSEN, or by means of a Geiger counter whose cylindrical electrode was of aluminum about 0.2 mm thick."[3] Five years after the discovery of carbon-11 the $^{14}N(p, \alpha)^{11}C$ reaction was described by BARKAS[4]. Thus, by 1939 the three still most commonly used reactions, the $^{10}B(d, n)^{11}C$, $^{11}B(p, n)^{11}C$ and $^{14}N(p, \alpha)^{11}C$, for producing carbon-11 had been discovered.

Practical methods for producing carbon-11 led almost immediately to the use of this nuclide in chemical and biological research. LONG[5] was the first investigator to publish work utilizing this nuclide in research in his study of the exchange between "chromioxalate" ion and oxalate ion. Shortly thereafter, RUBEN, HASSID and KAMEN[6] used $^{11}CO_2$ in a photosynthesis study (the first implicit biosynthesis of carbon-11 labeled compounds). A purely synthetic approach to using this nuclide appeared in an article by CRAMER and KISTIAKOWSKY[7,cf8] on lactic acids labeled in the 1, 2 and 3 positions. In order to accomplish these syntheses CN from $^{11}CO_2 + K + NH_3$, $HC{\equiv}^{11}CH$ from BaC_2—^{11}C and $^{(11)}CH_3{}^{11}CHO$ were prepared as precursors. These labeled lactic acids were subsequently used in biological studies.[9] KAMEN and RUBEN[10] reported on the use of bacteria to prepare carbon-11 labeled acetic, propionic, and succinic acids and *methane*!

A number of applications of carbon-11 to biological research followed[11–14] and the subject was reviewed by BUCHANAN and HASTINGS[15] in an article titled "The Use of Isotopically Marked Carbon in the Study of Intermediary Metabolism."

The first human experiments[16,cf also 17] using carbon-11 were carried out in 1945 by TOBIAS *et al.* and involved an investigation of the elimination of ^{11}CO by humans with specific emphasis on the possible conversion of ^{11}CO to $^{11}CO_2$. These authors, using about one μCi per experiment, were able to show that there was ["less than 0.1% (if any)"] essentially no conversion of ^{11}CO to $^{11}CO_2$ *in vivo* under their conditions, a fact confirmed by experiments[18] carried out 30 years later using a more sophisticated approach.

The study and use of carbon-11 entered a new and more diffuse phase after 1946. The classic

* By acceptance of this article, the publisher and/or recipient acknowledges the U.S. Government's right to retain a non-exclusive, royalty-free license in and to any copyright covering this paper.

† Research performed under the auspices of the U.S. Energy Research and Development Administration.

text of KAMEN[19] *Radioactive Tracers in Biology* is indicative of the change brought about by the increasing availability of carbon-14. The first and second editions[19] devoted full chapters to carbon-11. The third edition[19] called *Isotopic Tracers in Biology* combined carbon-11, carbon-13 and carbon-14 in one large chapter with most of the emphasis being on carbon-14. Carbon-11 disappeared from published use in purely biological experiments and began to find application in medical problems following the completion of the cyclotron at Hammersmith Hospital in London, England.[cf 20, 130, etc]

Basic research in the chemistry of carbon-11 atoms [Hot Atom Chemistry] began in the post war period and has been extensively reviewed by WOLF[21], WOLFGANG[22] and MACKAY[23]. This area will not be covered in this review except where it impacts directly on Nuclear Medicine and radiopharmaceuticals.

Carbon-11 began to receive increasing attention as a useful nuclide in medical application in the sixties with the greatest number of publications appearing between 1970 and the present.

This renewed effort is primarily due to the increasingly widespread installation of medical cyclotrons and the conversion of cyclotrons which had been used for research in physics and chemistry to medical application. In more recent times there has been renewed interest in photonuclear reactions, and the use of other accelerators such as the Van de Graaff is discussed in relation to carbon-11 production.

A list of the topics to be covered in this review follows. The main emphasis, however, will be on production of carbon-11 and synthesis of precursors and labelled organic compounds.

Sections:

(1) The literature.
(2) Nuclear reactions used for carbon-11 production.
(3) Preparation of simple precursors for synthesis: ^{11}CO, $^{11}CO_2$ and ^{11}CN.
(4) Synthetic and biosynthetic methods.
(5) "Non-synthetic" techniques.
(6) Miscellaneous topics.
(7) Medical uses.

1. THE LITERATURE

A review on an interdisciplinary subject such as this poses special problems. Literature references were obtained from Chemical Abstracts, Nuclear Science Abstracts, Nuclear Medicine Abstracts and Biological Abstracts. One of the peculiarities of this multidisciplinary area is the large number of reviews containing reference to carbon-11. The total literature of this area comprises less than 200 references at the present time, yet one runs into the paradoxical situation of reviews quoting mainly . . . reviews. The reviews containing information on carbon-11 fall into four categories: (1) Medical applications of accelerator produced nuclides, including reviews on imaging devices. (2) Accelerators in nuclear medicine. (3) Nuclide production by accelerators. (4) Misc. areas such as nuclear chemistry, labeling, etc.

A review by TER-POGOSSIAN[24] covers the general literature through 1965. The following year an overview of cyclotrons and short-lived nuclides was published by TER-POGOSSIAN and WAGNER[25]. A review that can serve as a primer on cyclotrons, nuclide production and target systems has been written by LAUGHLIN, TILBURY and DAHL[26]. A complete listing of carbon-11 compounds through 1973 can be found in the review of WOLF, CHRISTMAN, FOWLER and LAMBRECHT[27]. Reviews not referenced in the text but which mention carbon-11 are listed in Appendix I. Another problem arises with the large number of abstracts, the majority of which appear in the *Journal of Nuclear Medicine*. In some cases provocative new material appears in these abstracts which never seems to find its way into the chemical literature. Hence the work cannot be carried on by others or even repeated because of lack of sufficient detail. Another problem concerns reports referenced primarily in Nuclear Science Abstracts or Nuclear Medicine Abstracts. While many of these reports contain interesting new approaches and ideas they do not constitute publication in the open literature because of limited access to them by the public. They usually are available only at large universities or research centers in North America, Western Europe and Japan. As a consequence these reports will be referenced in Appendix II and the subject matter noted, but they will not be reviewed in the text.

This review covers the literature from the discovery of carbon-11 through 1975. A few publications that appeared in 1976 are included.

2. NUCLEAR REACTIONS USED FOR CARBON-11 PRODUCTION

Carbon-11 decays by positron emission with a $\beta+$ end point of 0.968 MeV[28] a half-life of 20.40 ± 0.04 min[29] and a $K\text{cap}/\beta^+$ ratio = 0.0023.[29] The positron can be detected directly or the nuclide may be assayed by utilizing the two 511 KeV photons created by annihilation of the positron.

The number of nuclear reactions[21,29] which result in the production of carbon-11 are indeed large and new reactions continue to appear in the literature, however, only the three listed below are in routine use for production purposes.

$$^{10}\text{B}(d, n)^{11}\text{C} \qquad ^{11}\text{B}(p, n)^{11}\text{C}$$
$$^{14}\text{N}(p, \alpha)^{11}\text{C}$$

Two other reactions deserve serious consideration, $^{12}\text{C}(^3\text{He}, ^4\text{He})^{11}\text{C}$ and $^{12}\text{C}(\gamma, n)^{11}\text{C}$, but as yet their potential remains to be developed. Spallation reactions could be used for production of very large quantities of carbon-11, however no experimental work has been published. The threshold energies for the routinely used reactions are low and all have appreciable cross sections up to particle bombardment energies of 30 MeV. [$^{10}\text{B}(d, n)^{11}\text{C}$[30,31] $^{11}\text{B}(p, n)^{11}\text{C}$[32,33] $^{14}\text{N}(p, \alpha)^{11}\text{C}$[34–39]]. The integrated cross sections for the energy ranges accessible on most multi-particle cyclotrons used in medical application are such that Curies of carbon-11 can be easily produced.

Yields are not given in Table 1 because of the considerable inconsistencies in published cross sections and integrated cross sections. It is nevertheless clear that the two most useful reactions, all things considered, and assuming availability of a multiparticle machine, are the $^{11}\text{B}(p, n)^{11}\text{C}$ and $^{14}\text{N}(p, a)^{11}\text{C}$ reactions. It is unfortunate that precisely determined cross sections are not available for the boron reactions since a careful reading of the literature indicates considerable confusion as to what the maximum carbon-11 yield might be under a given set of conditions, however, the situation with respect to the $^{14}\text{N}(p, \alpha)^{11}\text{C}$ reaction is considerably better. The first direct measurement of the excitation function by BLASER, MARMIER and SEMPERT[34] covered a narrow range of energies. Some ten years later, CACACE and WOLF[35] reported a crude excitation function in their attempts to locate the maximum of the function. Several groups continued efforts to determine the shape of the function[38,36,37] but it was not until the work of JACOBS *et al.*[38] that a definitive structure was obtained. CASELLA *al.*[39] have modified the structure slightly. Thus the calculated yields based on the work reported in [36,38,39] are within $\pm 10\%$ of the absolute yield. The relative merits of the boron and nitrogen reactions from the practical viewpoint are described in section 4.

Table 1 also lists the $^{12}\text{C}(^3\text{He}, ^4\text{He})^{11}\text{C}$ and $^{12}\text{C}(\gamma, n)^{11}\text{C}$ reactions because they are high yield reactions. Both of these reactions result in a product isotopic with the target. As yet no good way has been devised to obtain an essentially carrier-free product. The helium-3 reaction[40] gives an excellent yield and has been successfully applied in a recoil labeling[cf 66] experiment. However, the disadvantages of working with helium-3 and with a carbon containing target outweigh the slightly higher

TABLE 1. Carbon-11 production reactions

Reaction	Natural abundance	Threshold	Maximum cross section	Particle energy at maximum cross section
$^{10}\text{B}(d, n)^{11}\text{C}$	19.7%*	0	~200 mb	3.0–5.4 MeV†
$^{10}\text{B}(p, n)^{11}\text{C}$	80.3%	3.02 MeV	~360 mb	8.7–10 MeV†
$^{14}\text{N}(p, \alpha)^{11}\text{C}$	99.6%	3.13 MeV	~290 mb	~7.5 MeV
$^{12}\text{C}(^3\text{He}, ^4\text{He})^{11}\text{C}$	98.89%	0	~340 mb	~8.5 MeV
$^{12}\text{C}(\gamma, n)^{11}\text{C}$	98.89%	18.7 MeV	—	—

* There is a small contribution to the ^{11}C yield from the d, $2n$ reaction $^{11}\text{C}(d, 2n)^{11}\text{C}$, but the maximum cross section is only ~40 mb at ~10 MeV.

† Spread in published functions.

potential yield. The use of the $^{12}C(\gamma, n)^{11}C$ reaction first discussed by THEARD and GROCE[41] in the production sense and carefully delineated by WELCH[42] will be discussed in greater detail later in this review. The major difficulty here again is the necessity of using a carbon production target and the cumbersome nature of using a tungsten target bombarded with high energy electrons in order to effect the photonuclear reaction.

3. PREPARATION OF SIMPLE PRECURSORS FOR SYNTHESIS: ^{11}CO, $^{11}CO_2$ AND ^{11}CN

^{11}C-carbon monoxide, ^{11}C-carbon dioxide and ^{11}C-cyanide as $H^{11}CN$ or $Na^{11}CN$ are the three most commonly used carbon-11 precursors in the synthesis of compounds containing carbon-11. Other simple compounds such as ^{11}C-formaldehyde, ^{11}C-methanol, ^{11}C-acetylene etc. will be discussed in section 4.

The bombardment of B_2O_3 targets containing natural abundance boron or isotopically enriched boron (an advantage only if one is using the $^{10}B(d, n)^{11}C$ reaction) leads to the facile production of ^{11}CO and $^{11}CO_2$. A basic target for this purpose was described by KAMEN[19] for deuteron bombardment. Essentially the same target can be used for proton bombardment with a minor change in the window. BUCKINGHAM and FORSE[43, cf 44] improved this target design by coating a stepped wedge with B_2O_3 so that the B_2O_3 melts during bombardment. The rate of disappearance from the beam is slowed and bombardment times can be extended.

The yield of ^{11}CO and $^{11}CO_2$, appearing in the flushing gas in a dynamic system, can be optimized by regulating the flow or evaporation of the B_2O_3 in the incident beam. This is done by balancing the rate of cooling against the energy deposited by the incident beam. Construction materials are particularly important in targets used in deuteron bombardments because of activation problems. Another version of the stepped target appears in VONBERG *et al.*[45,46,47] Further comment on target problems can be found in WELCH and TER-POGOSSIAN[48] and HOOP, LAUGHLIN and TILBURY.[49] WELCH and TER-POGOSSIAN[48] using a conventional target and deuteron bombardment studied the effect of carrier gas and other parameters on obtaining a high yield of ^{11}CO. A similar study for both ^{11}CO and $^{11}CO_2$ production was later carried out by CLARK and BUCKINGHAM[46,47].

The primary product resulting from carbon-11 atom reactions in oxygen containing media is ^{11}CO.[50,51,21,22,23] The $^{11}CO_2$ which is observed in varying degree and described in the many articles on the B_2O_3 target is formed almost exclusively by radiolysis of the ^{11}CO in the presence of oxygen containing substrates.[52,53] Indeed ELIAS and WOLF[54] were able to show that the $^{11}CO_2/^{11}CO$ ratio observed in gases from an oxygen containing target, when properly calibrated can be used as an internal dosimeter. Thus, the yield of $^{11}CO_2$ appearing in the gases from a target is a secondary product affected by the radiolytic conditions which obtain during bombardment. However, it is a simple matter to obtain either ^{11}CO, by passing the effluent target gases over heated Zn at 400°C as the reducing agent[cf 48] or $^{11}CO_2$ by oxidizing the gas mixture over CuO held at 750°C.[cf 19,43,44,49] The targets are frequently run with traces of CO or CO_2 as carrier. Thus, the product $^{11}CO/^{11}CO_2$ is not "carrier free" (*vide infra*). The B_2O_3 target can be run using a helium sweep and without added carrier gases if ultra high specific activity is required. Little, if any, loss in efficiency of recovery is observed. The use of B_2O_3 is troublesome in that the target materials must be constantly renewed and careful control of target temperature must be maintained. In our experience daily carbon-11 production necessitates at least two targets being on hand in order to assure reliability of delivery, with reworking of the B_2O_3 in the target required after roughly 1 hour of running time.

The production of $^{11}CO/^{11}CO_2$ is considerably less complex if one uses the $^{14}N(p, \alpha)^{11}C$ reaction in N_2 gas.[55,56,cf 52] The oxygen contaminant ($\sim$1 ppm) in commercially available nitrogen is sufficient to oxidize the carbon atoms generated in the nuclear reaction because of the efficient reaction between oxygen and carbon relative to nitrogen plus carbon. Addition of excess oxygen generates unwanted oxides of nitrogen. Nitrogen-13 generated in the target appears as $^{13}N_2$ and is not trapped. The advantages of this target include high yield, minimal maintenance, and simplicity of design and operation. A number of methods of preparation of ^{11}CN as $H^{11}CN$ including recoil synthesis from $N_2 + H_2$ gas, direct bombardment of NaC≡N and chemical synthesis from $^{11}CO_2 + K + NH_3$ have been evaluated by FINN *et al.*(57) and recoil synthesis via direct bombardment of $LiNH_2$ or $N_2 + H_2$ gas has been described by LAMB, JAMES and WINCHELL(58). The direct bombardment of $N_2 + H_2$ gas mixtures does indeed yield $H^{11}CN$, however, radiolysis of the product in the target gas reduces the yield to less than 50% of the carbon-11 produced in the target when a large scale run is carried out.(56, cf 52) Nevertheless, hundreds of millicurie of $H^{11}CN$ are obtainable by this direct method.(58, cf 57) CHRISTMAN *et al.*(59, cf 56) have utilized the radiolytic destruction of $H^{11}CN$ to $^{11}CH_4$ in the target during proton bombardment to devise a method for quantitative production of $H^{11}CN$. The reaction sequence involves bombardment of $N_2 + H_2$ to give ^{11}C which reacts with N_2 to give ^{11}CN (+ ^{11}C=N=N)(60) followed by proton induced radiolysis of ^{11}CN (+ ^{11}C=N=N and any ^{11}CO present) to give $^{11}CH_4$. ^{11}C-methane is quantitatively (95–100% radiochemical yield) removed from the target and allowed to react with NH_3 (either added or produced radiolytically in the target) over Pt at 1000°C to give a >95% overall yield of $H^{11}CN$. The advantages of this approach are near quantitative production of H^{11}C≡N, simplicity of targetry, and the fact that the $H^{11}CN$ can be conveniently prepared from either a N_2–H_2 target as carbon-11 source or from $^{11}CO_2$ as carbon-11 source. Thus multi-curie amounts of $H^{11}CN$ can also be prepared at those institutions where boron targets are used for carbon-11 production.

This laboratory has been using the nitrogen gas target for routine ^{11}CO, $^{11}CO_2$ and $H^{11}CN$ production in Curie quantities for years. The production methods for these basic intermediates are given in Table 2.

A problem which was first noted in the production of radioactive carbon-14 was the difficulty in producing a truly carrier-free nuclide. The problem in carbon-14 production was readily pinpointed. Target materials (e.g. nitrides) used in nuclear reactors ($^{14}N(n, p)^{14}C$) could not be cleansed of minute quantities of stable carbon. Furthermore, other sources of contaminating stable carbon were introduced in the conversion process to obtain carbonate, the usual form in which carbon-14 is sold. Many of the same difficulties also apply in the case of carbon-11 production.(cf 56) Nitrogen gas cannot be made "carbon" free. Similar problems exist for B_2O_3 targets. Processing of ^{11}CO, $^{11}CO_2$ and H^{11}C≡N prior to their use in synthesis can result in further adulteration with carbon-12. While the slight reduction in specific activity may be of no real consequence in the above case, larger reductions do occur in practice especially if one needs intermediates such as ^{11}C-formaldehyde. This point is of consequence if one remembers that one of the advantages of carbon-11 involves preparing labeled compounds whose concentration is "sub-physiological", i.e. no detectable deviation from normal function can be observed in the short term after administration of the labeled material. One relatively minor consequence of this problem is whether the term carrier free in quotes i.e. "carrier free" or "CF" should be used, or the term no carrier added, or NCA, should be used. At present, both are in common use.

TABLE 2. Production of basic intermediates

Target material	Nuclear reaction	Product	Conversion method	Final product
B_2O_3	$^{10}B(d, n)^{11}C$	$^{11}CO + ^{11}CO_2$	Reduction	^{11}CO
	$^{11}B(p, n)^{11}C$		Oxidation	$^{11}CO_2$
N_2(+1–10 pp, O_2)	$^{14}N(p, \alpha)^{11}C$	$^{11}CO + ^{11}CO_2$	Reduction	^{11}CO
			Oxidation	$^{11}CO_2$
N_2(+5% H_2)	$^{14}N(p, \alpha)^{11}C$	$^{11}CH_4$	See Text	$H^{11}CN$
N_2(+1% H_2)	$^{14}N(p, \alpha)^{11}C$	$H^{11}CN$	See Text	$H^{11}CN$
B_2O_3	$^{10}B(d, n)^{11}C$	$^{11}CO + ^{11}CO_2$	1. Reduction to $^{11}CH_4$	$H^{11}CN$
	$^{11}B(p, n)^{11}C$		2. $^{11}CH_4 + NH_3$ over Pt	

In any event the specific activities attainable by the use of ^{11}CO, $^{11}CO_2$ and $H^{11}C{\equiv}N$ by the methods currently in use make possible the preparation of compounds whose specific activities are only an order of magnitude to a few orders of magnitude removed from the theoretical maximum specific activity of 9.22×10^9 Ci/mol.

4. SYNTHETIC AND BIOSYNTHETIC METHODS

The use of carbon-11 in preparing a wide variety of organic compounds receives frequent mention in the literature particularly in the post 1969 period. However, sufficiently detailed methodology is not always given. To some extent the reason for this is the obvious nature of the chemical reaction. With one exception(77) no new synthetic reactions have been discovered by individuals reporting on labeled compound syntheses. For example the carboxylation of a Grignard or the use of an S_N2 reaction involving cyanide ion and an alkyl halide is hardly new. Nevertheless, the methodology and techniques required to successfully use a simple carbon-11 labeled precursor in the preparation of a more complex product is not always a direct copy of the parent synthetic approach. Indeed, the problems posed by the constraints of time, design relative to immediate utilization of a cyclotron produced product, problems relative to the "carrier free" nature of the precursor, radiation and radiation chemistry problems, technology of handling, peculiarities of assay, etc. make the detailed reporting of procedures not only useful but in many cases, absolutely necessary. The further constraint, namely the end use of these compounds in Nuclear Medicine, is perhaps the most important of all. Generally speaking the compounds must be prepared in high chemical and radiochemical yield and the specific activity and total activity delivered must equal or exceed the demands of clinical application. It is fair to say that we are a long way from that goal with respect to most of the compounds discussed in this review. Nevertheless, some compounds are available in sufficient quantity to equal or exceed the capacity for procedures of the cameras in common use.

A listing of carbon-11 compounds with specific yield and assay data is given in Table 3. Table 4 lists compounds where detailed data are not readily available.

The preparation of ^{11}C-formaldehyde has been described by CHRISTMAN *et al.*(61). The procedure involves the lithium aluminum hydride reduction of $^{11}CO_2$ to $^{11}CH_3OH$ which is then catalytically oxidized to $^{11}CH_2O$ over a ferric molybdenum catalyst. The ^{11}C-methanol is another useful intermediate product. The use of a silver catalyst in place of the ferric molybdenum catalyst has been described by MAZIERE *et al.*(62,cf63,64) COMAR, MAZIERE and CROUZEL(65) have used the ^{11}C-methanol to prepare ^{11}C-methyl iodide.

^{11}C-acetylene was prepared by CRAMER and KISTIAKOWSKY(7) by hydrolyzing ^{11}C–BaC_2 which they had prepared from $Ba^{11}CO_3$. MYERS(66) used a recoil synthesis (*vide infra*) to prepare ^{11}C–CaC_2 which on hydrolysis also leads to ^{11}C-acetylene. ^{11}C-acetylene can be obtained by the bombardment of alkanes especially methane, ethane, propane and cyclopropane(67,cf 27,68,69) using the $^{12}C(p, pn)^{11}C$ reaction.(21) The total activity that can be produced by this latter method is low, but the material is "carrier free". Most other one and two carbon intermediates are readily accessible via simply reactions.

One of the most extensively used reactions has been the carbonation of a Grignard, organolithium compound or other carbanionoid reagent in order to prepare carboxylic acids. As previously noted it was first used to prepare simple acids such as acetic and propionic.(12,19) Extensive use of these reactions was made by WINSTEAD *et al.* to produce a large number of aliphatic, aromatic and heterocyclic acids.(70,71,72) A novel synthesis of carboxyl labeled ^{11}C-acetoacetic acid has been described by STRAATMANN, HORTMANN and WELCH(73). Carbonation of 3-pyridyl lithium was used by MACHULLA, LAUFER and STÖCKLIN to prepare ^{11}C-nicotinic acid.(74)

The conversion of carboxyl labeled compounds to other compounds by reduction or other reactions has as yet not been extensively explored although reduction to alcohols is occasionally mentioned.

More versatility in synthetic approach is apparent in the application of ^{11}C-cyanide.

Both ^{11}C-dopamine[75,76] and ^{11}C-norepinephrine[77] and ^{11}C-iododopamine[80] have been prepared by the reduction of the nitrile precursor. ^{11}C-dopamine has in turn been used to prepare ^{11}C-salsolinol and 6,7-dihydroxy-1,2,3,4 tetrahydro isoquinoline-8-^{11}C,[78] by utilizing ring closure condensation reactions. Aliphatic amines are also readily prepared by reduction of the nitrile.[79,129]

A large number of carbon-11 labeled amino nitriles have been prepared using the addition of ^{11}C-cyanide to substituted amines.[81]

The reaction between carbonyl compounds, potassium ^{11}C-cyanide and $(NH_4)_2CO_3$ has been utilized to prepare labeled hydantoins.[82–84] Extensive experimental detail can be found in reference.[84]

The chemical synthesis of amino acids has not as yet been investigated in depth. β ^{11}C-alanine has been prepared by reducing the nitrile resulting from the reaction of the alpha haloacid and recoil labeled ^{11}C-cyanide.[85] The *dl*-α-alanine has been prepared from carboxyl labeled propionic acid.[86] The complex *dl*-aspartic acid synthesis of ATKINSON[87] was modified to effect a rapid synthesis of carboxyl labeled *dl*-aspartic acid.[88] The carboxylation of α-lithioisocyanides has been used to prepare *dl*-^{11}C-α-phenylglycine and *dl*-^{11}C-α-phenylalanine.[89,90] A promising approach to the synthesis of carbon-11 labeled amino acids (the authors used a carbon-14 model) involving the Bucherer modification of the Strecker synthesis has been reported in the fast synthesis of 1-aminocyclopentane carboxylic acid and *dl*-tryptophan.[91]

Chlorpromazine methiodide-methyl-^{11}C was prepared by methylation of chlorpromazine with ^{11}C-methyl iodide.[65] COMAR *et al.* subsequently prepared chlorpromazine itself, labeled with carbon-11 in a methyl group by using the Eschweiler-Clarke modification of the Leuckart-Wallach reaction (also known as reductive formylation).[62,63] The required carbon-11 reagent is ^{11}C-formaldehyde (*vide supra*). This same group has used reductive formylation to prepare a wide variety of carbon-11 labeled drugs including ^{11}C-thioproperazine, ^{11}C-imipramine and ^{11}C-nicotine.[92] ^{11}C-diazepam and ^{11}C-caffeine were prepared using $^{11}CH_3I$. STRAATMANN and WELCH[64] have developed a general method for labeling proteins with carbon-11 by reductive formylation using albumin and fibrinogen as models. MARAZANO *et al.*[93] have labeled ovine luteinizing hormone by reductive formylation. Both groups[64,93] used $NaBH_4$ as the reducing agent and modified the general procedure so as not to denature the protein. WELCH *et al.*[94] have prepared a putrescine analog, N-methyl-^{11}C-1,4 diaminobutane, by this method.

Additional comment on synthesis with carbon-11 can be found in references.[95–97]

The advantages of biosynthetic techniques for compound preparation are twofold; rapidity and the labeling of natural products in their optically active form wherever pertinent. The major disadvantages are scaling problems, isolation and purification. If the product is to be administered by injection in humans, apyrogenicity and sensitivity to impurities pose special problems. The labeling is usually non-specific. If it is necessary to know the quantity, the location and relative amounts of the label, work with carbon-13 or carbon-14 is usually necessary.

The early history of biosynthesis with carbon-11 has been noted in the introduction and is discussed in KAMEN's book.[19] However, unlike other aspects of research in and use of carbon-11, biosynthetic techniques fell into disuse after 1942. It was not until 1971 that interest in the use of biosynthesis for the preparation of carbon-11 labeled compounds reappeared. Methods for the preparation of ^{11}C-glucose[98,99] and L-aspartic acid-4-^{11}C[100] were published. In each case $^{11}CO_2$ was the feed material. Modification of the method of LIFTON and WELCH[98] has been reported by GOULDING and PALMER[101]. These authors also reported on biosynthetic methods for the preparation of ^{11}C-galactose, ^{11}C-mannitol and ^{11}C-glycerol.[102] STRAATMANN and WELCH[103] have described a greatly improved technique for purifying the biosynthetic ^{11}C-glucose.[103] These methods utilize intact organisms or plant materials for the source of enzymatic activity.

A crystalline enzyme, thymidylate synthetase, was used in the preparation of ^{11}C-thymidine labeled with carbon-11 in the methyl group.[61] The source of carbon-11 was $H^{11}CHO$.

COHEN *et al.*[104] have discussed the use of immobilized enzymes in labeled compound

TABLE 3. Carbon-11 labeled radiopharmaceuticals

Compound	Chemical yield, %	Radiochem yield, %	Synthesis time (min)	Activity EOB total	Activity EOB S.A.	Activity A.D. total	Activity A.D. S.A.	Ref.
Alcohols								
methanol			5–10				2 mCi/ml	128
ethanol			10–60				1–5 mCi/ml	128, 120
isopropanol			10–60				1–3 mCi/ml	128
glycerol			75			72 μCi	26.8 μCi/mg	101
mannitol			75			910 μCi	77 μCi/mg	101
hexadecanol	16		55				2.9 mCi/mg	88
Carboxylic acids								
aliphatic and aromatic (26 acids)		40–98%	40			76 mCi		72
acetoacetic acid		52–55	40			1 mCi		73
nicotinic acid (carrier-free)		60–85%	25			10 mCi		74
urocanic acid	45						0.092 μCi/mMol	110
Amino acids								
glycine				20–80 μCi			0.28 μCi/mMol	117, 110
β-alanine	35	4.5	90					89
aspartic acid	29						35 μCi/mg	88
methionine				20–80 μCi				117
DL-α-phenylglycine	45	6	40			25 mCi	1 mCi/mg	89, 90
DL-α-phenylalanine		1	70			4 mCi		89, 90
Amines								
aliphatic amines C_4–C_8*		20–40					2000 Ci/mMol	79
N-methyl-1,4-diaminobutane			15			10 mCi		94
dopamine: HCl*	10–15	52	65				30 μCi/mg	75
		30	60			20 mCi		76, 80
norepinephrine: HCl*	10–15	10	40				40 mCi/mg	59, 77
iododopamine*			60				20000 Ci/mMol	80

Nitriles							
α-N-alkylaminophenylacetonitriles	6–53	11–65	49–82		16 mCi		81
α-N-arylaminoarylacetonitriles	20–66	19–61	35–75		40 mCi		81
Sugars							
glucose-fructose		6	45–50		350 μCi		98
glucose						100 μCi/mg	103
galactose			75		135 μCi	25.8 μCi/mg	101
Hydantoins							
dialkylhydantoins	11–50	7–36	70–90	22–163 mCi	1–15 mCi		82, 84
diarylhydantoins	37–61	24–60	71–83	173 mCi	10–59 mCi		82, 84
alkylarylhydantoins		11–36	106	58 mCi	1–7 mCi		82, 84
spirohydantoins	30–81	41–60	73–81	253–932 mCi	16–36 mCi		82, 84
p-hydroxyphenylphenylhydantoin		40–58	70		60 mCi		83
Nucleotides							
thymidine	77	5	110			5 mCi/μMol	61
Macromolecules							
albumin	70	38.6			3–5 mCi		64
fibrinogen		33			3–5 mCi		64
ovine luteinizing hormone			15			20 Ci/mMol	93
blood cells			10			250 mCi/mol	138,139,141
Miscellaneous							
acetyl phosphate		20–35	10		1–2 mCi		105
chlorpromazine*		10–20	30		100 mCi	1.5 Ci/mMol	62, 65

* carrier-free

TABLE 4. Other carbon-11 compounds

Compound	Reference
methane	10
methanol	154, 155
ehthaol	154, 155
octanol	129
acetic acid	10
propionic acid	10
octanoate	154, 155
palmitate	154, 155
oleic acid	154, 155
succinic acid	10
lactic acid	7
salicylic acid	68
anthranilic acid	68
α-alanine	68
dl-trytophan	91
l-aminocyclopentanecarboxylic acid	91
p-cyanophenylalanine	88
6,7-dihydroxy-1,2,3,4-tetrahydroisoquinoline	78
salsolinol	78
blood cells	111
inulin	114
α-chymotrypsin	112, 113
caffeine	92
diazepam	92
imipramine	92
nicotine	92
thioproperazine	92

biosynthesis. The application of immobilized enzymes should be an important advance in terms of simplicity and the ability to carry out continuous reaction. Recently SPOLTER *et al.*[105] have prepared ^{11}C-acetyl phosphate by incubating $CH_3{}^{11}COOH$ and acetate kinase.

While biosynthetic techniques are not as yet a significant part of the effort in carbon-11 labeling of biologically active compounds, the development of rapid and simple methods, for example the combination of specific immobilized enzymes, simple carbon-11 precursors and rapid purification by high pressure liquid chromatography holds much promise in making complex substances of known stereochemistry available for research and application.

It cannot be expected that much in the way of wholly new synthetic and biosynthetic methods will emerge from the area of carbon-11 labeling since the major emphasis in this field is on the use of known reactions. Nevertheless, a great deal needs to be done in devising techniques for rapid and efficient syntheses of these compounds. The pressures of demand may perhaps permit new and better syntheses or biosyntheses to be found.

5. "NON-SYNTHETIC" TECHNIQUES

"Non-synthetic" techniques cover a multitude of approaches to labeling compounds other than the usual synthetic and biosynthetic methods. Included are recoil labeling, excitation labeling (also called decay induced labeling), radiation labeling, discharge labeling (i.e. the labeling species is produced in a microwave or other discharge) accelerated ion labeling, Wilzbach labeling, etc.

Recoil labeling has been reviewed by WOLF,[106] and STÖCKLIN.[108] Irradiation and non-synthetic methods have been discussed by WOLF.[109] For example ^{11}CO and ^{11}CN are formed by recoil labeling during bombardment.[50–52,57] In each case it is a combination of the excess kinetic energy of the carbon atom and its spin state at the time of reaction that determine its reactivity. However the only true hot reaction (i.e. one which does not occur when the reacting atomic species is in thermal equilibrium with its surroundings with a Boltzmann distribution of kinetic energies) is the formation of $^{11}CN(^{11}CN_2)$. It cannot form in the reaction between a thermal ground state carbon atom and a nitrogen molecule. In contrast carbon monoxide can form when a thermal carbon atom collides with an oxygen molecule.

There are relatively few papers in the literature on the direct activation of organic compounds. Curiously the first paper in which an organic compound labeled with carbon-11 was used in a medical experiment involved recoil labeling, apparently without the people involved being aware that they were using a recoil method. FROST, MICHEL and SCHWARZ[110] used the $^{12}C(\gamma, p)^{11}C$ reaction in bombarding urocanic acid, glycine and glucose. Indeed, they show a scintigram of a stomach obtained by the use of the bombarded glucose. There is no question they produced carbon-11, but there is also no question that the material in the stomach pictured was a mixture of many carbon-11 labeled compounds with ^{11}C-glucose being at best a minor component. In the same year LARSSON, GRAFFMAN and JUNG[111] reported labeling blood by direct 185 MeV proton irradiation. KASCHE[112,113] attempted to label α-chymotrypsin by using the $^{16}O(p, 3p3n)^{11}C$ reaction in an aqueous solution of this material. His approach was based on the work of STENSTRÖM[114] who did an extensive study on the recoil chemistry of carbon-11 in water and in aqueous solutions and who did an interesting study on the attempted labeling of inulin with carbon-11. TRAMPISCH and SATTLER[115,116] have reported on recoil labeling of methionine with carbon-11 by utilizing the γ, n reaction on the parent substance for carbon-11 atom generation. DONNERHACK, SATTLER and TRAMPISCH[117] have extended this work and labeled aminobutyric acid and glutathione. They have investigated the distribution of the unpurified recoil activated methionine in rats. 20–80 μCi per injection were used. MOERSDORF *et al.*[118] claim to have labeled methionine and glucose by recoil, but no proof of radiochemical purity is provided.

The recoil labeling of organic molecules, while giving reasonable yields in a few cases, is not a generally applicable method for the preparation of radiopharmaceuticals. The major difficulty is caused by the fact that large numbers of carrier-free products which are closely related in structure and reactivity to the parent molecule are formed by the reaction of the recoiling carbon atom with the substrate during the formation of these atoms by charged particle or photon bombardment. The more complex the substrate the more varied the products. Purification and clearly demonstrable radiochemical purity is thus made very difficult. This means that the preparation of a pure product, unless it is one that is fortuitously easy to separate from its contaminants, is very difficult. Nevertheless, an understanding of the method can make it useful in some cases as, for example ^{11}CO and ^{11}CN preparation. Aspects of recoil labeling which have not been explored include the possibility of utilizing to advantage the very complexity of products, by isolating a desired carrier free product which is produced, rather than purifying the labeled substrate itself. Another approach might be a screening of compounds for efficacy. A particular set of compounds could be bombarded and activated and tested in animals. The difficulty, however, when a positive test is obtained, would be to determine the source of efficacy, since there would be no *a priori* way of telling if the labeled substrate or one of the carrier free contaminants had been responsible for the observed response.

Ease of labeling can be contrasted with difficulty in purification. Activation of substrates with the carbon-11 distributed among many products including the labeled substrates, may prove to be a viable technique. Recoil labeling and indeed other non-synthetic methods not discussed here because they have not been applied to the preparation of carbon-11 labeled radiopharmaceuticals, may yet prove to be valuable in providing materials for Nuclear Medicine.

6. MISCELLANEOUS TOPICS

6.1 *Photonuclear Reactions*

In the previous section on recoil labeling a number of the articles referenced described the use of the $^{12}C(\gamma, n)^{11}C$ reaction for carbon-11 production in the sample. The use of a photonuclear reaction[41,42] for carbon-11 production is not generally applicable because of the scarcity of appropriate accelerators and the complex targetry needed to produce the necessary high energy bremsstrahlung. Nevertheless, large quantities of carbon-11 can be made in this way. Carrier-free production of ^{11}CO–$^{11}CO_2$ is probably not possible with a carbon target, yet a heated porous carbon target over which a steady stream of oxygen is flowing would allow the isolation of high specific activity ^{11}CO–$^{11}CO_2$ by an on-line method. The article by WELCH[42] places the problem in correct perspective. A more recent article by MCNEIL[119] is a good introduction to the photoactivation process, but is less accurate in dealing with labeling. Photospallation of oxygen-16, $^{16}O(\gamma, 2p3n)^{11}C$, has recently been described for the preparation of $^{11}CO_2$.[120] These authors also describe a synthetic method for preparing ^{11}C-ethanol.

Photonuclear reactions for carbon-11 production have as yet not been adequately explored. At the present moment, it seems unlikely that these reactions will be competitive with production by charged particle reactions from "medical cyclotrons." The technology of use is certainly more complex and the cost of the bremsstrahlung targets is not a minor factor. However, electron accelerators can be utilized wherever they exist for the study of such reactions and there is sufficient compelling reason to exploit this use, in order to make carbon-11 more generally available.

6.2 *The van de Graaff Accelerator*

The van de Graaff accelerator has also been suggested[121–125] for carbon-11 production. The optimistic view of COHEN and CANADA[122] was challenged by PHELPS and WIELAND[123]. What should of course be apparent is that all three production methods for carbon-11, $^{10}B(d, n)^{11}C$, $^{11}B(p, n)^{11}C$ and $^{14}N(p, \alpha)^{11}C$ are in principle possible, by using protons or deuterons from van de Graaff accelerators. It is simply a question of the maximum energy and current the particular accelerator can provide. Activities of 80 μCi[121] and 50 mCi[124] have been obtained. There is little doubt that in some places sufficient activity can be produced for research and in cases where a reasonably high current is available (energies usually do not exceed 6 MeV) sufficient carbon-11 activity for radiopharmaceutical preparation for human use could be generated. However, a realistic assessment of carbon-11 requirements for most Nuclear Medicine purposes would indicate that the use of presently available van de Graaff accelerators is marginal. In any event, carbon-11 yields can be estimated from available cross sections (*vide supra* under Nuclear Reactions used for Carbon-11 Production).

6.3 *Carbon-11 in Basic Research*

This is an area that in the view of this author has been grossly neglected. The value of carbon-11 in basic research was recognized very shortly after its discovery (cf. Introduction), however, the cost and ease of using carbon-14 washed out the advantages of carbon-11 at that time. Carbon-11 is superbly suited to probing the dynamics of rapid physiological processes. Furthermore, these processes can be studied in living systems and non-invasively. The property of non-invasive use is one of the factors that holds such a promising future for carbon-11 labeled compounds in Nuclear Medicine, as function and metabolism are important factors in the diagnosis of disease. However, carbon-11 can also be used in basic research without a necessary medical connotation. Several examples of this can be found in the recent literature.

MORE and TROUGHTON[126] used $^{11}CO_2$ to study translocation in living plants. RAICHLE *et al.*[127] have studied brain glucose transport and metabolism with ^{11}C-glucose and blood brain barrier permeability with ^{11}C labeled alcohols.[128]

FOWLER *et al.*[129] have studied MAO mediated amine metabolism in the lung by utilizing ^{11}C labeled amines. While the studies by RAICHLE *et al.* and FOWLER *et al.* had their motivation in Nuclear Medicine application, they nevertheless represent basic research that can stand independent of any medical context.

This area of research should again become active because of the recent almost exponential growth of carbon-11 availability. Many new insights into the dynamics of enzyme action and other processes in living systems hopefully will be illuminated by the application of carbon-11 as a tracer.

Carbon-11 labeled compounds have not been used in the traditional pre-clinical testing of new pharmaceutical products. Studies with compounds labeled with stable and long lived nuclides are the standard for organ distribution and metabolic studies. Use of carbon-11 labeled new drugs would not only provide organ distribution data in a few hours, but in addition would provide the dynamics of distribution in the animal being studied. If the drug is not metabolized then influx and efflux of critical organs can also be determined. A breakdown product spectrum if that is required would be more convenient with a longer lived or stable tracer.

7. MEDICAL USES

While the major portion of this review is on the production and utilization of carbon-11 in research and radiopharmaceutical production it would hardly be complete without mention of the uses to which it has been put in the practice of Nuclear Medicine. A number of these uses have already been apparent in the discussion presented in previous sections. The emphasis in most of the papers referenced to this point has been on the physical, chemical and biochemical aspects. The papers referenced in this section concern themselves primarily with medical aspects, yet many of them also contain valuable comment on methods of preparation and other pertinent information. No attempt at comprehensive cross-referencing is made.

The first use of carbon-11 in humans was reported in 1945.[16] However, following this report no published work appeared until the early sixties. A number of papers were published on the use of ^{11}CO and $^{11}CO_2$ in the lung.[130–136] These papers concern themselves primarily with clearance rates, gas exchange, blood flow and pulmonary function. Extravascular lung water has also been studied using carbon-11.[137] Red cell labeling with ^{11}CO[138] has been used for measurement of blood volume[139,140] measurement of splenic red blood cell mass[141] and red blood cell survival studies.[142] Placental localization by labeling of red cells by the simple expedient of ^{11}CO inhalation has been studied[143,144] and has been used by Hammersmith Hospital for the diagnosis of placenta praevia. Comparisons of the ^{11}CO method and the ultrasound method have been made.[145,146] Johnson and King[146] concluded the methods were of comparable value, but preferred ultrasonography because of the presumed lower risk to the patient and fetus.

Exchange between blood bicarbonate and bone carbonate was suggested as a mechanism for carbon-11 deposition in bone via $^{11}CO_2$ administration.[147] While it was possible to demonstrate good localization in bone[148] extensive work on the use of carbon-11 for bone scanning and tumor detection has not continued[149,150] quite possibly because of the greater ease of using technetium compounds (or fluorine-18) for similar purposes.

Organic acids have been studied in depth for use as organ visualization agents.[70,151–153] The ability of fatty acids to localize in the heart[cf 154–156] presents intriguing possibilities for precise infarct localization and evaluation. While Poe, Robinson and MacDonald[154] concluded that potassium-43 was more desirable because of extraction efficiency and access, the use of ^{11}C-palmitate[156] allows one to apply positron emission transaxial tomography (PETT, which is just now becoming generally available). Thus, ^{11}C-palmitate or some other positron emitting labeled compound and the PETT approach may ultimately prove to be the superior technique.

Uses of ^{11}C-aminonitriles,[157,cf 81] hydantions,[158,cf 82–84] psychotropic drugs e.g. chlorpromazine, imipramine, diazepam etc.[159,160,cf 62,63,65,92] ^{11}C-cyanide as such,[161,162] catecholamines or aliphatic amines[76–80] and alcohols[163] have been also detailed in the literature.

The study of dose delivered by carbon-11 labeled compounds has not been addressed in any detail in the literature.[164,165] The calculation of dose in terms of energy deposited by the positron and the annihilation photons (contribution by EC is very small) can be carried out using currently available methods. However, carbon-11 presents another aspect that has not been explicitly studied as yet. As carbon-11 labeled organic compounds are catabolized and metabolized some of the degradation products can be incorporated into biologically active compounds. For example when carbon-11 labeled thymidine is administered a sizeable fraction[61] ends up in the DNA. As the carbon-11 decays, DNA chains may be damaged beyond repair from the decay event occurring in the molecule. Whether or not this poses a problem is not known at present. Hopefully basic research in this area will be able to answer this question.

8. CONCLUSION

Studies involving carbon-11 have been carried out in essentially three areas of interest: (1) The nuclide has been studied for its own sake. These studies have involved the physics and decay properties of carbon-11, the chemical reactivity of the excited and ground state carbon atom and the fundamental nature of the electronic and kinetic state of the atom in the broader context of chemical dynamics. (2) Carbon-11 has been used as a tracer tool in basic chemical, biochemical, biological and medical research. (3) Carbon-11 has been used as a tool in the diagnosis of disease and increasing interest is being generated in its ability to be applied to the non-invasive and dynamic assessment of metabolism and function.

The unique positron of carbon in the scheme of living systems makes carbon-11 the nuclide of choice for probing the action of living systems. The potential of this nuclide need not be belabored. It has been detailed time and again, as a sampling of the references listed in this review will show. However, a great deal needs to be done. The successful application to the diagnosis of human malfunction depends on many factors: (1) Much basic research still needs to be done in order to understand how to use these compounds effectively. The question of the biochemistry and action at the cellular level of "carrier free" compounds is an area in which little has been done and less is known. The effective design of organic agents as radiopharmaceuticals is very different from what has been required for many of the technetium labeled compounds now in use. New principles and techniques such as using enzyme mediated response, affinity labeling, chelate carrier labeling, etc. are already being used to design highly specific agents. Unfortunately, the natural pressures to prepare agents which are immediately applicable in a medical environment have a tendency to force the neglect of the basic research required to understand what is occurring at the cellular and subcellular level, knowledge which might ultimately be more important in the long run. (2) Another factor is ease of accessibility to accelerators capable of producing carbon-11. While methods now exist for producing not only multicurie amounts of carbon-11, but also multicurie amounts of certain compounds, these materials are only available in areas where medical cyclotrons are located. Should the broad scale efficacy and utility of carbon-11 ultimately be demonstrated, the construction of "inexpensive desk top" accelerators becomes a necessity. (3) As yet very little attention has been paid to converting those materials which do have promise, to "kit" form. It is unlikely that every accelerator installation can have, or indeed should have a custom synthesis group attached to it. Thus, the facile conversion of research compounds to ready availability must become a necessary part of the effort in future work.

In preparing this review in what is very obviously a multi-disciplinary application the authors were struck by the not infrequent total lack of adequate referencing. This can be underlined by the opening sentence of an article which appeared quite recently "A potential new biophysical tracer is described: ^{11}C, . . ."! We cannot help but wonder in retrospect how inadequate our referencing and cross-referencing may be. We would be most grateful if lapses and inaccuracies were brought to our attention.

While the focus in this paper has been on radiopharmaceuticals and Nuclear Medicine, basic research in carbon-11 remains an active and exciting area and the application of carbon-11 to the study of basic biological and biochemical processes is again becoming viable. Much rewarding research remains to be done.

Acknowledgement—The authors wish to express their thanks for the invaluable help provided by the extensive literature search of Dr. D. Christman.

REFERENCES

1. Lauritsen C. C., Crane H. R. and Harper W. W. *Science* **79**, 234 (1934).
2. Crane H. R. and Lauritsen C. C. *Phys. Rev.* **45**, 430 (1934); Crane H. R. and Lauritsen C. C. *Phys. Rev.* **45**, 497 (1934).
3. Yost D. M., Ridenour L. N. and Shinohara K. *J. Chem. Phys.* **3**, 133 (1935).
4. Barkas W. H. *Phys. Rev.* **56**, 287 (1939).
5. Long F. A. *J. Am. Chem. Soc.* **61**, 570 (1939).
6. Ruben S., Hassid W. Z. and Kamen M. D. *J. Am. chem. Soc.* **61**, 661 (1939).
7. Cramer R. D. and Kistiakowsky G. B. *J. biol. Chem.* **137**, 549 (1941).
8. Solomon A. K. *J. appl. Phys.* **12**, 310 (1941).
9. Hastings A. B. and Kistiakowsky G. B. *J. appl. Phys.* **12**, 322 (1941).
10. Kamen M. D. and Ruben S. *J. appl. Phys.* **12**, 310 (1941).
11. Conant J. B., Cramer R. D., Hastings A. B., Klemperer F. W., Solomon A. K. and Vennesland B. *J. biol. Chem.* **137**, 557 (1941).
12. Nahinsky P., Rice C. N., Ruben S. and Kamen, M. D. *J. Am. Chem. Soc.* **64**, 2299 (1942).
13. Vennesland B., Solomon A. K., Buchanan J. M., Cramer R. D. and Hastings A. B. *J. biol. Chem.* **142**, 371 (1942).
14. Vennesland B., Solomon A. K., Buchanan J. M. and Hastings A. B. *J. biol. Chem.* **142**, 379 (1942).

15. BUCHANAN J. M. and HASTINGS A. B. *Physiol. Rev.* **26,** 120 (1946).
16. TOBIAS C. A., LAWRENCE J. H., ROUGHTON F. J. W., ROOT W. S. and GREGERSEN M. I. *Am. J. Physiol.* **145,** 253 (1945).
17. LAWRENCE J. H. and TOBIAS C. A. *Air Surg. Bull.* **2,** 452 (1945).
18. WEINREICH R., RITZL F., FEINENDEGEN L. E., SCHNIPPERING H. G. and STÖCKLIN G. *Radiat. and Environm. Biophys.* **12,** 271 (1975).
19. KAMEN M. D. *Radioactive Tracers in Biology,* 1st Edition, Academic Press, New York (1947); KAMEN M. D. *Radioactive Tracers in Biology,* 2nd Edition, Academic Press, New York (1951); KAMEN M. D. *Radioactive Tracers in Biology,* 3rd Edition, Academic Press, New York (1957).
20. WOOD C. A. P. and NEWBERY G. R. *Nature, Lond.* **173,** 233 (1954); GALLOP J. W., VONBERG D. D., POST R. J., POWELL W. B., SHARP J. and WATERTON P. J. *Proc. IEE* Part B, vol. **104,** 452 (1957); NEWBERY G. R. *Br. J. Radiol.* **XXXII,** No. 382, 633 (1959).
21. WOLF A. P. *Advances in Physical Organic Chemistry* (Edited by GOLD V.), Vol. 2, p. 201. Academic Press, New York (1964).
22. WOLFGANG R. *Progress in Reaction Kinetics* (Edited by Porter G.), p. 97. Pergamon Press, Oxford (1965).
23. MACKAY C. In *Carbenes* (Edited by JONES M. and MOSS R. A.), Vol. II, Wiley, New York (1975).
24. TER-POGOSSIAN M. M. In *Progress in Atomic Medicine* (Edited by LAWRENCE J. H.), p. 107. Grune & Stratton, New York (1965).
25. TER-POGOSSIAN M. M. and WAGNER H. N., JR. *Nucleonics* **24,** 50 (1966).
26. LAUGHLIN J. S., TILBURY R. S. and DAHL J. R. In *Progress in Atomic Medicine* (Edited by LAWRENCE J. H.), p. 39. Grune & Stratton, New York (1971).
27. WOLF A. P., CHRISTMAN D. R., FOWLER J. S. and LAMBRECHT R. M. In *Radiopharmaceuticals and Labelled Compounds,* Vol. 1, p. 345. IAEA, Vienna (1973).
28. WONG C. *Phys. Rev.* **95,** 761 (1954).
29. AJZENBERG-SELOVE F. *Nucl. Phys.* **A248,** 30 (1975).
30. BRILL O. D. and SUMIN L. V. *Atom. Energ.* **7,** 377 (1959).
31. WOHLLEBEN K. and SCHUSTER E. *Radiochim. Acta* **12,** 75 (1969).
32. FURUKAWA M., ISHIZAKI Y., NAKANO Y., NOZAKI T., SIEJI Y. and TANAKA S. J. *Phys. Soc. Japan* **15,** 2167 (1960).
33. LEGGE G. J. F. and BUBB I. F. *Nucl. Phys.* **26,** 616 (1961).
34. BLASER J. P., MARMIER P. and SEMPERT M. *Helv. Phys. Acta* **25,** 442 (1952).
35. CACACE F. and WOLF A. P. *Chimica e L'Industria* **44,** 752 (1962).
36. NOZAKI T., OKUO T., AKUTSU H. and FURUKAWA M. *Bull. Chem. Soc. Japan* **39,** 2685 (1966).
37. EPHERRE M. and SEIDE C. *Phys. Rev.* C **3,** 2167 (1971).
38. JACOBS W. W., BODANSKY D., CHAMBERLIN D. and OBERG D. L. *Phys. Rev.* C **9,** 2134 (1974).
39. CASELLA V. R., CHRISTMAN D. R., IDO T. and WOLF A. P. *Int. J. appl. Radiat. Isotopes* **27** (1976).
40. COCHRAN D. R. F. and KNIGHT J. D. *Phys. Rev.* **128,** 1281 (1962).
41. THEARD L. M., GROCE D. E. In *Progress in Biomedical Engineering* (Edited by FOGEL L. J. and GEORGE F. W.), pp. 29–41. Spartan Books, Washington (1967)
42. WELCH M. J. CONF-730301-P2, 1179.
43. BUCKINGHAM P. D. and FORSE G. R. *Int. J. appl. Radiat. Isotopes* **14,** 439 (1963).
44. DOLLERY C. T., FOWLER J. F., HUGH-JONES P., MATTHEWS C. M. E. and WEST J. B. *Strahlentherapie* **5,** 88 (1963).
45. VONBERG D. D., BAKER L. C., BUCKINGHAM P. D., CLARK J. C., FINDING K., SHARP J. and SILVESTER D. J. In *The Uses of Cyclotrons in Chemistry, Metallurgy and Biology* (Edited by AMPHLETT C. B.), p. 258. Butterworths, London (1970).
46. CLARK J. C. and BUCKINGHAM P. D. *Radiochem. Radioanal. Lett.* **6,** 281 (1971).
47. CLARK J. C. and BUCKINGHAM P. D. *Int. J. appl. Radiat. Isotopes* **22,** 639 (1971).
48. WELCH M. J. and TER-POGOSSIAN M. M. *Radiat. Res.* **36,** 580 (1968).
49. HOOP B. JR., LAUGHLIN J. S. and TILBURY R. S. In *Instrumentation in Nuclear Medicine* (Edited by HINE G. and SORENSEN J.), Vol. II, ch. 13. Academic Press, New York (1973).
50. DUBRIN J., MACKAY C., PANDOW M. L. and WOLFGANG R. *J. inorg. nucl. Chem.* **26,** 2113 (1964).
51. STÖCKLIN G and WOLF A. P. *Chemical Effects of Nuclear Transformations* Vol. I, pp. 121–132. IAEA, Vienna (1964).
52. ACHE H. J. and WOLF A. P. *J. phys. Chem.* **72,** 1988 (1968).
53. ANDERSON A. R., BEST J. V. F. and WILLETT M. J. *Trans. Farad. Soc.* **62,** 595 (1966).
54. ELIAS H. and WOLF A. P. *Radiochim. Acta* **10,** 111 (1968).
55. FINN R. D. and WOLF A. P. *J. nucl. Med.* **13,** 429 (1972).
56. CHRISTMAN D. R., FINN R. D., KARLSTROM K. I. and WOLF A. P. *Int. J. appl. Radiat. Isotopes* **26,** 435 (1975).
57. FINN R. D., CHRISTMAN D. R., ACHE H. J. and

Wolf A. P. *Int. J. appl. Radiat. Isotopes* **22,** 735 (1971).
58. Lamb J. F., James R. W. and Winchell H. S. *Int. J. appl. Radiat. Isotopes* **22,** 475 (1971).
59. Christman D. R., Finn R. D., Karlstrom K. I. and Wolf A. P. *J. nucl. Med.* **14,** 864 (1973).
60. Ache H. J. and Wolf A. P. *Radiochim. Acta* **6,** 32 (1966).
61. Christman D. R., Crawford E. J., Friedkin M. and Wolf A. P. *Proc. Nat. Acad. Sci. U.S.A.* **69,** 988 (1972).
62. Comar D., Maziere M. and Raynaud C. (Hopital d'Orsay, France). In *Radioactive Isotopes in Clinical Medicine and Research* (Edited by Hoefer R.), pp. 81–89. Urban & Schwarzenberg, Munich (1975).
63. Maziere M., Sainte-Laudy J. L., Crouzel M. and Comar D. In *Radiopharmaceuticals* (Edited by Subramanian, G., Rhodes B. A., Cooper J. F. and Sodd V. J.), p. 189. Society of Nuclear Medicine, Inc., New York (1975).
64. Straatmann M. G. and Welch M. J. *J. nucl. Med.* **16,** 425 (1975).
65. Comar D., Maziere M. and Crouzel C. In *Radiopharmaceuticals and Labelled Compounds,* Vol. 1, p. 461. IAEA, Vienna (1973).
66. Myers W. G. *J. nucl. Med.* **13,** 699 (1972).
67. Finn R. J., Machulla H. J., Stöcklin G. and Wolf A. P. Unpublished work (1973).
68. Machulla H. J. and Stöcklin G. *Proc. 13th Int. Ann. Meeting Soc. Nucl. Med.* Copenhagen, September 10–13 (1975).
69. Weinreich R., Bräutigam W., Machulla H. J., Probst H. J., Qaim S. M. and Stöcklin G. *7th Int. Conf. Cyclotrons and Their Applications*, Zurich, August (1975).
70. Winstead M. B., Winchell H. L. and Fawwaz R. *Int. J. appl. Radiat. Isotopes* **20,** 859 (1969).
71. Winchell H. S. In *Frontiers of Nuclear Medicine* (Edited by Horst W.), p. 306. Springer-Verlag, Berlin, Heidelberg, New York (1971).
72. Winstead M. B., Lamb J. F. and Winchell H. S. *J. nucl. Med.* **14,** 747 (1973).
73. Straatmann M. G., Hortmann A. G. and Welch M. J. *J. Labelled Compds.* **10,** 175 (1974).
74. Machulla H. J., Laufer P. and Stöcklin G. *Radiochem. Radioanal. Lett.* **18,** 275 (1974).
75. Christman D. R., Hoyte R. M. and Wolf A. P. *J. nucl. Med.* **11,** 474 (1970).
76. Fowler J. S., Ansari A. N., Atkins H. L., Bradley-Moore P. R., MacGregor R. R. and Wolf A. P. *J. nucl. Med.* **14,** 867 (1973).
77. Fowler J. S., MacGregor R. R., Ansari A. N., Atkins H. L. and Wolf A. P. *J. med. Chem.* **17,** 246 (1974).
78. Fowler J. F., Wolf A. P., Christman D. R., MacGregor R. R., Ansari A. and Atkins H. In *Radiopharmaceuticals* (Edited by Subramanian G., Rhodes B. A., Cooper J. F. and Sodd V. J.), p. 196. Society of Nuclear Medicine, Inc., New York (1975).
79. Ansari A., Atkins H., Briggs J., Christman D., Fowler J., Gallagher B., MacGregor R. and Wolf A. *J. nucl. Med.* **16,** 511 (1975).
80. Fowler J. F., Gallagher B. M., MacGregor R. R., Wolf A. P., Ansari A. N. and Atkins H. L. *J. med. Chem.* **19,** 356 (1976).
81. Winstead M. B., Widner P. J., Means J. L., Engstrom M. A., Graham G. E., Khentigan A., Lin T. H., Lamb J. F. and Winchell H. S. *J. nucl. Med.* **16,** 1049 (1975).
82. Winstead M. B., Parr S., Lin T. H., Khentigan A., Lamb J. and Winchell H. S. *J. nucl. Med.* **13,** 479 (1972).
83. Tilbury R. S., Stavchansky S., Ting C. T., MacDonald J. M., Maugham E., Freed B. R., Russ G. R., Helson L., Benua R. S., Kostenbauder H. B. and Laughlin J. S. *J. nucl. Med.* **16**, 575 (1975).
84. Winstead M. B., Parr S. J., Rogal M. J., Brockman P. S. and Lubcher R. *J. med. Chem.* **19,** 279 (1976).
85. Elias H., Lotterhos H. F. and Weimer K. *Chem. Ber.* **105,** 3754 (1972).
86. cf. 68.
87. Atkinson R. J. *J. chem. Soc.* 3317 (1952).
88. Weimer K. Development of fast syntheses of amino acids and fatty alcohols labelled with carbon-11. Dissertation, Univ. Darmstadt, Germany (1973).
89. Vaalburg W. Cyclotron produced short-lived isotopes in nuclear medicine. 81 pp. Thesis. Rijksuniversiteit, Groningen, Netherlands (1974).
90. Vaalburg W., Beerling-Van Der Molen H. D. and Woldring M. G. *Nuclear-Medizin* **14,** 60 (1975).
91. Washburn L. D., Sun T. T., Wieland B. W. and Hayes R. L. *J. nucl. Med.* **16,** 579 (1975).
92. Comar D., Maziere M., Marazano C. and Raynaud C. *J. nucl. Med.* **16,** 521 (1975).
93. Marazano C., Maziere M., Morgat J. L., De La Llosa P., Comar D. and Fromageot P. *Radiochem. Radioanal. Lett.* **21,** 53 (1975).
94. Welch M. J., Coleman R. E., Straatmann M. G., Asberry B. E., Primeau J. L., Fair W. R. and Ter-Pogossian M. M. Carbon-11 labeled N-methyl 1,4-diaminobutane: a putrescine analog for prostate and tumor localization, private communication (1976).
95. Lathrop K. A., Harper P. V., Rich B. H., Dinwoodie R., Krizek H., Lembares N. and Gloria I. In *Radiopharmaceuticals and Labelled*

Compounds, Vol. 1, p. 471. IAEA, Vienna (1973).
96. WINSTEAD M. B., LIN T. H., KHENTIGAN A., LAMB J., FLEISCHER A., WINCHELL H. S. and MYERS W. In *Radiopharmaceuticals and Labelled Compounds,* Vol. 1, p. 451. IAEA, Vienna (1973).
97. GELBARD A. S., HARA T., TILBURY R. S. and LAUGHLIN J. S. In *Radiopharmaceuticals and Labelled Compounds,* Vol. 1, p. 239. IAEA, Vienna (1973).
98. LIFTON J. F. and WELCH M. J. *Radiat. Res.* **45,** 35 (1971).
99. WOLF A. P. *Radioisotopy* **12,** 499 (1971).
100. HARA T., TAYLOR C., LEMBARES N., LATHROP K. A. and HARPER P. V. *J. nucl. Med.* **12,** 361 (1971).
101. PALMER A. J. and GOULDING R. W. *J. Labelled Compds.* **10,** 627 (1974).
102. GOULDING R. A. and PALMER A. J. *Int. J. appl. Radiat. Isotopes* **24,** 7 (1973).
103. STRAATMANN M. G. and WELCH M. J. *Int. J. appl. Radiat. Isotopes* **24,** 234 (1973).
104. COHEN M. B., MACDONALD N., CHANG C. C. and BOBINET D. D. The 17th Nat. Meeting of the APhA Academy of Pharmaceutical Sciences, p. 21. New Orleans, Louisiana, November 10–14 (1974).
105. SPOLTER L., CHANG C. C., BOBINET D., COHEN M. B., MACDONALD N. S., FLESHER A. and TAKAHASHI J. *J. nucl. Med.* **16,** 572 (1975).
106. WOLF A. P. *Annual Review of Nuclear Science* (Edited by SEGRE E., FRIEDLANDER G. and MEYERHOF W. E.), Vol. X, pp. 259–290. Annual Reviews, Inc., Palo Alto (1960).
107. WOLF A. P. *Proc. Symp. on Preparation and Biomedical Application of Labelled Molecules* 423–235 EUR2200e. Venice (1964).
108. STÖCKLIN G. *Proc. Symp. Preparation and Biomedical Application of Labelled Molecules* 437–457. EUR2200e. Venice (1964).
109. WOLF A. P. *Proc. 2nd Int. Conf. Methods of Preparing and Storing Labelled Molecules* pp. 907–934 (EUR3746 (1968) 755). Brussels (1966).
110. FROST D., MICHEL L. and SCHWARZ E. *Strahlentherapie* **127,** 510 (1965).
111. LARSSON B., GRAFFMAN S. and JUNG B. *Nature, Lond.* **207,** 543 (1965).
112. KASCHE V. *Radiochem. Radional. Lett.* **3,** 51 (1970).
113. KASCHE V. *Biophys. Biochem. Res. Comm.* **38,** 875 (1970).
114. STENSTROM T. On the chemical fate of nascent ^{11}C atoms induced by irradiation of water and aqueous solutions with 185 MeV protons. Dissertation Gustaf Werner Inst. Univ. of Uppsala (1970).
115. TRAMPISCH W., SATTLER E. L. and SCHRAUB A. *Biophysik* **7,** 355 (1971).
116. TRAMPISCH W. and SATTLER E. L. *Proc. Int. Conf. Photonuclear Reactions and Applications, Asilomos Calif.* (Edited by BERMAN B. L.) (CONF 730301), p. 1013. Lawrence Livermore Lab., Livermore, Calif. (1973).
117. DONNERHACK A., SATTLER E. L. and TRAMPISCH W. *Strahlentherapie* **151,** 250 (1976).
118. MOERSDORF D., WUNDER S., SCHWALL H., BEHSCHAD H. and SCHENK P. In *Frontiers of Nuclear Medicine* (Edited by HORST W.), pp. 41–45. Springer-Verlag, Berlin, Heidelberg, New York (1971).
119. MCNEIL K. G. *Phys. Today* **27,** 75 (1974).
120. DE GRAZIA J. A., RODDEN A. F., TERESI J. D., BUSICK D. D. and WALZ D. R. *J. nucl. Med.* **16,** 73 (1975).
121. MORE R. D. and TROUGHTON J. H. *Int. J. appl. Radiat. Isotopes* **23,** 344 (1972).
122. COHEN B. L. and CANADA T. R. *Phys. Med. Biol.* **17,** 416 (1972).
123. PHELPS M. E. and WIELAND B. W. *Phys. Med. Biol.* **18,** 284 (1973).
124. PERRIS A. G., LANE R. O., TONG J. Y. and MATTHEWS J. D. *Int. J. appl. Radiat. Isotopes* **25,** 19 (1974).
125. ENGLER R. A. and RYDIN R. A. *Radioisotope Medical Applications* **19,** 55 (1974).
126. MORE R. D. and TROUGHTON H. *Photosynthetica* (Prague) **7,** 271 (1973).
127. RAICHLE M. E., LARSON K. B., PHELPS M. E., GRUBB R. L. JR., WELCH M. J. and TER-POGOSSIAN M. M. *Am. J. Physiol.* **228,** 1936 (1974).
128. RAICHLE M. E., EICHLING J. O., STRAATMANN M. G., WELCH M. J., LARSON K. B. and TER-POGOSSIAN M. M. *Am. J. Physiol.* **230,** 543 (1976).
129. FOWLER J. S., GALLAGHER B. M., MACGREGOR R. R. and WOLF A. P. *J. Pharm. Exp. Ther.* **198,** 133 (1976).
130. WEST J. B. and DOLLERY C. T. *J. appl. Physiol.* **17,** 9 (1962).
131. WEST J. B., HOLLAND R. A. B., DOLLERY C. T. and MATTHEWS C. M. E. *J. appl. Physiol.* **17,** 14 (1962).
132. WEST J. B. In *Dynamic Clinical Studies with Radioisotopes,* p. 213. USAEC Report ORNL TID-7678 (1964).
133. MATTHEWS C. M. E., LASZLO G., CAMPBELL E. J. M., KIBBY P. M. and FREEDMAN S. *Resp. Physiol.* **6,** 29 (1968).
134. WEST J. B. In *Progress in Atomic Medicine.* Vol. 2, pp. 39–64. Grune and Stratton, New York (1968).
135. ARNOT R. N., GLASS H. I., CLARK J. C., DAVIS J. A., SCHIFF D. and PICTON-WARLOW C. G. In *Radioisotopen in Klinik und Forschung.* Vol.

9 p. 60. Urban & Schwarzenburg, Munich-Berlin (1970).
136. WEST J. B. In *Pulmonary Investigations with Radionuclides* (Edited by GILSON A. J.), pp. 118–122. Chas. C. Thomas, Springfield, Ill. (1970).
137. JONES T., HUGHES M. B., BUCKINGHAM P. D., RHODES C. G., JONES H., ASHTON P. and FORSE G. R. *J. nucl. Med.* **14,** 414 (1973).
138. CLARK J. C., GLASS H. I. and SILVESTER D. J. *Proc. 2nd Int. Conf. Methods Prep. Stor. Labelled Compounds* (Edited by SIRCHIS J.), p. 603. EURATOM, Brussels, Belgium (1966).
139. GLASS H. I., BRANT A., CLARK J. C., DE GARRETA A. C. and DAY L. G. *J. nucl. Med.* **9,** 571 (1968).
140. GLASS H. I., EDWARDS R. H. T., DE GARRETA A. C. and CLARK J. C. *J. appl. Physiol.* **26,** 131 (1969).
141. GLASS H. I., DE GARRETA A. C., LEWIS S. M., GRAMMATICOS P. and SZUR L. *Lancet* **1,** 669 (1968).
142. SZUR L. In *Radioisotopes in Medical Diagnosis,* p. 342. Appleton-Century-Crofts, New York (1971).
143. GLASS H. I., JACOBY J., WESTERMAN B., CLARK J. C., ARNOT R. N. and DIXON H. G. *J. nucl. Med.* **9,** 468 (1968).
144. JACOBY H. E. and MCCLURE BROWN J. C. *Lancet* **2,** 1323 (1968).
145. HAKIN C. A. *J. Obst. & Gynac. Br. Commonwealth* **77,** 625 (1970).
146. JOHNSON P. M. and KING D. L. *Semin. Nucl. Med.* **4,** 75 (1974).
147. MYERS W. G. and HUNTER W. W., JR. *J. nucl. Med.* **8,** 305 (1967).
148. MYERS W. G. and HUNTER W. W., JR. *Proc. IAEA Symp. Med. Radioisotope Scintigraphy.* Vol. 11, p. 43. Salzburg (1969).
149. GALASKO C. S. B. In *Frontiers of Nuclear Medicine* (Edited by HORST W.), pp. 46–58. Springer-Verlag, New York (1971).
150. GALASKO C. S. B., COOMBS R. R. and CLARK J. C. *Br. J. Radiol.* **43,** 670 (1970).
151. WINSTEAD M. B., WINCHELL H. S., FAWWAZ R. and LAWRENCE J. H. *J. nucl. Med.* **10,** 382 (1969).
152. WINCHELL H. S. and WINSTEAD M. B. In *Frontiers of Nuclear Medicine* (Edited by HORST W.), pp. 161–170. Springer-Verlag, New York (1971).
153. WINSTEAD M. B., WINCHELL H. S., FAWWAZ R. A. and LAWRENCE J. H. *J. nucl. Med.* **11,** 378 (1970).
154. POE N. D., ROBINSON G. D. and MACDONALD N. S. *J. nucl. Med.* **14,** 440 (1973).
155. ROBINSON G. D., JR. and MACDONALD N. S. *J. nucl. Med.* **14,** 446 (1973).
156. WELCH M. J., TER-POGOSSIAN M. M. and PHELPS M. E. unpublished work.
157. WINSTEAD M. B., WIDNER P. J., MEANS J. L., ENGSTROM M. A., GRAHAM G. E., KHENTIGAN A., LIN T. H., LAMB J. F. and WINCHELL H. S. *J. nucl. Med.* **16,** 582 (1975).
158. STAVCHANSKY S. A., TILBURY R. S., LAUGHLIN J. S., MCDONALD J. M., TING C., FREED B. and KOSTENBAUDER H. B. *The 17th National Meeting of the APhA Academy of Pharmaceutical Sciences,* New Orleans, Louisiana, November 10–14 Vol. 4 (1974).
159. RAYNAUD C., TODD-POKROPEK A. E., COMAR D., PIZER S. M., KACPEREK A., MAZIERE M., MARAZANO C. and KELLERSHOHN C. In *Dynamic Studies with Radioisotopes in Medicine 1974.* Vol. 1, pp. 45–59. IAEA, Vienna (1974).
160. cf. Ref. 63.
161. MYERS W. G., LAMB J. F., JAMES R. W. and WINCHELL H. S. *J. nucl. Med.* **11,** 637 (1970).
162. MYERS W. G., LAMB J. F., JAMES R. W. and WINCHELL H. S. *Nuclear-Medizin* **12,** 154 (1973).
163. ROBINSON G. D., JR., USZLER J. M. and BENNETT L. R. *J. nucl. Med.* **16,** 561 (1975).
164. BROWNELL G. L. *Strahlentherapie* **60,** 13 (1965).
165. IZAWA M. and KASHIDA Y. (National Inst. of Radiological Sciences, Shiba, Japan). *Genshiryoku Kogyo* **20,** 20 (1974).

APPENDIX 1

Reviews not referenced in text.

1. SARGENT T. W. *Whole Body Counting.* IAEA, Vienna 447 (1962).
2. MATTHEWS C. M. E. *J. nucl. Med.* **6,** 155 (1965).
3. CLARK J. C., MATTHEWS C. M. W., SILVESTER D. J. and VONBERG D. D. *Nucleonics* **25,** 54 (1967).
4. STICKLEY E. E. *Radiol. Clin. No. Am.* **7,** 207 (1969).
5. SCHEER K. E. *Br. J. Radiol.* **42,** 641 (1969).
6. SILVESTER D. J. In *Radioactive Isotopes in the Localization of Tumors* (Edited by MC CREADY V. R.), Wm. Heinman Medical Books Ltd., London (1969).
7. GLASS H. I. and SILVESTER D. J. *Br. J. Radiol.* **43,** 589 (1970).
8. TER-POGOSSIAN M. M. In *Pulmonary Investigation with Radionuclides* (Edited by GILSON A. J.), pp. 205–209. Chas. C. Thomas, Springfield, Ill. (1970).
9. TILBURY R. S., MAMACOS J. P. and LAUGHLIN J. S. In *The Uses of Cyclotrons in Chemistry, Metallurgy and Biology* (Edited by AMPHLETT C. B.), p. 117. Butterworths, London (1970).
10. TER-POGOSSIAN M. M. In *Pulmonary Investigations with Radionuclides,* Miami, Florida. (Edited by GILSON A. J.), pp. 65–68. Chas. C. Thomas, Springfield, Ill. (1970).
11. BEAVER J. In *Central Nervous System Investigation*

with Radionuclides. Second Annual Nuclear Medicine Seminar, 1970. (Edited by GILSON A. J. and SMOAK W. M., III), p. 136. Springfield, Ill. (1971).
12. BROWNELL G. L., BURNHAM C. A., HOOP B., JR. and BOHNING D. E. In *Dynamic Studies with Radioisotopes in Medicine*, pp. 161–175. IAEA, Vienna (1971).
13. IYA V. K., MANI R. S. and DESAI C. N. In *Radioisotope Production and Quality Control*, p. 823. Tech. Rep. Ser. 128, IAEA, Vienna (1971).
14. SVOBODA K. In *Radioisotope Production and Quality Control*, p. 747. Tech. Rep. Series 128, IAEA, Vienna (1971).
15. DAVIES J. W. L. In *Radioisotopes in Medical Diagnosis*, p. 319. Appleton-Century-Crofts, New York (1971).
16. FAWWAZ R. A. In *Progress in Atomic Medicine*, pp. 1–18. Grune & Stratton, New York (1971).
17. GLASS H. I. In *Medical Radioisotope Scintigraphy* 1972, Vol. 2, p. 299. Vienna (1973).
18. SILVESTER D. J. In *Radiopharmaceuticals and Labelled Compounds*, Vol. 1, p. 197. IAEA, Vienna (1973).
19. COHEN B. L. *Phys. Med. Biol.* **18,** 286 (1973).
20. MONAHAN W. G., BEATTIE J. W., POWELL M. D. and LAUGHLIN J. S. In *Medical Radioisotope Scintigraphy* **1,** 285. IAEA, Vienna (1973).
21. GÖTTE H. and KLOSS G. *Angew. Chem. Int. Ed. Engl.* **12,** 712 (1973).
22. PETERS R. E. and TER-POGOSSIAN M. M. In *Radioisotopen in Klinik und Forschung* **10,** 79 Urban & Schwarzenberg, Munich–Berlin (1973).
23. TILBURY R. S. and LAUGHLIN J. S. *Semin. Nucl. Med.* **4,** 245 (1974).
24. UMEGAKI Y. (National Inst. of Radiological Sciences, Chiba, Japan). *Rinsho Kagaku* **10,** 1229 (1974).
25. FOWLER J. S. and WOLF A. P. *17th Nat. Meeting APhA Acad. Pharmaceutical Sciences*, New Orleans, Lousiana, Nov. 10–14, 4, 18 (1974).
26. TER-POGOSSIAN M. M., PHELPS M. E., HOFFMAN E. J. and MULLANI N. A. *Radiology* **114,** 89 (1975).

APPENDIX 2

Reports Containing Information on Carbon-11 and Its Compounds.
Key:

AD:	Defense Document Center, Alexandria, VA.
AEC:	Atomic Energy Commission.
BNL:	Brookhaven National Laboratory, Upton.
CEA:	Commissariat a L'Energie Atomique, France.
COO:	Chicago Operations, AEC.
CONF:	Technical Information Center, Oak Ridge.
FMI:	Franklin McLean Memorial Institute, Chicago.
NYO:	New York Operations, AEC.
SAN:	San Francisco Operations, AEC.
SUNI:	Southern Universities Nuclear Institute, Faure, South Africa.
UCLA:	University of California, Los Angeles.
URCL:	University of California, Berkeley.

1. STENSTROM T. Report AD652514 (1976). Chemical distribution of ^{11}C after proton irradiation of inulin in dilute water solution.
2. LAUGHLIN J. S., TILBURY R. S. and MAMACOS J. P. Report NYO-910-124 (1969). Biological Effects of Radiation and Related Biochemical and Physical Studies. Proposal No. 4. Cyclotron facility and radionuclide production. Progress Report May 1, 1968–May 1, 1969.
3. LAUGHLIN J. S., TILBURY R. S. and MAMACOS J. P. Report AT(30-1) 910 AEC Sept. 30, 1969. Cyclotron Facility and Radionuclide Production.
4. WINCHELL H. S. WINSTEAD M. B. URCL-19420 (1969). Visualization of radioactivity in the dog following administration of various ^{11}C-carboxylates.
5. BROWNELL G. L. (COO-3333-1) 1971 Technical Progress Report January 1, 1971–October 31, 1971. Positron scanning with computer interface.
6. LAMBRECHT R. M. and WOLF A. P. Proc. Panel of IAEA, Amsterdam (1971). Accelerator produced nuclides and radiopharmaceutical production.
7. LAUGHLIN J. S., TILBURY R. S. and KUO T. Y. T. Sloan-Kettering Inst. for Cancer Research, N.Y. (1971) Contract AT(30-1)-910. Biological effects of radiation and related biochemical and physical studies.
8. HARPER P. V., SKAGS L. S., PORTER J., LEMBARES N., FORSTHOFF H., KRIZEK H., LATHROP K. A. and KUCHNIR F. Conf. Proc. No. 9 (1972) 627. See CONF-720717. Argonne Cancer Research Hospital Cyclotron.
9. LATHROP K. A., HARPER P. V., RICH B. K., DINWOODIE R., KRIZEK H., LEMBARES N. and GLORIA I. FMI 1000 450 (1972) 16. Rapid incorporation of short lived cyclotron-produced radionuclides into radiopharmaceuticals.
10. MORRISON R. T. and MINCEY E. K. Conf. Proc., No. 9, 650 (1972). See CONF-720717. Nuclear Medicine uses of TRIUMF.
11. PRETORIUS R., BAIN C. A. R., PEISACH M., KRITZINGER J. J., VERBRUGGEN R. and SCHMITT H., SUNI-23 (1972) 35. Production of Carbon-11 in the Form of CO and CO_2.
12. ATKINS H. L. (CONF-730119-3) from American Physical Society Meeting, New York, New York (29 Jan. 1973). Radioisotopes in medicine.

13. FOWLER J. S., WOLF A. P., CHRISTMAN D. R., MACGREGOR R. R., ANSARI A. and ATKINS H. (BNL-18379)(CONF-740203-2) 1973, 23. Carrier-free ^{11}C-labelled catecholamines in nuclear medicine and biochemical research.
14. LIN T. H. and WINCHELL H. S. (SAN-849-2) March 1973 (Contract AT904-3)-849, p. 12. Synthesis and evaluation of ^{11}C labeled organic compounds for use in nuclear medicine.
15. LAUGHLIN J. S., TILBURY R. S. and KUO T. Y. T. AEC Progress Report AT (11-1) 3521, April 30, 1973. Biological effects of radiation and related biochemical and physical studies.
16. *Scientific Seminars of Nuclear Medicine*—Laboratory Dedication. (Edited by VAN TUINEN R. and SCHOLZ K. L.), November, 1973. DHEW Publication (FDA) 74-8012.
17. SODD V. J. In *Scientific Seminars of Nuclear Medicine*. FDA-74-8012, pp. 1–31. Production of radionuclides and preparation of radiopharmaceuticals: Panel discussion. Nov. 1973.
18. LAUGHLIN J. S., TILBURY R. S. and KUO T. Y. T., Sloan-Kettering Inst. for Cancer Research N.Y. Oct. 31, 1974. (COO-3521-142). Contract AT(11-1)-3521. Report Period May 1, 1973–October 31, 1974. Biological effects of radiation and related biochemical and physical studies. Production of radionuclides, labeled compounds and fast neutron beams.
19. LAUGHLIN J. S. and MYERS W. P. L., Sloan-Kettering Inst. for Cancer Research, New York (1974). (COO-3521-144) Biological Effects of Radiation and Related Biochemical and Physical Studies. Contract AT(11-1)-3521. Report Period May 1, 1973–October 31, 1974. Metabolic studies in cancer with radioactive isotopes.
20. LATHROP K. A., HARPER P. V., RICH B. H., DINWOODIE R., KRIZEK H., LEMBARES N. and GLORIA I., Mar. (1974) in Semiannual Report to the Atomic Energy Commission. (FMI-41, pp. 68–76). Rapid incorporation of short-lived cyclotron-produced radionuclides into radiopharmaceuticals.
21. MAZIERE M., MARAZANO C. and COMAR D. (CEA Centre d'Etudes Nucleaires de Saclay, 91-Gif-sur-Yvette (France). Dept. de Biologie). 1974. (CEA-CONF-2757) Labelling of drugs with carbon-11 of interest in nuclear medicine.
22. Annual Progress Report for AEC, Contract No. AT(04-1)GEN-12 for period ending June 30, 1974. UCLA-12-966, pp. 71–111 (1974). Clinical and Basic Nuclear Medicine.
23. COMAR D., CROUZEL C. and KELLERSOHN C. (CEA-Conf-2758) (CEA, 91-Orsay, France). Conf-74069-1). From 16th French Colloquim of Nuclear Medicine, Clermont-Ferrand, France (1974). Production of and medical interest in the radioisotopes ^{11}C, ^{13}N, ^{15}O and ^{18}F.
24. COMAR D., CEA-CONF-2959 Vienna, Dec. 9–13, 1974. Tumor localization using compounds labelled with cyclotron produced short lived radionuclides.
25. LIN T. H. and WINCHELL H. S. (Medi-Physics, Inc., Emeryville, Calif.) Jan. 1975. Contract AT(04-3)-849. (SAN-849-3). pp. 3–10. Synthesis and evaluation of ^{11}C-labeled organic compounds for use in nuclear medicine.
26. The Franklin McLean Memorial Research Institute, Annual Report (19[5) (Edited by LANZL E. F.), FMI-42, Contract E(11-1)69. pp. 1–61.

International Journal of Applied Radiation and Isotopes, 1977, Vol. 28, pp. 49 – 52. Pergamon Press. Printed in Northern Ireland

Synthesis of Methyl Iodide-^{11}C and Formaldehyde-^{11}C

CHRISTIAN MARAZANO, MARIANNICK MAZIERE, GÉRARD BERGER and DOMINIQUE COMAR

Service Hospitalier Frédéric Joliot, Département de Biologie, Commissariat à l'Energie Atomique, 91406 Orsay, France

(*Received* 8 *March* 1976)

A fast 2-stage method of preparing formaldehyde-^{11}C and methyl iodide-^{11}C is described. $^{11}CO_2$ is reduced to methanol-^{11}C by $LiAlH_4$ in THF, this product leading to formaldehyde-^{11}C by oxidation on a silver catalyst, or to methyl iodide-^{11}C by reaction with hydriodic acid. By means of these two molecules it is easy to methylate certain chemical groups and the method has been used to label substances of biological interest for research *in vivo*. The specific activity at the moment of use is 20 to 30 mCi/μmole, which indicates a high isotopic dilution.

INTRODUCTION

THE PROPERTIES of carbon-11 make it suitable as a tracer for the dynamic study of organic molecules in a living organism, but complicate its use in organic synthesis. Its short half-life (20.4 min) limits the chemical reaction time, while its very high specific activity (10^4 Ci/μmole) necessitates working in small volumes with small quantities of reagents. The first problem was to produce the radioisotope in a form suitable for rapid incorporation into a given organic molecule. Two reagents, formaldehyde and methyl iodide, were chosen for their ability to methylate groups present in very many natural substances, the amino, thiol or alcohol groups for instance.[1] Since a direct synthesis of formaldehyde-^{11}C from $^{11}CO_2$ hardly seemed possible under our experimental conditions it was decided, like CHRISTMAN *et al.*[4] and STRAATMANN *et al.*[5], to use a two-stage process with methanol as intermediate product, the latter also forming the precursor of methyl iodide-^{11}C.[6]

MATERIAL AND METHOD

All stages take place in a single system chosen for its small size and simplified as much as possible to minimize the dangers of irradiation (Fig. 1.). A variable speed current of nitrogen is used as carrier gas for the different transfers.

1. *Preparation of* $^{11}CO_2$

$^{11}CO_2$ is obtained by 20 MeV proton irradiation of nitrogen (80 cm-long target under 2 bars pressure), according to the nuclear reaction $^{14}N(p, \alpha)^{11}C$, then collected at $-20°C$ in a glass tube containing a molecular sieve (5 Å, 60/80 mesh).

2. *Preparation of methanol-*^{11}C

The $^{11}CO_2$ trapped on the molecular sieve is released by heating to 220°C and carried by a current of nitrogen (100 ml/min), after passage over dehydrating agent (phosphoric anhydride), into a three-necked flask (125 cm^3) containing 50μl of a 2 M lithium aluminium hydride solution in tetrahydrofuran (THF) diluted in 200 μl of the same solvent and cooled to $-20°C$ to avoid evaporation. The temperature of the solution is afterwards brought to 60°C, then the THF is distilled and eliminated in a current of nitrogen or under vacuum. The residue is immediately hydrolysed by 0.2 ml water, releasing at least 90% of the initial activity in the form of methanol-^{11}C.

3. *Preparation of methyl iodide-*^{11}C

The methanol-^{11}C released by hydrolysis from its complex with the hydride is carried by a current of nitrogen (100 ml/min) into a three-necked flask containing 1 ml hydriodic acid

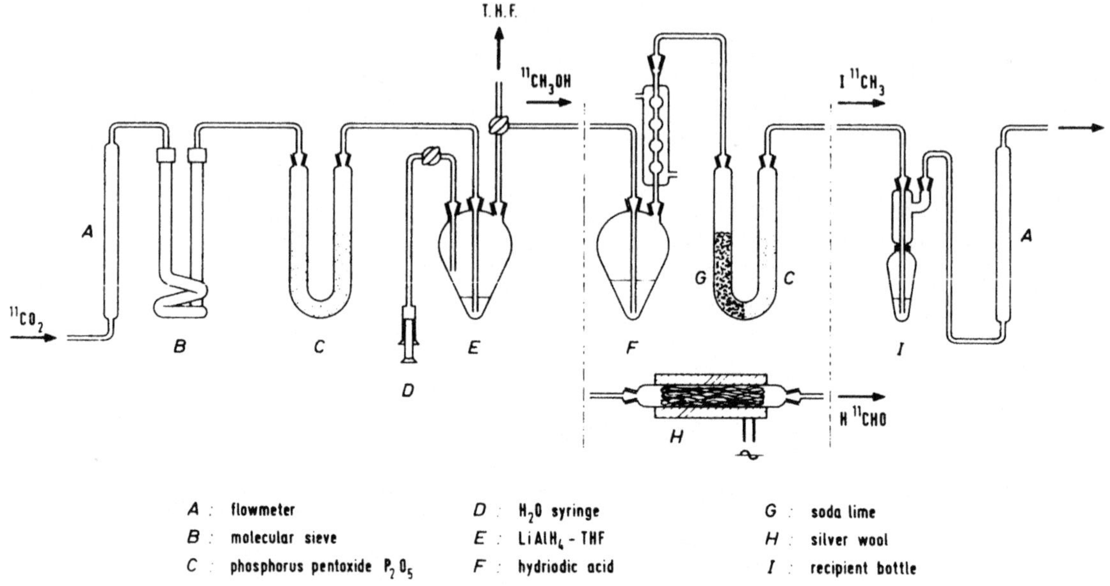

FIG. 1. Apparatus for production of $I^{11}CH_3$ and $H^{11}CHO$.

boiled under reflux conditions.[7,8] The methyl iodide-^{11}C formed is carried over soda lime then phosphoric anhydride and finally collected at -20°C in a miscible solvent, generally acetone. The yield is about 80% in 5 min. The methyl iodide can be determined by gas chromatography on a 68–80 mesh firebrick 20% reoplex column. Results of about 2–3 μmoles are obtained.

4. *Preparation of formaldehyde-^{11}C*

Methanol-^{11}C is carried by a current of nitrogen containing 4% oxygen (100 ml/min) and oxidized by passage through a quartz tube (length: 18 cm; dia.: 2 cm) containing 15 g silver wool (99.9% silver—PROLABO) heated to 450°C. The products formed are collected in 0.5 cm^3 water at 0°C or in a medium suitable for later synthesis. This stage takes 3 min. The formaldehyde-^{11}C yield is evaluated after addition of carrier by selective precipitation with dimedone. The yields found by this method range from 50 to 60%, obtained on about 50 preparations. Secondary products consist of non-oxidized methanol-^{11}C and $^{11}CO_2$ which can be detected by trapping in molecular sieve at 0°C. The quantity of formaldehyde produced, determined by Nash's colorimetric method,[9] falls regularly between 1 and 2 μmoles.

RESULTS

From 100 mCi $^{11}CO_2$ it is thus possible to obtain routine preparations of about 30 mCi $H^{11}CHO$ in 8 min and 50 mCi $I^{11}CH_3$ in 10 min, the specific activity in both cases lying in the region of 15–30 mCi/μmole.

DISCUSSION

These results call for a first comment. A considerable isotopic dilution is observed since the specific activity of the labelled molecules obtained is 15 to 30 mCi/μmole, nowhere near that of carbon-11 (10^4Ci/μmole). This means that carbon is introduced from the atmosphere, which is not surprising since this latter contains at least 14 μmoles CO_2/l. (330 ppm). At what stage does this occur? First of all it is clear that, given the affinity of lithium aluminium hydride solutions for CO_2, it is difficult to avoid atmospheric contamination during their preparation. However, if this operation is carried out under very pure nitrogen and in a small volume (serving for both dissolution and filtration) it is possible to obtain a clear 2M hydride solution in

THF (previously redistilled on $LiAlH_4$ then kept under nitrogen, away from light, on 4 Å molecular sieve) which by hydrolysis releases no more than 2 μM methanol/ml (the methanol is determined by gas chromatography on a Porapak P. column). Such a solution can be kept for several months under nitrogen and away from the light. To obtain the best specific activity, it is therefore necessary to use small quantities of hydride to reduce the $^{11}CO_2$, although under our experimental conditions it is not possible to use less than 20μmoles in 100 μl THF since the reduction is then too slow or incomplete. It may be noted that solvents of high boiling point (diethylene glycol diethyl ether,[10] solvent T[11]) have been used to prepare hydride solutions. Being non-volatile they trap the $^{11}CO_2$ even at 60°C, and also avoid the evaporation stage necessary when THF is used. Unfortunately on heating they give off undesirable products, particularly diethylene glycol diethyl ether which produces ethanol. Diethyl ether must be avoided under the conditions employed since it has been found to give rise by passage over the silver catalyst at 450°C to the formation of formaldehyde, acetaldehyde, methanol and ethanol.

Reduction of 0.5–6 μmoles CO_2 (given off from potassium carbonate) in the system used (50 μl 2M $LiAlH_4$ in THF and nitrogen carrier) was shown to take place with a constant yield of at least 80%.

These tests also showed that the hydride is responsible to only a small extent (0.2 μmole) for the isotopic dilution observed. Most of the contamination therefore comes from the irradiation circuit and trapping system although the causes are not yet clearly defined.

Finally, some comments are necessary on the technical problems raised by the use of silver wool to catalyse the methanol–formaldehyde conversion.[12] In fact this method, though simple, needs adjusting before use, and systematic tests were therefore carried out in order to optimize the yields. This is achieved by oxidizing 100 μg methanol diluted in 100 μl water and measuring the quantity of formaldehyde produced by colorimetry.[9] The first problem is to obtain reproducible yields because the catalyst is found to become more active with use, the result being an increase in the CO_2 production at the expense of formaldehyde. This disadvantage is eliminated if the silver is activated beforehand by oxidation of 2 ml methanol, carried in air at 550°C. The catalyst is then at its optimum efficiency and remains stable enough to allow a number of operations (more than 50) before being deactivated once more. It is necessary however to use only a nitrogen–oxygen mixture as carrier gas since nitrogen alone (oxidation by dehydrogenation) leads to deactivation phenomena and random yields. Several parameters have been studied. The proportion of oxygen has little effect, concentrations from 1 to 20% giving much the same yields. On the other hand for a given amount of silver the yields depend on the gas flow rate and above all on the temperature of the catalyst. It is convenient to choose a flow rate first (40–300 ml/min) and then to determine the adequate temperature (from 400 to 550°).

CONCLUSION

Formaldehyde-^{11}C and methyl iodide-^{11}C produced routinely in this way can be used to methylate either amino, amide or thiol groups. Since these chemical groups are fairly common in biologically active molecules a number of drugs or chemical mediators have thus been successfully labelled, 10 to 20 mCi of labelled product usually being obtained from 100 mCi $^{11}CO_2$. However in certain cases the method is limited by the specific activity which, though much higher than that of carbon-14, is still far from the theoretical specific activity of ^{11}C. Improvements are still necessary therefore, especially for the labelling of peptide hormones at physiological doses[13] or of toxic products, injectable at doses well below the pharmacological activity threshold.

REFERENCES

1. Maziere M., Marazano C. and Comar D. 16° Colloque de Médecine Nucléaire de Langue Française. CLERMONT-FERRAND (France) 6–9 juin (1974).
2. Weygand F. and Linden H. *Angew. Chem.* **66,** 174 (1954).
3. Lane A. C., McCoubrey A. and Peaker R. *J. Labelled Compounds* **2,** 284 (1966).
4. Christman D., Crawford E. J., Friedkin M. and Wolf A. P. *Proc. Nat. Acad. Sci. USA* **69,** 988 (1972).

5. Straatmann M. G. and Welch M. J. *J. nucl. Med.* **16,** 425 (1975).
6. Comar D., Maziere M. and Crouzel C. *Radiopharmaceuticals and Labelled Compounds,* vol. 7, p. 461. IAEA, Vienna (1973).
7. Harman D., Stewart T. O. and Ruben S. *J. Am. Chem. Soc.* **64,** 2293 (1942).
8. Murray A. and Ronzio A. R. *J. Am. Chem. Soc.* **74,** 2408 (1952).
9. Nash T. *Biochem. J.* **55,** 416 (1953).
10. Nystrom F., Yanko W. H. and Brown W. G. *J. Am. Chem. Soc.* **70,** 441 (1948).
11. Cox J. D., Turner H. S. and Warne R. J. *J. Chem. Soc.* 3167 (1950).
12. Weygand F. and Schaefer G. *Ber* 307 (1952).
13. Marche P., Marazano C., Maziere M., Morgat J. L., de la Llosa D., Comar D. and Fromageot P. *Radiochem. Radioanal. Lett.* **21,** 53 (1975).

Since this paper was first proposed the specific activity of $HCHO^{11}C$ was increased to 100 mCi/μM by replacing the molecular sieve trap by liquid oxygen and by degassing the target under radiation. Irradiation conditions routinely used are 15 μA/15 min 18 MeV protons which gives rise to 300 mCi $^{11}CO_2$.

International Journal of Applied Radiation and Isotopes, 1977, Vol. 28, pp. 53–65 . Pergamon Press. Printed in Northern Ireland

The Preparation of Fluorine-18 Labelled Radiopharmaceuticals

A. J. PALMER, J. C. CLARK and R. W. GOULDING
MRC Cyclotron Unit, Hammersmith Hospital, Ducane Road, London W12 0HS, U.K.

(*Received* 3 *April* 1976)

Progress in the preparation of fluorine-18 labelled radiopharmaceuticals is reviewed. The pharmacology and design of ^{18}F-labelled organic compounds of biomedical interest, production of the radionuclide and methods of labelling which have been used to date are discussed. Emphasis has been placed on the practical chemical problems encountered, together with the important fields of product purification and quality control.

INTRODUCTION

IN THE past three decades the study of fluorine compounds has become a major branch of organic chemistry, concerning both highly fluorinated systems and those in which fluorine may replace only one or two hydrogen atoms in an organic molecule. Many fluorine-containing compounds have found use in medicine and in recent years some research has been directed towards the preparation of radiopharmaceuticals incorporating the only radionuclide of fluorine with a useful half life, ^{18}F ($T_{1/2}$ 110 min β^+). This article is concerned with the preparative aspects of incorporating ^{18}F into organic molecules of potential biomedical interest, together with the design of such compounds. ROBINSON has recently reviewed their potential applications[(1)] while the use of ^{18}F as fluoride ion for bone scanning and dental studies is well known.

For labelling organic molecules ^{18}F is usually considered in conjunction with three other common cyclotron produced β^+ emitters, ^{11}C, ^{13}N and ^{15}O. The *in vivo* detection of these nuclides requires specialized detectors and electronics if their full potential is to be realized, but using modern positron imaging devices[(2–4)] it is possible to obtain more detailed positional information than is obtainable from a γ-emitter. From the biochemical point of view ^{11}C, ^{13}N and ^{15}O are ideal for labelling an organic radiopharmaceutical since no "foreign" atom is introduced. However, they all have very short half-lives, and if this factor proves unacceptable then ^{18}F may offer an alternative. Since, with minor exceptions fluorocarbon compounds do not occur in nature, they are wholly artificial and consequently the electronic properties and size of fluorine relative to hydrogen (which it usually replaces in an organic molecule) must be considered together with the related pharmacological effects.

Superficially ^{18}F would appear to be an ideal label from the radiochemical point of view. It may be produced in a reactor or on a cyclotron using a variety of nuclear reactions. Times of up to 4 hr ($2 \times T_{1/2}$) have been considered acceptable for chemical syntheses. As stated previously fluorine usually replaces hydrogen, consequently a large number of possible positions of labelling are frequently available, and the very high dissociation energy of the C–F bond means that these compounds may be expected to have improved *in vivo* stability when compared for example with similar iodine compounds.

Many fluorination reactions give a low yield based on fluorine, and classical methods of introducing fluorine use extreme reaction conditions and very reactive or corrosive reagents. Often simple compounds have been fluorinated and then transformed into more biologically interesting compounds by a series of time consuming reactions. The aim in ^{18}F chemistry is to avoid such reactions and procedures and where possible adopt the most direct and rapid high yield method. It is always desirable to optimize the yield with respect to the ^{18}F in the starting material. Biosynthetic techniques,[(5,6)] which are frequently used for preparing other radiopharmaceuticals and have the advantage that they yield high specific activity products, cannot be used for the direct introduction of inorganic ^{18}F although they may of course be used for subsequent transformations.

Up to the present time one of the principal problems in the preparation of ^{18}F labelled compounds has been that it has proved difficult to prepare them in specific activities comparable with those of other radiopharmaceuticals. This may lead to toxicity problems and a general lack of pharmacological data may be an additional handicap in this respect.

Pharmacology of fluoro-compounds

1. C–F *in place of* C–H. Fluorine is of comparable size to hydrogen and both form strong bonds to carbon, but the chemical reactivities are very different. The enzymic effect of this is that the analogue with fluorine at the reaction centre will very often act as inhibitor rather than substrate. With a fluorine adjacent to the reaction centre the "normal" reaction may still occur but at a greatly changed rate. A remote fluorine atom usually has no effect on the process. Thus the introduction of a halogen atom into an aromatic ring system often prevents biochemical reaction at that position, an effect described as "obstructive halogenation".[7] The effect is very pronounced for fluorine especially at the para position of simple phenyl groups when the metabolically important enzymatic hydroxylation can be severely inhibited.[8] If hydroxylation cannot occur then neither can subsequent conjugation, and the tissue distribution is affected. As an example, Fig. 1 outlines the possible *in vivo* fates of three selected fluoroamino acids. The tissue distribution of ^{18}F at any time may be seen to be due to a range of metabolites in addition to the administered compound (I, II, III).

These considerations are important when it is desired to produce γ-emitting labelled molecules which have known biochemical properties and may tend to localize in specific tissues. For this purpose a single fluorine atom should be placed in a site remote from functional groups. It is useful to synthesize as many isomers of the ^{18}F-labelled compound as possible so that the differences can be evaluated.

2. CF_3 *in place of* CH_3, Br *or* NO_2. When a CF_3 group replaces a CH_3 group in a molecule the change in properties is usually fairly pronounced. The former has significant steric crowding effects and is very strongly electron withdrawing. In aliphatic systems a remote CF_3 group can be reasonably inert (mimicking CH_3) for example in the leucine analogue (IV)[10].

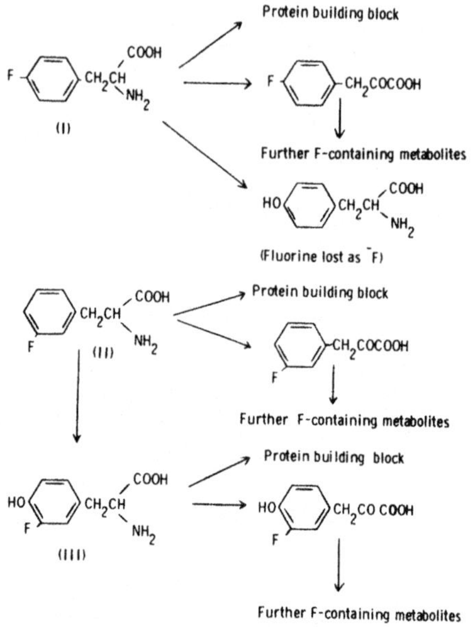

FIG. 1. *In vivo* transformations of *p*-fluorophenylalanine (I) *m*-fluorophenylalanine (II) and 3-fluorotyrosine (III).[8,9]

In aromatic systems the CF_3 group normally resembles the NO_2, CN and SO_2–R groups in electronic effects, and in particular activities ortho and para positions to nucleophilic attack. In more complex aromatic systems the effect of the CF_3 group is usually not so pronounced, when it may resemble Br, Cl or even CH_3 in biological behaviour.[8,11]

Therefore a labelled CF_3 group could be introduced to make a ^{18}F-labelled analogue of a CH_3, Cl, Br or NO_2 parent compound under favourable circumstances.

3. C–F *in place of* C–OH. A fluorine atom can

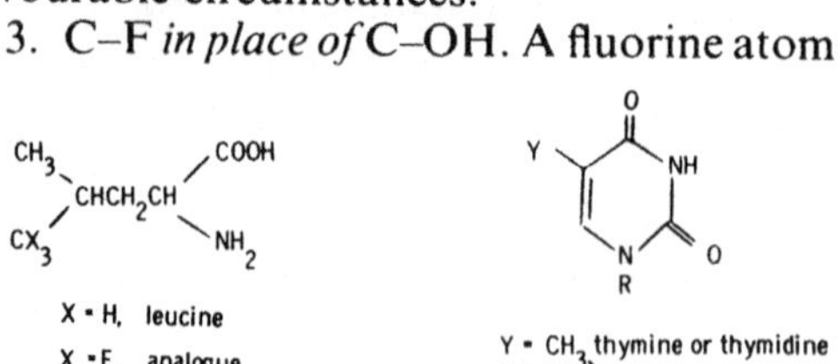

FIG. 2. Trifluoromethyl analogues of leucine (IV)[10] and thymine (V).[12]

also be introduced as the analogue of a hydroxyl group. Pharmacological application has been concentrated on polyhydroxy compounds such as carbohydrates. The fluorine atom can only act as a hydrogen bond acceptor, whereas the hydroxyl group can act as donor and acceptor. It is best to introduce a single fluorine atom because polyfluoro compounds have a different pharmacology. It has been shown that one monofluoro analogue of glucose out of all the possible isomers is transported in the same manner as glucose itself.[13] This is because only in this isomer can the same H-bonding occur in the receptor site of the transport molecule.

4. $-C_6H_4{}^{18}F$ *as a protein label.* Various hapten groups have been used for tagging proteins and enzymes for non-radioactive tracer work. Proteins have been labelled directly with radioiodine, or with ^{111}In *via* covalently bound EDTA groups[14] (these are often not very stable *in vivo*). It would appear that a perfect radio-label would have a half-life of 2–24 hr and would be covalently bound to the protein *via* a spacer group. Proteins have been usefully labelled with ^{11}C-methyl groups despite the short half life.[15,16] ^{18}F labelled fluorophenol, fluoroaniline and fluorobenzoic acid are readily prepared (see under reaction scheme 1) and could be used to label proteins by established procedures.

Up to the present time only ^{18}F-labelled compounds of the type described in section (1) have been reported.

Stereochemical effects

Biological activity is closely related to the stereochemistry of the substrate molecule, and when such a molecule exists in two or more isomeric forms (geometrical or optical) the biological activity of each of these forms can be very different. This isomerism may already be present in the molecule or it may be generated when a foreign atom (the label) is introduced.

In those cases where a single stereo-isomer of a labelled compound is required it may be possible to use a synthetic sequence in which the precursor to be labelled is available as a single stereo-isomeric form and can be converted to the final product via stages which proceed with effectively 100% stereospecificity. Alternatively an epimeric or racemic mixture may be generated during or after the labelling stage (cf. the amino acids below) and this may be followed by a procedure for resolution. The consequent loss of time and material must be offset against the fact that this may be an easier technique than a stereospecific synthesis.

Strategy of ^{18}F *labelled synthesis*

Because of the short half life it is essential to introduce the ^{18}F label at the latest stage possible. Usually a suitable precursor is prepared in bulk and stored for future use. Most reactions used for ^{18}F labelling have to be completed in less than the optimum times and using other than the optimum relative amounts of reactants. Consequently mixtures of labelled and/or unlabelled products are often obtained. Preparative gas (gc) or liquid chromatography (lc) are especially useful separation techniques since they can be carried out rapidly. Solvent extraction and sublimation may also be useful. Conversely, preparative thin layer chromatography is not so useful because spotting, zone identification and elution are relatively lengthy processes.

^{18}F labelled compounds for *in vivo* use are preferably administered in approximately isotonic solutions. Because of the short half-life, lengthy de-salting or solvent removal stages are not practicable. The best approach is to either be able to isolate a non-polar compound from a volatile organic solvent such as ether, or to carry out reactions in aqueous solutions of low ionic strength. Immobilized reagents are especially useful (e.g. H^+-form ion exchange resin in place of acids, or F^--form resin for fluorination[17]). If any final liquid chromatographic purification is required the preferred eluant is water or an approximately isotonic, non-toxic buffer.

Analytical considerations

It is necessary to establish the identity and homogeneity of any starting materials one may synthesize for subsequent ^{18}F-labelling. Infrared and nuclear magnetic resonance (nmr) are useful for the former, and thin layer (tlc) or gas (gc) chromatography for the latter.

Since the range of compounds that have been labelled with ^{18}F is not very extensive new reactions attempted in future will present new problems to be solved. Development work normally requires a lengthy series of trial reactions which are followed by various standard radio-chromatographic techniques e.g. tlc/autoradiography. The technique of coupled glc-mass spectrometry is valuable for the analysis and identification of mixtures that are volatile or can be made volatile by derivatization.

When a complete synthesis of the required ^{18}F-compound has been achieved it is often not possible to perform an extensive chromatographic analysis before the *in vivo* use because of the short half-life of the label. Therefore, in a series of preliminary experiments employing stable or active fluorine and using the projected procedure, the identity and homogeneity of the product should be ascertained. Then a standardized experimental procedure is established and the final product may be analysed by a shortened procedure concurrently with its utilization. In order to determine the specific activity of the solution to be administered, measurement of the u.v. absorbance at a selected wavelength is often a valuable and rapid method.

The points mentioned in this section may appear to be self evident, but failure to appreciate them will invariably give misleading results. Much information in the literature concerning short-lived labelled compounds is unsatisfactory because adequate product analysis is not described. Certain claimed preparations do not quote any method of analysis and the phrase "analysed by tlc" has appeared without a description of the stationary or mobile phase used. Even when exchange processes are used compound degradation may occur and so the identity of the labelled "product" can never be assumed without proof.

PRODUCTION OF THE RADIONUCLIDE

Some nuclear reactions which lead to the production of fluorine-18 are listed in Table 1. The excitation curves for these nuclear reactions together with the thick target saturation activities for deuterons and tritons of up to 24 MeV, and ^3He and ^4He up to 42 MeV have been measured by NOZAKI[18].

TABLE 1. Some nuclear reactions for the production of Fluorine-18

No.	Target	Projectile	Reaction
1	O	^3H	$^{16}O(^3H, n)^{18}F$
2	O	^4He	$^{16}O(^4He, pn)^{18}F$
3	O	^3He	$^{16}O(^3He, p)^{18}F$
4	Ne	d	$^{20}Ne(d, \alpha)^{18}F$
5	Ne	^3He	$^{20}Ne(^3He, \alpha p)^{18}F$

Reaction 1 has been in use for many years using the 2.73 MeV (max) tritons generated in a nuclear reactor by the $^6Li(n, \alpha)^3H$ reaction.[19] The target material is usually Li_2CO_3 with a ^6Li enrichment of at least 90%. Fluorine-18 is recovered as $^{18}F^-$ without the addition of carrier by HF distillation into aqueous alkali,[20–22] or by ion exchange.[23,24] Partial removal of the undesirable impurity tritium (typically 750 mCi ^3H to 2.5 mCi ^{18}F[23]) as water and $^3H_2$ is achieved by repeated evaporations to dryness. The traces of tritium that remain after this treatment have been identified as ^3H-acetate and ^3H-formate.[25]

The recovery of anhydrous $H^{18}F$ from reactor irradiated Li_2CO_3 by distillation of HF onto ion exchange resin which is subsequently dried and eluted with HF has been described.[26] However, as up to 1.5 g of HF was used to achieve a 90% recovery from the resin, the resulting $H^{18}F$ was of low specific activity.

Improvements relating to the Li_2CO_3 target design have been described which enable high neutron fluxes to be used more effectively and batches of 65–75 mCi of fluorine-18 to be prepared.[22]

Reactions 2 and 3 employing oxygen[26] or water[27–30] as target materials have been extensively used for ^{18}F production. Aqueous solutions of ^{18}F from water targets have been used to label aromatic diazonium fluoroborates by exchange.[31,32] Fluoride has been recovered from aqueous solutions using a fluoride form anion exchange resin column which after careful drying has been used in anhydrous interhalogen exchange reactions to label a variety of aliphatic compounds with fluorine-18.[17] The use of an oxygen gas target to produce anhydrous HF has been demonstrated.[26] Here the ^{18}F trapped heterogenously by the silver plated target walls during irradiation is subsequently recovered by exchange with ~1 g of anhydrous HF carrier.

Reactions 4 and 5 have received a great deal of attention in recent years, particularly under anhydrous conditions. Here the aim has been to produce a variety of high specific activity fluorinating agents. The investigations have followed three main approaches.

1. *Homogeneous scavenging of ^{18}F with recovery during irradiation*

The mixing of nitric oxide, hydrogen or chlorine with neon before irradiation has yielded $NO^{18}F$[33], $H^{18}F$[34] and $Cl^{18}F$[34] reportedly carrier-free from nickel targets preconditioned with F_2. A recovery of 60% is reported for $NO^{18}F$ but the data for $H^{18}F$ and $Cl^{18}F$ suggest large losses of ^{18}F to the target walls. The addition of F_2 to the neon resulted in the recovery of $^{18}F_2$ with 30% efficiency.[34] However, these preliminary results provide encouragement for further development of the technique.

2. *Heterogeneous scavenging with post irradiation recovery*

The losses of ^{18}F in the homogeneous scavenging approach are largely to the walls of the target vessel. There are several reports of the recovery of carrier-free ^{18}F by the washing of glass-lined target vessels with normal saline[35] or water[36,37] with recovery efficiencies of around 60–90%. The washing of a glass-lined target with dilute solutions of AgF, BF_3 and aromatic fluoroborates in anhydrous organic solvents with 50–80% recoveries described by Ido (1974) appears to have some attractive features although the specific activity of the fluorinating agent is reduced by the introduction of carrier.[38]

A further heterogeneous approach to ^{18}F recovery has been to coat the target walls with a thin layer of the organic compound to be labelled. After irradiation the organic material is recovered in solution by washing the target walls. Inorganic ^{18}F is removed by ion exchange and the labelled product separated from the starting material by ion exchange chromatography.[39]

The fluorinating agents AgF, KF and SbF_3 (100 mg–1 g) have been labelled by a similar technique.[18] A variation of the technique has been described where a target lined with AgF or AgF_2 was used. After irradiation the neon was removed and the vessel used as a reactor for the synthesis of CCl_3F and CCl_2F_2.[40] This technique is in principle applicable to many fluorinations involving volatile organic substrates.

3. *Heterogeneous scavenging during irradiation remote from the target*

The recovery of ^{18}F by exchange labelling of a small quantity of a simple or complex fluoride placed remote from the target has been demonstrated.[41] The neon is circulated during irradiation through a glass-lined target vessel, and the effluent gas is led through PTFE tubing to a filter holder where a glass fibre element, previously coated with a 5–50 mg of an ionic fluoro-compound, e.g. KF, $NaBF_4$, SbF_3, or an aromatic diazonium fluoroborate is supported. A reactor containing dry fluoride-form ion exchange resin (Dowex 1, ~2 gm) may also be used.

The volatile ^{18}F-labelled intermediate which is continuously swept from the target during irradiation is removed almost quantitatively by the coated filter. The filter element or ion exchange resin can often be used directly in the chemical syntheses that follow. This technique is useful for fluorinations involving non-volatile organic substrates.

Methods of ^{18}F Labelling

The Balz–Schiemann reaction has been used extensively for making aryl fluoro-substituted compounds. Nucleophilic displacement reactions have been used for the preparation of aliphatic fluoro-compounds. Additionally various other methods have been employed in special cases.

PREPARATION OF AROMATIC FLUORO-CARBONS VIA THE BALZ–SCHIEMANN REACTION

Compounds which are monofluorinated in an aromatic ring system often show biological activity related to the corresponding unfluorinated or phenolic compound. Hence methods of preparing such compounds in labelled form are of immediate interest. Direct mono-fluorination of aromatic rings is only possible in very special circumstances, but the Balz–Schiemann reaction[42] allows the ready conversion of an aryl nitro compound, via the amine and diazonium salt, to the corresponding fluoro-compound (Fig. 3, general reaction scheme).

The tretrafluoroborate anion of the diazonium salt may be labelled by exchange and the C–^{18}F bond is formed when the diazonium fluoroborate is decomposed thermally. Many simple ^{18}F-aryl fluoro-compounds have been prepared by this method[43] and of these, amines, phenols and carboxylic acids could be used as starting materials for the preparation of other labelled compounds. The chemical yield in the Balz–Schiemann reaction is often low, and a large amount of side-products may be produced. Also, since the fluorine is introduced as the labelled fluoroborate anion, the maximum possible radiochemical yield is 25% and in practical examples it is ~2–15%.

A further complication is introduced because other reactive groups present in the molecule can prevent any of stages I, II or III from proceding. In such cases (in fact with almost all the compounds of biological interest) it is necessary to use "protected" derivatives of such groups, prepared either by a direct protection reaction or by synthesis from simpler compounds. After stage III the protecting group(s) are removed by suitable deprotection reactions. The reported chemistry of protecting groups is very extensive especially in the peptide, steroid and carbohydrate fields,[44] but so far only the simplest protecting groups have been employed during ^{18}F-syntheses.

The diazonium fluoroborates may be labelled by exchange with carrier-free ^{18}F in water[32] or water/acetone[31] or by heterogenous exchange between the solid supported in an inert matrix and carrier-free fluorine-18 extracted from a recirculatory neon gas target.[41] The last method is far superior since exchange is instantaneous and much higher specific activities may be achieved. The labelled diazonium fluoroborate may be decomposed dry or in an inert solvent. A mixture of products is always obtained as shown by tlc (silica gel, 10% ethyl acetate in chloroform) and the ratio of various components varies with the decomposition solvent used. Biphenyl, tetrahydronaphthalene and phenanthrene give good yields, whereas dichlorobenzene and malononitrile give poor yields of the desired protected fluoro-compound.[41,45] Purification of the mixture may be effected by preparative tlc or lc when the diazonium salt is decomposed dry, but lc is the best method when a solvent is used.

1. *Aromatic amino acids and catecholamines*

Fluorine-18 labelled analogues of four naturally occurring aromatic amino acids have been prepared. These are phenylalanine, tyrosine, tryptophan and 3,4-dihydroxyphenylalanine (DOPA). The first three are utilized in the production of enzymes and proteins and the ^{18}F-labelled analogues are being investigated as potential pancreas scanning agents[46] since this organ is a site of rapid utilization of amino acids for protein production. DOPA is associated with melanin-formation and Parkinsonism. All of the ^{18}F-amino acids have been evaluated as potential melanoma localising agents[47] and fluoro-DOPA.

Abnormalities in the quantities and metabolism of catecholamines are associated with hypertension, Parkinsonism, chromaffin tissue tumours and other pathological conditions.[49] ^{18}F-6-Fluoro-dopamine is being evaluated as an adrenal scanning agent.[50] Fluorine-18 labelled derivatives of amino acids and catecholamines which have been prepared are listed in Tables 2–4.

Labelled *p*- and *m*-fluorophenylalanine may be prepared by a published method[51] used for the inactive compounds essentially without modification. A diethyl acetomidomalonate derivative of the corresponding diazonium tetrafluoroborate (Table 2) is first prepared and labelled. After the decomposition the protecting groups (ester and amide) may be removed by vigorous acid hydrolysis when the *dl*-amino acid is obtained.[31,52] Alternatively it is possible to prepare the *dl*-acylamino acid which may be stereoselectively deacetylated using the fungal enzyme amino acylase to give the *l*-amino acid.[53,53]

A mixture of the enzyme, *d*-acylamino acid and *l*-amino acid is obtained and the required *l*-amino acid can be recovered efficiently by 1c

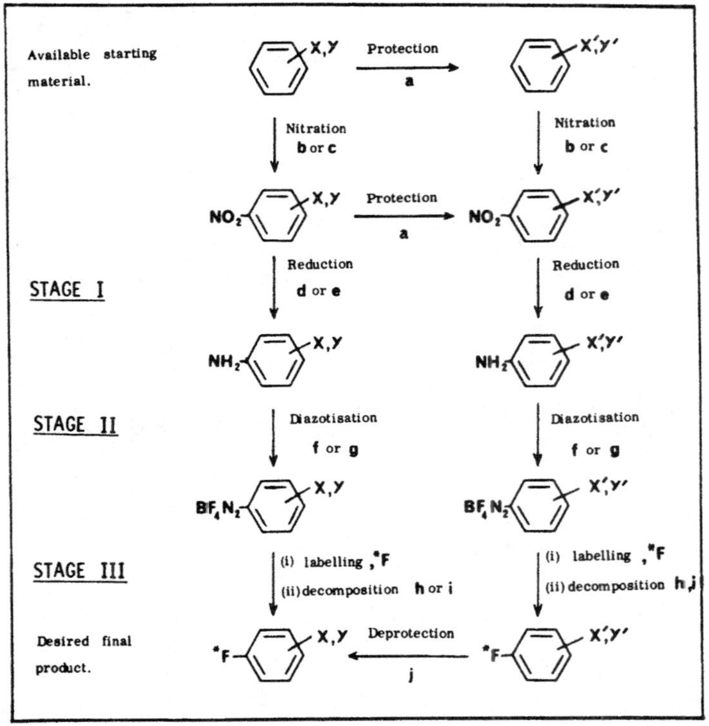

GENERAL REACTION SCHEME.

X,Y general substituents. X',Y' protected substituents.

Reagents : a see table below. b HNO_3/H_2SO_4. c HNO_3/CH_3COOH. d $Na_2S_2O_4$. e H_2, Pd/C, CH_3COOH. f HNO_2/HBF_4. g $NOBF_4$. h Δ(130-200°, heat alone). i Δ(130-200°, in inert solvent). j see table below.

PROTECTING GROUPS:

Group (X,Y)	Protected Group (X',Y')	Deprotection
$-NH_2$ (aryl, alkyl)	acetamido-, formamido- phthalimido-	strong acid or alkali hydrazine
-OH (aryl)	methyl ether	HBr, HI or BBr_3
-COOH (alkyl)	ethyl ester	alkali
amino acid	diethyl acetamido -malonate	strong acid or alkali OR (i) alkali (ii) acid (iii) enzyme

^{18}F-LABELLED COMPOUNDS PREPARED:

(a) Potential Synthetic Precursors.

Fluoro-aniline (ex $FC_6H_4NHCOCH_3$)
Fluorophenol (ex $FC_6H_4OCH_3$)
Fluorobenzoic acid
o-Fluorobenzyl chloride

References 43, 52.

(b) Biologically Important Compounds.

p-Fluorohippuric acid
o/m/p-Fluorophenylalanine (L, DL)
3-Fluorotyrosine (DL)
5-Fluoro-DOPA (DL)
5/6-Fluorotryptophan (L, DL)
6-Fluorodopamine
Haloperidol (TM)

For references see tables 2 - 4.

FIG. 3. Application of the Balz–Schiemann reaction.

5

TABLE 2. O/m/p-Fluorophenylalanines

Compound	Starting Material	Protected NO_2 - Compound	Ref
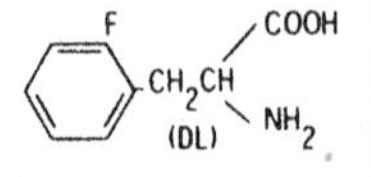	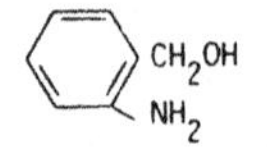	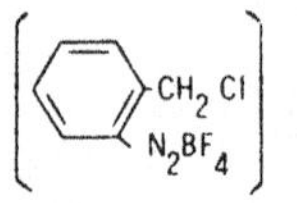	52
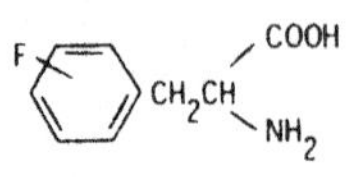 m, p (L, DL)	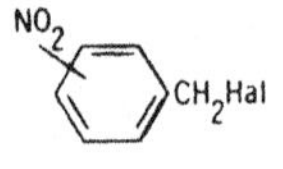	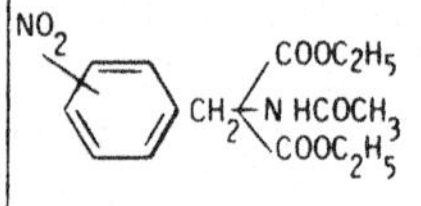	31, 41, 52, 53

TABLE 3. 3-Fluorotyrosine, 5-Fluoro-DOPA and 6-Fluorodopamine

Compound	Starting Material	Protected NO_2 - Compound	Ref
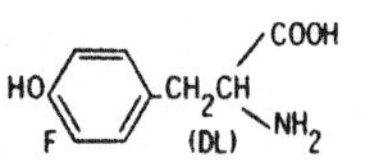	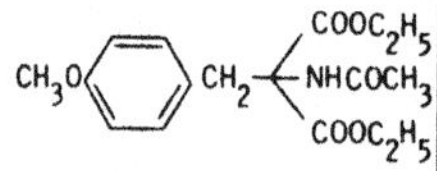	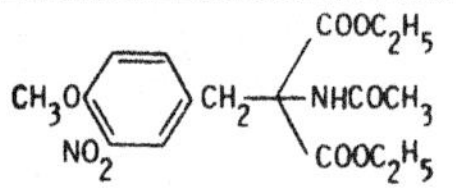	36, 41
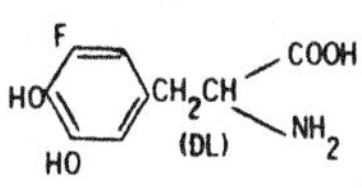	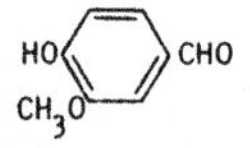	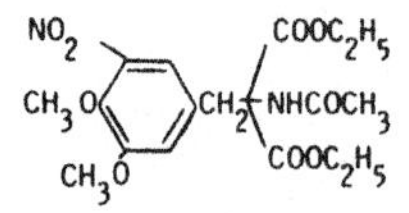	48, 55
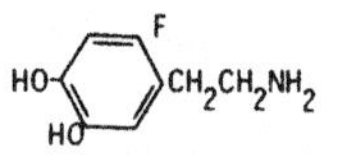	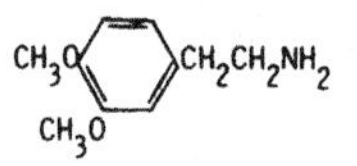	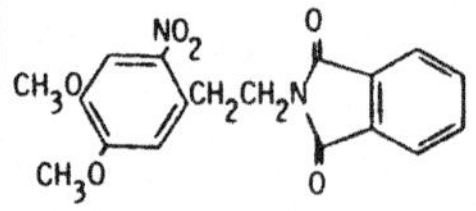	50

TABLE 4. 5/6-Fluorotryptophan

Compound	Starting Material	Protected NO_2 - Compound	Ref
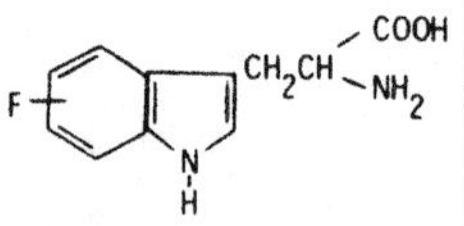 (L, DL)	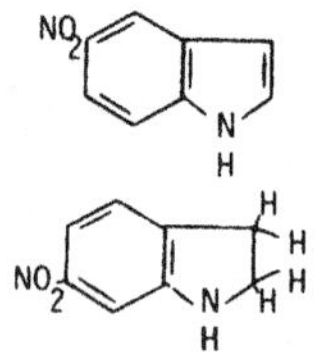	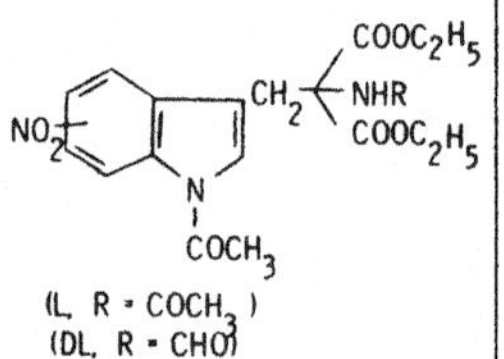	32, 41, 45

using Biogel-P2.[53] Fluorine-18 labelled *dl*-*o*-fluorophenylalanine cannot be prepared by the above route but has been synthesized by a longer method.[52]

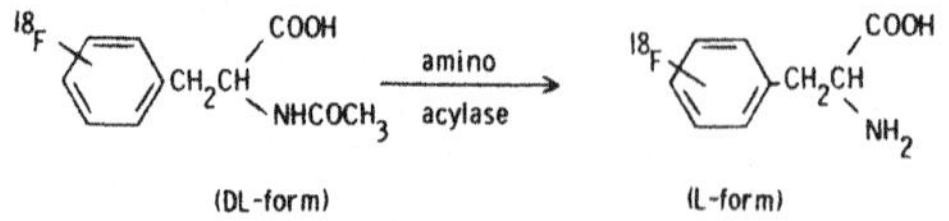

FIG. 4. Preparation of ^{18}F-labelled *L*-*p*- and *m*-fluorophenylalanine.

In the case of *dl*-3-fluorotyrosine[36] and *dl*-5-fluoro-DOPA[55] the amino acid function is generated from the diethyl acetamidomalonate group, and the phenolic OH function from methyl ether groups. The nitro-compounds shown (Table 3) can be converted into diazonium fluoroborates, labelled and decomposed to give the corresponding fluoro-compounds. These in turn (after purification) may be completely deprotected to give the *dl*-amino acid by treatment with constant boiling HBr or HI. In the case of fluoro-DOPA care is required to avoid loss of the product through oxidation and polymerization. So far it has not been possible to prepare these amino acids in the *l*-form because of chemical problems encountered.[45]

The preparation of ^{18}F-labelled *dl*-5- and 6-fluorotryptophan has been reported.[32] The 4- and 7-isomers could be prepared by similar procedures if they aroused any interest in the future. Indoles in which a free ring (1 –) NH group is present cannot be diazotized without self-reaction occurring. However successful diazotizations and subsequent decompositions have been effected through the use of ring (1 –) N-acetylindoles,[56,57] and the nitro-compounds shown (Table 4) may thus be employed. After the thermal decomposition reaction, purification is again carried out by preparative tlc or lc.

Indoles with a free ring (1 –) NH group present such as tryptophan are also unstable to to mineral acids and air oxidation. To avoid all of these problems the diethyl formamido-malonate (R = CHO, Table 4) is used in the preparation of labelled *dl*-fluorotryptophans, when the two stage (dilute alkali followed by very dilute acid) hydrolysis gives the free *dl* amino acid directly[32,58] (compare with fluoro-phenylalanine above). If the diethyl acetamido malonate ($R = COCH_3$, Table 4) is used the product is the α-*N*-acetyltryptophan derivative[58] which can be treated with the enzyme to give the labelled *l*-amino acid.[45] It should be noted the acetyl group is a very good protecting group for the ring (1 –)NH group because it is readily lost in the hydrolyses.

^{18}F-labelled 6-fluorodopamine has been prepared from the corresponding protected nitro-compound[50] by the standard stages. The catechol group was protected as the bis-methyl ether, and the amino group by phthaloylation. Deprotection was effected by hydrazinolysis followed by treatment with hydrobromic acid or boron tribromide (Table 3).

2. *Other compounds of biomedical interest*

The syntheses of ^{18}F-labelled 4-[4-(*p*-chlorophenyl)-4-hydroxypiperidino]-4′ - fluorobutyrophenone (VI)[59] (the neuroleptic drug "Haloperidol") and ^{18}F-*p*-fluorohippuric acid (VII)[38] have been reported. The former was for use in investigations of its tissue distribution and pharmacokinetics. Both these compounds were prepared directly from the corresponding diazonium tetrafluoroborates, protecting groups being unnecessary.

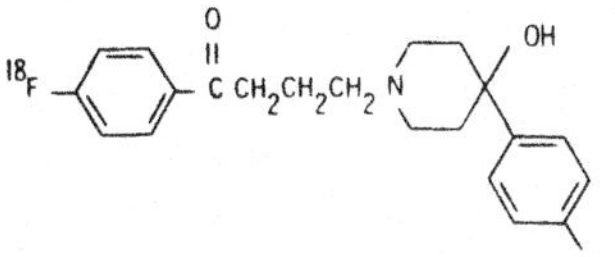

VI

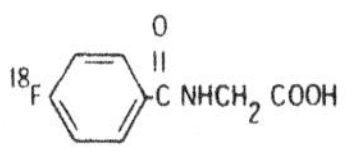

VII

FIG. 5. ^{18}F-labelled "haloperidol" (VI)[59] and *p*-fluorohippuric acid (VII).[38]

METHODS EMPLOYING NUCLEOPHILIC DISPLACEMENT

1. *Alkali metal and related fluorides*

A wide range of aliphatic fluoro-compounds have been prepared by nucleophilic displacement of a good leaving group, i.e. bromide, iodide, tosylate (– OTs) or an ammonium salt ($-NMe_3{}^+X^-$). Often a very large molar excess of the fluoride is necessary to achieve reasonable yields of organic fluoride.

The nature of the solvent used is of critical importance. It should be highly polar and ionizing, and have very low nucleophilicity[60] (in order not to compete with fluoride). It is desirable to have good solvation of the inorganic cation but poor or (preferably) no solvation of the anion. In the aprotic solvents DMA, DMSO or HMPA* the large cations K^+, Rb^+, Cs^+, Tl^+ and Bu_4N^+ are highly solvated but the fluoride ion is essentially "naked". Thus compared with other solvents, e.g. glycol, the rate of reaction is greatly increased (up to 10^6 times). A further improvement is to use catalytic amounts of a crown ether[61] (very effective solvation of cations) or a phase transfer catalyst e.g. cetyl tributyl phosphonium bromide.[62]

In order to obtain useful amounts of ^{18}F-labelled compounds (say > 500 μCi at > 1 mCi/mg) it is essential to take full account of the above considerations. The salts KF, Bu_4NF etc. are extremely difficult to render completely anhydrous especially on the small scale. The presence of traces of water will cause failure of the reaction. In view of the problems involved it

* Dimethyl acetamide, dimethyl sulphoxide and hexamethyl phosphoramide, ($PO(NMe_2)_3$).

is not surprising that relatively little use has been made of these reagents for ^{18}F- labelling.

ROBINSON[17] has developed a novel alternative technique in which fluoride-form ion exchange resin (Dowex-1) is used as the fluorinating agent

$$\text{Resin}\,[CH_2NMe_3+F^-]_n,$$

which is labelled by exchange with carrier-free fluorine-18 in neon[45] or water.[17] The success of this particular reagent may be due to the ease with which it can be rendered anhydrous (1 hr/150°C *in vacuo*). It has a certain formal similarity to the phase transfer catalysts, and, as an immobilized reagent, facilitates very rapid work up after the reaction.[63] The mixture of alkyl bromide and fluoride obtained can be separated by preparative gas chromatography. Various ^{18}F-2-fluorocarboxylic acids (protected as ethyl esters) have been prepared by the method, and from these, by reduction, the corresponding 2-fluoroalkanols. These compounds have been used for brain and heart studies and also investigation of ethanol metabolism.[64]

2. *Silver fluorides*

Silver (I) fluoride is much more versatile than the alkali metal fluorides in that it can be used to prepare *pri*-, *sec*- and *tert*-alkyl fluorides[65] and heteroaromatic compounds. The reactions are relatively rapid and the reagent is often used without solvent with both liquid and gaseous substrates (at 25–200°C). High yields of the organic fluoro-compound are obtained even with ^{18}F-labelled material (up to 50% radiochemical yield, X = Br or I):

$$RX+2AgF \rightarrow RF+AgF\cdot AgX$$

However silver fluoride is awkward to handle because it is extremely hygroscopic and corrosive (it normally cannot be used in glass). It has been employed to prepare labelled freon-11 (CCl_3F),[41] 6-fluoro-9-benzyl purine[38] and 3-fluoro-cholestene[38] (Table 5).

Silver difluoride, AgF_2, can occasionally act as nucleophilic fluorinating agent and is more reactive than the monofluoride. It is instantly hydrolysed by water. See also electrophilic fluorinations below. Mercury (I) and (II) fluorides have been used in similar way.[66,67] One fluorine-18 reaction has been reported (Table 5)

TABLE 5. Preparation of ^{18}F-labelled compounds by halogen displacement

Reagent	Substrate	Product	Ref
$^{18}F^-$ Ion exchange resin	$BrCH_2CH_2OH$	$^{18}FCH_2CH_2OH$	74
$^{18}F^-$ Ion exchange resin	Ethyl halocarboxylates	Ethyl ^{18}F-fluorocarboxylates, acids and alcohols (by subsequent transformation)	17, 74, 75
$Ag^{18}F$	CCl_4	$C^{18}FCl_3$	40
$Ag^{18}F_2$	$CFCl_3$	$C^{18}F_2Cl_2$	40
$Ag^{18}F$	3-Iodocholestene	^{18}F-3-Fluorocholestene	38
$Ag^{18}F$	6-Chloro-9-benzyl purine	^{18}F-6-Fluoro-9-benzyl purine	38
$K^{18}F$ /COOH	2, 4, Di-iodoestrone	^{18}F-2, 4-Difluoroestrone	76
$K^{18}F$ /COOH	3, 5-Di-iodotyrosine	^{18}F-3, 5-Difluorotyrosine	76
$K^{18}F$	21 - Iodopregnolone - 3 - acetate	^{18}F-21-Fluoropregnolone-3-acetate	77

with $Ag^{18}F_2$. Pratical details concerned with the small scale handling of fluorine gas and silver fluorides have been published.[40]

3. *Antimony and bismuth fluorides*

SbF_3, BiF_3, SbF_3Cl_2, SbF_5 and BiF_5, have been used to affect the transformation $RCCl_3 \rightarrow RCF_3$. The group R may be an alkyl, aryl or heteroaromatic group, but very few compounds of type $RCCl_3$ have been prepared and hence are available to be used as substrates. This is unfortunate since good yields of product are obtained even with 10–100 mg amounts of ^{18}F-labelled SbF_3 ($R{=}CH_3$, C_6H_5).[45] Antimony trifluoride may be labelled by exposure to carrier-free fluorine-18 in neon.[18,41] It undergoes partial hydrolysis by exposure to moisture during handling but may afterwards be completely reactivated (as SbF_3Cl_2) by treatment with $SbCl_5$.[66]

Some compounds which have been reported to have been prepared by methods employing nucleophilic displacement are listed in Table 5.

PREPARATION BY DIRECT FLUORINATION AND OTHER ELECTROPHILIC FLUORINATING AGENTS

The past few years has seen the introduction of some novel methods of fluorination which seem likely to lend themselves to the preparation of ^{18}F labelled radiopharmaceuticals. Perchloryl fluoride ($FClO_3$) and fluoroxytrifluoromethane (CF_3OF) have been used extensively as a source of electrophilic fluorine to fluorinate activated alkenes. CF_3OF and

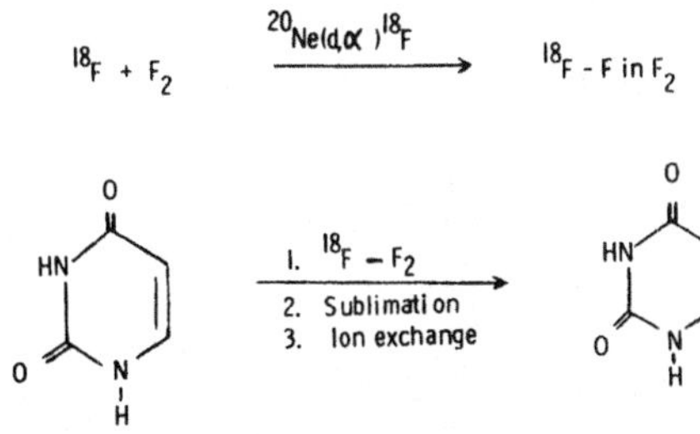

FIG. 6. Preparation of ^{18}F-5-fluorouracil.(70)

molecular fluorine have also been used for the direct electrophilic fluorination of 2,4-dioxopyrimidines to produce the 5-fluoro-derivatives.(68,69) This reaction has been employed in ^{18}F-labelling—when 5-fluorouracil was prepared from uracil and $^{18}F-F_2$ 5-fluorouracil is a cytoxic analogue of uracil which finds some application in cancer chemotherapy,(71) while the active material is a potential tumour localizing agent and may also be used to investigate the pharmacokinetics of the stable compound.(72)

This is a rapid procedure which is well suited to labelling, being capable of yielding high specific activity material. Although specialized targetry is required, elemental fluorine is prepared quite easily in the laboratory.(40) Syntheses using labelled CF_3OF have not so far been reported. Other reactions of this type which have been reported in the chemical literature include the preparation of 5-fluorocytosine, 5-fluoro-orotic acid and 5-fluorobarbituric acid.(68)

MISCELLANEOUS REACTIONS

Several labelled fluorinating agents are currently available but there have been no reports of their applications. These include nitrosyl fluoride, fluoroxytrifluoromethane and chlorine monofluoride. Labelled hydrogen fluoride, which may be prepared by several different methods, has been available for some time but appears so far to have found only very limited use. Cleavage of an epoxide by this reagent or boron trifluoride is a promising method in view of its rapidity and possible use on unprotected substrates, particularly in the preparation of labelled fluoro-steroids.(38, 73) STRAATMANN and WELCH(78) have recently prepared ^{18}F-labelled diethylamino sulphur trifluoride (DAST) a reagent which can be used to effect the following transformations:(79, 80)

$$RR'CO \rightarrow RR'CF_2.$$
$$RCOOH \rightarrow RCF_3$$
$$ROH \rightarrow RF.$$

CONCLUSION

Several ^{18}F-labelled organic compounds have been prepared and are being evaluated in animals with the ultimate aim of clinical application. However, as this article has attempted to show, the difficulties encountered in designing and preparing ^{18}F-labelled compounds of radiopharmaceutical quality are considerable.

The Balz–Schiemann reaction is relatively easy to apply but is only applicable to aromatic compounds and gives low yields. Fluorinations using silver or antimony fluorides give high yields but there are special problems to be overcome, while the F^- form ion exchange method appears to offer promise for certain aliphatic compounds. The use of elemental fluorine is at present limited to one class of compound, and other methods are a long way from regular utilization. However, radiopharmaceuticals labelled with ^{18}F may offer many potential advantages of *in vivo* stability and improved detection, both in those cases where the pharmacology is already known, for example fluorosteroids and fluorocarbohydrates, and also other classes of compounds in which development is at a very early stage.

Acknowledgement—The authors wish to express their thanks for Dr. D. A. WIDDOWSON, Dept. of Chemistry, Imperial College, London SW7 for many helpful discussions.

REFERENCES

1. ROBINSON G. D. Prospects for ^{18}F radiopharmaceuticals. In *Radiopharmaceuticals*, p. 141. Soc. Nucl. Med., New York (1975).
2. BROWNELL G. L., BURNHAM C. A., HOOP B. and KAZEMI H. Positron Scintigraphy with short-lived cyclotron produced radiopharmaceuticals and a multi-crystal positron camera. In *Medical Radioisotope Scintigraphy*, p. 313. IAEA, Vienna (1973).

3. Ter-Pogossian M. M., Phelps M. E., Hoffman E. J. and Mullani N. A. *Radiology* **114**, 89 (1975).
4. Muehllehner G. *J. nucl. Med.* **16**, 653 (1975).
5. Lifton J. F. and Welch M. J. *Radiat. Res.* **45**, 35 (1971).
6. Cohen M. B., Spolter L., Chang C. C., MacDonald N. S., Takahashi J. and Bobinet D. D. *J. nucl. Med.* **15**, 1192 (1974).
7. Chenoweth M. D. and McCarty L. P. *Pharmacol. Rev.* **15**, 673 (1963).
8. Smith F. A. Chem. Tech. U.S. **3**, 422 (1973).
9. Fowden L. Fluoroamino acids and protein synthesis. In *Carbon-Fluorine Compounds* (A Ciba Foundation Symposium), p. 141. Associated Scientific Publishers, Amsterdam (1972).
10. Rennert O. M. and Anker H. S. *Biochemistry* **2**, 471 (1963).
11. Smith F. A. Handbook of Experimental Pharmacology, (Edited by Smith F. A.), Vol. 20, Part 2, p. 351. Springer-Verlag, New York (1970).
12. Heidelberger C. Fluorinated pyridine nucleotides. In *Carbon-Fluorine Compounds*. (A Ciba Foundation Symposium), p. 125. Associated Scientific Publishers, Amsterdam (1972).
13. Barnett J. E. G. Fluorine as a substitute for oxygen in biological systems: examples in mammalian membrane transport and glycosidase action. *ibid* p. 95.
14. Goodwin D. A., Sundberg M. W., Diamanti C. I. and Meares C. F. ^{111}In-labelled radiopharmaceuticals and their clinical use. In *Radiopharmaceuticals*, p. 80. Soc. Nucl. Med., New York (1975).
15. Straatmann M. G. and Welch M. J. *J. nucl. Med.* **16**, 425 (1975).
16. Marche P., Marazano C., Maziere M., Morgat J. L., de la Losa P., Comar D. and Fromageot P. *Radiochem. Radioanal. Lett.* **21**, 53 (1975).
17. Robinson G. D. Biologically active ^{18}F-fluoro-organic compounds: Rapid high activity synthesis of ^{18}F-fluorocarboxylates and derivatives. In *Radiopharmaceuticals and Labelled Compounds*, Vol. 1, 423. IAEA, Vienna (1973).
18. Nozaki T., Iwamoto M. and Ido T. *Int. J. appl. Radiat. Isotopes* **25**, 393 (1974).
19. *Radioisotope Production and Quality Control*, IAEA Technical Report Series No. 128, p. 589. IAEA, Vienna (1971).
20. Thomas C. C., Sondel J. A. and Kerns R. C. *Int. J. appl. Radiat. Isotopes* **16**, 71 (1965).
21. Helus F., Krauss D. and Maier-Borst W. *Radiochem. Radioanal. Lett.* **15**, 225 (1973).
22. Chan P. K. H., Firnau G. and Garnett E. S. *ibid* **19**, 237 (1974).
23. Beg K. and Brown F. *Int. J. appl. Radiat. Isotopes* **14**, 137 (1963).
24. Scholz K. L. and Sodd V. J. *ibid* **23**, 465 (1972).
25. Collins C. H., Collins K. E., Ackerhalt R. E. and Blau M. *ibid* **26**, 571 (1975).
26. Nozaki T., Tanaka Y., Shimamura A. and Karasawa T. *ibid* **19**, 27 (1968).
27. Clark J. C. and Silvester D. J. *ibid* **17**, 151 (1966).
28. Tilbury R. S. and Dahl R., Mamacos J. P. and Laughlin J. S. *ibid* **21**, 277 (1970).
29. Hinn G. M., Nelp W. B. and Weitkamp W. G. *ibid* **22**, 699 (1971).
30. Lindner L., Suer T. H. G. A., Brinkman G. A. and Veenboer J. T. *ibid* **24**, 124 (1973).
31. Goulding R. W. and Palmer A. J. *ibid* **23**, 137 (1972).
32. Atkins H. L., Christman D. R., Fowler J. S. Hauser W., Hoyte R. M., Klopper J. F., Linn S. S. and Wolf A. P. *J. nucl. Med.* **13**, 713 (1972).
33. Welch M. J., Lifton J. F. and Gasper P. P. *ibid* **12**, 405 (1971) Abstract.
34. Lambrecht R. M. and Wolf A. P. Cyclotron and short lived halogen isotopes for radiopharmaceutical applications. In *Radiopharmaceuticals and Labelled Compounds*, Vol. 1, p. 275. IAEA, Vienna (1973).
35. Harper P. V., Lembares N. and Krizek H. *J. nucl. Med.* **12**, 362 (1971) Abstract.
36. Palmer A. J., Clark J. C., Goulding R. W. and Roman M. The preparation of ^{18}F-labelled DL-3-fluorotyrosine. In *Radiopharmaceuticals and Labelled Compounds*, Vol. 1, p. 291. IAEA, Vienna (1973).
37. Wolber G., Hartmann G., Helus F., Hover K. H. and Lorenz W. J. Applications of the compact cyclotron of the German Cancer Research Centre in nuclear medicine, neutron therapy and radiation biophysics. *7th Int. Conf. Cyclotrons and their Applications*, p. 444. Birkhauser Verlag, Basel (1975).
38. Ido T. Synthesis and *in vivo* distribution patterns of ^{18}F-organic compounds. *Proc. 1st World Cong. Nucl. Med.*, p. 901. World Federation of Nuclear Medicine and Biology (1974).
39. Lebowitz E., Richards P. and Baranowsky J. *Int. J. appl. Radiat. Isotopes* **23**, 392 (1972).
40. Clark J. C., Goulding R. W. and Palmer A. J. Preparation of ^{18}F-labelled fluorocarbons for use in pharmacodynamic studies. In *Radiopharmaceuticals and Labelled Compounds*, Vol. 1, p. 411. IAEA, Vienna (1973).
41. Clark J. C., Goulding R. W., Roman M. and Palmer A. J. *Radiochem. Radioanal. Lett.* **14**, 101 (1973).
42. Suschitzky H. The Balz-Schiemann reaction In *Advances in Fluorine Chemistry* (Edited by Stacey M., Tatlow J. C. and Sharpe A. G.), Butterworths, Washington (1965).

43. Nozaki T. and Tanaka Y. *Int. J. appl. Radiat. Isotopes* **18,** 111 (1967).
44. McOmie (Editor) *Protective Groups in Organic Chemistry*. Plenum, London (1973).
45. Goulding R. W. Ph.D. Thesis, University of London (1976).
46. Taylor D. M. and Cottrall M. F. The evaluation of amino acids labelled with ^{18}F for pancreas scanning. In *Radiopharmaceuticals and Labelled Compounds*, Vol. 1, p. 405. IAEA, Vienna (1973).
47. Dampure H. J., Osman S. and Somaia S. To be published.
48. Garnet E. S. and Firnau G. ^{18}F-5-Fluoro-DOPA as a new brain scanning agent. In *Radiopharmaceuticals and Labelled Compounds*, Vol. 1, p. 405. IAEA, Vienna (1973).
49. Axelrod J. and Weinshilboum R. *N. Engl. J. Med.* **287,** 237 (1972).
50. MacGregor R. R., Ansari A. N., Atkins H. L., Christman D. R., Fowler J. S. and Wolf A. P. *J. nucl. Med.* **15,** 513 (1974) Abstract.
51. Okuda T. and Tatsumi S. *J. Biochem.* **44,** 631 (1957).
52. Hoyte R. M., Lin S. S., Christman D. R., Atkins H. L., Hauser W. and Wolf A. P. *J. nucl. Med.* **12,** 280 (1971).
53. Goulding R. W. and Gunasekera S. W. *Int. J. appl. Radiat. Isotopes* **26,** 561 (1975).
54. Tosa T., Mori T., Fuse N. and Chibata I. *Enzymologia* **31,** 214 (1968).
55. Firnau G., Nahmias C. and Garnett S. *Int. J. appl. Radiat. Isotopes* **24,** 182 (1973); *J. Med. Chem.* **16,** 416 (1973).
56. Hydorn A. E. *J. org. Chem.* **32,** 4100 (1967).
57. Ikan R., Hoffmann E., Bergman E. D. and Gahin A. *Israel J. Chem.* **2,** 32 (1964).
58. Bergman E. D. and Hoffman E. *J. Chem. Soc.* 2827 (1962).
59. Kook C. S., Reed M. F. and Digenis G. A. *J. med. Chem.* **18,** 533 (1975).
60. Schadt F. L. and Schleyer P. R. *Tetrahedron Lett.* 2335 (1974).
61. Liotta C. L., Harris H. P., McDermott M., Gonzales T. and Smith K. *Tetrahedron Lett.* 71 (1975).
62. Landini D., Quici S. and Dolla F. *Synthesis* 430 (1975).
63. Cainelli G. and Manescalchi F. *ibid.* 472 (1976).
64. Robinson G. D., Uszler J. M. and Bennett L. R. *J. nucl. Med.* **16,** 561 (1975) Abstract.
65. Bhandari K. S. and Pincock R. E. *Synthesis,* 655 (1975).
66. Hudlicky M. *Chemistry of Organic Fluorine Compounds*, pp. 51; 59. Pergamon Press, Oxford (1961).
67. Chambers R. D. *Fluorine in Organic Chemistry*, p. 39. Wiley, New York (1973).
68. Barton D. H. R., Bubb W. A., Hesse R. H. and Pechet M. M. *J. Chem. Soc. Perkin I.* 2095 (1974).
69. Barton D. H. R., Hesse R. H., Toh H. T. and Pechet M. M. *J. org. Chem.* **37,** 329 (1972).
70. Fowler J. S., Finn R. D., Lambrecht R. M. and Wolf A. P. *J. nucl. Med.* **14,** 63 (1973).
71. Busch H. *An Introduction to the Biochemistry of the Cancer Cell* p. 139. Academic Press, New York (1962).
72. Wolf W., Bermon J. A. and Shani J. *J. nucl. Med.* **16,** 582 (1975) Abstract.
73. Palmer A. J. Unpublished results.
74. Robinson G. D. *J. nucl. Med.* **14,** 446 (1973) Abstract; *ibid* **16,** 561 (1975) Abstract.
75. de Kleijn J. P., Meeuwissen H. J. and Van Zanten B. *Radiochem. Radioanal. Lett.* **23,** 139 (1975).
76. Mantescu C., Genunche A. and Simonescu L. ^{18}F-labelling of bioactive molecules using $K^{18}F$–CH_3COOH solutions as fluorinating agent. In *Radiopharmaceuticals and Labelled Compounds*, Vol. 1, p. 395. IAEA, Vienna (1973).
77. Eng R. *J. nucl. Med.* **16,** 526 (1975) Abstract.
78. Straatmann M. G. M.Sc. Thesis, Washington University, St. Louis (1976).
79. Markovskij L. N., Pashinnik V. E. and Kirsanov A. V. *Synthesis* 787 (1973).
80. Middleton W. J. *J. org. Chem.* **40,** 574 (1975).

International Journal of Applied Radiation and Isotopes, 1977, Vol. 28, pp. 67–82. Pergamon Press. Printed in Northern Ireland

Radiopharmaceuticals Labelled with Technetium

WILLIAM C. ECKELMAN*
Division of Nuclear Medicine, George Washington University, Washington, D.C. 20037, U.S.A.

and

STANLEY M. LEVENSON†
Department of Nuclear Medicine, National Institutes of Health, Bethesda, Maryland 20014, U.S.A.

(*Received* 12 *May* 1976)

The ready availability and ideal nuclear properties of ^{99m}Tc has led to its widespread use for imaging purposes. In general the localization of the present radiopharmaceuticals is based on the ability of an organ to remove foreign substances from the blood. Further application of this ideal nuclide seems related to the development of methods to label biologically active molecules or drugs in such a way as not to interfere with their desired *in vivo* behavior. However, directly binding the ^{99m}Tc to these molecules may prevent the expected distribution (except in the case of large molecular weight proteins and cells) by altering the critical functional groups. Therefore the synthesis of derivatives containing a chelating group to bind the technetium is suggested as a possible solution to the problems associated with direct labelling. This could result in a molecule with similar biological properties to the parent molecule. But before useful derivatives of biologically active molecules and drugs can be prepared, extensive study of the chemistry of ^{99m}Tc is needed.

INTRODUCTION

OF THE conveniently available radionuclides, technetium has by far the best nuclear properties for diagnostic imaging. With the advent of commercial generator systems, instant technetium, innovations in chelation, and new chelating agents, there has been a marked expansion in the use of ^{99m}Tc labelled compounds. Chemical forms of ^{99m}Tc are presently the most widely used radiopharmaceuticals for radionuclide imaging of the brain, liver, lung, and skeleton, and to a lesser extent in thyroid scintigraphy. Compounds for both static and dynamic evaluation of renal and cardiac pathology have gained popularity, particularly with increasing computer technology.

In general, the use of these radiopharmaceuticals is based on the ability of specific organs to remove foreign substances from the blood. The ^{99m}Tc chelates for kidney studies are low molecular weight, water soluble compounds which are rapidly excreted via the renal pathway, hence, images of the kidneys are obtained. For hepatic imaging, the most common ^{99m}Tc agent is radiolabelled colloid which is visualized following rapid phagocytosis by the reticuloendothelial system of the liver. The basis for pulmonary artery blood flow evaluation is the mechanical obstruction of an innocuous percentage of the arteriolar–capillary pulmonary circulation by technetium-labelled particles ranging in size from 10 to 50 μm.

For future progress in nuclear medicine, however, it appears that a more refined, specific approach to compound localization, which depends upon the use of radiolabelled biologically active compounds or synthetic drugs, will be needed. In spite of the potential rewards offered by this type of investigation, glaring difficulties have become evident in attempts to

*Reprint requests may be sent to: William C. Eckelman, Ph.D., Associate Professor of Radiology, Nuclear Medicine Research, The George Washington University Medical Center, 2300 Eye Street, N.W., Washington, D.C. 20037, U.S.A.

†Present Address: Division of Nuclear Medicine, Georgetown University Hospital, Washington, D.C. 20007, U.S.A.

directly radiolabel the functional groups of compounds such as hormones, enzymes, and drugs. Firstly, the native functional group(s) may be needed to interact with the active biological site responsible for compound localization; should the radiolabel interfere with this, the normal behavior of the molecule will be altered and tracer studies will be a failure. Secondly, the radionuclide may bind to the molecule with insufficient affinity to produce a stable chelate. Just as with the previous problem, the desired behavior of the labelled compound will not be achieved.

The importance of both of these factors is exemplified by radiolabelled bleomycin. Bleomycin is a mixture of closely related antibiotics that have been used successfully to treat a variety of malignancies. Chemically, this antibiotic acts as a chelating agent and has been shown to bind a number of divalent and trivalent cations, but with varying affinities.[1] Although chelates of indium, gallium and copper have not demonstrated the necessary *in vivo* and *in vitro* stability, the bond in cobalt bleomycin is more stable.[2] Ideally, a technetium labelled bleomycin would be the most efficacious form; however, inspection of the bleomycin structure[3] indicates that the disaccharide moiety is the most likely chelating group but unfortunately, this is a low affinity site for ^{99m}Tc. This statement is based on work by Richards and Steigman[4] who demonstrated that the sugar moiety has a high affinity for ^{99m}Tc at pH 10–12, but a weak affinity at neutral pH. Therefore, the use of the native functional groups of bleomycin to bind a tracer such as technetium, results in a weak chelate with poor stability.

Another factor for consideration is the change in biological activity of a drug or biological derivative secondary to the addition of a radiolabel. It has been shown that the chelation of copper to bleomycin destroys its ability to cleave strands of DNA.[5] When labelled with cobalt, the antibacterial activity of bleomycin is deleteriously affected and becomes negligible when tested against the usually responsive *Bacillus subtilis* ATC 6633.[6] In this instance, the cobalt appears to alter the biological effectiveness of the bleomycin because of its bond to the functional groups responsible for maintaining the antibiotic integrity of this drug.

Another approach (for which there has been much precedent) is derivatization.[7] Drug derivatives have been prepared to increase absorption, eliminate bitterness and odor, diminish gastric upset, increase or decrease metabolism, and improve stability of the parent compound both *in vivo* and *in vitro*.

More recently, another area of investigation dealing with specific site directed synthetic derivatives has begun to unfold. Among those molecules which have potential value due to sites of increased concentration secondary to their physiological action are: (1) steroid hormones, (2) peptide hormones, (3) adrenergic substances, (4) vitamins, and (5) certain synthetic drugs.[8] The "ideal" properties for any site directed derivative were outlined by Paul Ehrlich some seventy years ago and recently have been enumerated by Sinkula and Yalkowsky as follows:[7]

(1) Exclusive and complete transport to the diseased tissue or target organ, including high binding affinity and interaction with these cell systems and tissues.

(2) Absence of binding by the derivative to protein or tissue not specifically diseased and absence of degradation or metabolism of the derivative prior to contact with the diseased bioenvironment.

(3) Lack of toxicity for normal tissue in the body.

(4) Complete elimination from the body of the non-localized pharmaceutical.

For the simple reason that these criteria are so stringent, the "ideal" site directed drug derivative has yet to be synthesized.

Several attempts at derivatization will be cited as examples of the approach. In efforts to develop more effective chemotherapy for malignancy, a number of agents have been designed to interfere with nucleic acid metabolism; the nonspecificity of action against both normal and neoplastic cells remain a problem and little specificity is obtained. In another type of approach, Tsou *et al.*[9] prepared a nitrogen mustard derivative of propionamide to take advantage of the increased levels of enzymatic amidase produced by neoplastic cells in tumor-bearing animals. Some specificity for tumor cells was noted in that leukopenia was not as predominant with this as with other cytotoxic agents, and neoplastic cells were destroyed. This type of effort utilizes the specific characteristics of neoplasm at the cellular level.

Similar attempts to destroy tumors in specific organs stimulated the use of breast cancer of various steroid hormones bound to nitrogen mustard through an ester linkage.[10–12] Although there was some encouragement from the initial clinical trials,[13] subsequent studies demonstrated no specific binding of the estradiol derivative to the estradiol receptor in breast tissue.[14]

The utility of anti-tumor antibodies has also been reported. The coupling product of daunomycin and antibodies directed against tumor-associated surface antigens demonstrated immunological specificity against different murine lymphoid tumors, with cytotoxicity against the tumor bearing the appropriate surface antigen.[15] There is potential for much more work in this area.

Recently a derivative of the beta adrenergic antagonist, pindolol, has been synthesized.[16] This derivative, termed HYP (hydroxybenzylpindolol), has been iodinated in the phenol moiety and shown to bind the beta adrenergic receptor with the same affinity as pindolol itself. This is but another example of practical efforts at derivatization.

From the preceding, it can be seen how the derivative approach has already been applied with varying success in clinical research, particularly in the field on oncology. Just as the previously described "ideal derivative properties" have not been achieved with non-radioactive substances, these properties are not found with present-day medical radioisotopes. Even with radioiodine, perhaps the best example of a physiologic site directed radiopharmaceutical, the usual 24-hr thyroid uptake of 15–25% is but a fraction of the administered dose. Localization also occurs in the stomach and salivary glands, with the primary portion excreted via the kidneys as the major competitive pathway.

In regard to the use of radiopharmaceuticals, certain other factors must be taken into consideration. One of these is the time frame during which a radiopharmaceutical must maintain its integrity. When evaluating pharmacokinetics, no absolutes exist, but rather, a medical tracer must remain stable *in vivo* only for the duration of the study to be performed. As an example, if RBCs or human serum albumin (HSA) labelled with technetium accurately reflect the vascular pool size at times up to 30 min post-injection, then these agents fulfill the criteria for stable *in vivo* derivatives and their use is valid during this period of time.

Because further discussion will concentrate on the application of ^{99m}Tc to biological derivative formation, the importance of the *in vivo* stability of technetium chelates must be stressed. Without this crucial component, site directed studies would be unobtainable. With this in mind, some knowledge of those radiopharmaceuticals in which the ^{99m}Tc is directly bound to the molecule of choice is essential to predict the types of chelate containing derivatives possible. Radiochemical purity, defined as the proportion of the total activity present in the stated or desired chemical form in the final product, will be an important concept to remember.[17] As a general rule, any chelating agent derivative to be discussed which does not avidly bind ^{99m}Tc both *in vitro* and *in vivo*, cannot be regarded as a true drug tracer and the synthetic effort will be considered as ill-conceived.

In order to better understand the potential for derivatization with technetium, the authors would feel remiss in not providing some background information concerning the chemistry of ^{99m}Tc.

CHEMISTRY OF ^{99m}Tc

The choice of ^{99m}Tc for use in nuclear medicine imaging procedures rests upon its favorable nuclear properties ($T_{1/2}$–6 hr, 140 keV gamma emission, absence of β decay) and ready availability.[18] As the decay produce of ^{99}Mo, ^{99m}Tc is eluted with saline from a generator consisting of ^{99}Mo adsorbed by an alumina column. In the nuclear transformation, TcO_4^- (VII), the stable chemical state of technetium in aqueous solution is formed.[19] However, this pertechnetate generator product will not bind to chelating agents nor adsorb to particles necessary for bone, pulmonary, or renal imaging procedures. Consequently, to perform these studies, a less stable, positively charged reduced state of ^{99m}Tc must be produced.[20,21]

Concerning the chemical states of technetium, the anion pertechnetate in aqueous solution is the most stable form. When ^{99m}Tc is bound to various chelating agents, the reduced states, Tc(III), Tc(IV), Tc(V), predominate. Reduced states of technetium can be achieved by treatment with a variety of reducing substances. The most frequently used are: (1) stannous ion,[22] (2) ferric chloride and ascorbic acid,[23] (3) ferrous ion,[24] (4) sodium borohydride,[25] and (5) concentrated HCl.[26] Pertechnetate can also be reduced electrolytically, although with the zirconium electrode, other reducing species are present.[21] In the reduced state, technetium readily binds to chelating agents forming compounds commonly used in diagnostic imaging; among these are ^{99m}Tc-diethylenetriamine-pentacetic acid (DTPA), 1-hydroxy-ethylidene-1, 1-diphosphonic acid (HEDP), and -glucoheptonate.

While it is desirable to know the exact oxidation state of technetium in these radiopharmaceuticals, present methodology makes this determination with nanomolar quantities impossible.

Experiments with millimolar amounts have indicated that ^{99}Tc is reduced by stannous ion to the V state and then slowly to Tc(IV) at pH 7 in citrate buffer. In HCl, ^{99}Tc is also reduced by stannous ion to the IV state. With a DTPA buffer at pH 4, the ^{99}Tc(III) state prevails.[27] Unknown however, is whether these millimolar determinations can be extrapolated to the nanomolar quantities used in diagnostic imaging procedures.

Investigators have also attempted to define the reduced valence state of ^{99m}Tc in the DTPA chelate.[28] It is known that macroscopic concentrations of ^{99}Tc (10^{-3} M), for which *in vitro* analysis can be made, form ^{99}Tc(IV) in the presence of the following reducing agents: (1) stannous ion, (2) ferric chloride and ascorbic acid, (3) ferrous ion, and (4) concentrated HCl–HI. In each case, and over a wide range of concentrations, binding efficiency of the reduced technetium to DTPA was greater than 85%. When the same measurements are performed with concentrated HCl alone as the reducing agent, only 10% of the ^{99}Tc was bound to the DTPA. Differences in the efficiency of compound formation almost certainly reflect a difference in valence state produced by the reducing agent. Since the same binding trend is evident for ^{99m}Tc and ^{99}Tc, it can be assumed that the same valence state which occurs with ^{99}Tc will also be found with ^{99m}Tc (10^{-9} M) utilized in nuclear medicine laboratories. Although Tc(IV) is produced with the first four reducing agents in the absence of DTPA, any conclusions regarding the valence state of technetium labelled to DTPA would be unwise in view of the potential changes affected by the addition of DTPA.

A recent effort to extrapolate from millimolar to nanomolar chemistry by observing the biological behavior of the resulting radiopharmaceuticals has shown that ^{99m}Tc HEDP and ^{99m}Tc glucoheptonate have the same biological distribution as the ^{99}Tc compounds and therefore should be in the same oxidation state, namely, Tc(IV) in both cases.[29]

All experiments assuming a similarity of chemical behavior between ^{99}Tc and ^{99m}Tc have considered the concentration of technetium as the only variable in the reaction. Unfortunately, variations exist in other parameters such as the quantity of reducing agent employed in chelate formation. The ratio of stannous and stannic ion to the amount of technetium has been considered responsible for reports of both ^{99}Tc(III) and ^{99}Tc(IV) in the chelate.[27] The quantity of the reducing agent used in compound formation is therefore crucial. These findings also strongly suggest that more than one valence state of technetium can bind to the same chelate, although it is uncertain as to the effect this has on the *in vivo* distribution of the compound. Another important consideration is the method used to determine the oxidation state of ^{99}Tc at the millimolar level. As an example of this, iodine titration could theoretically yield misleading information concerning the oxidation state of technetium if the iodine oxidizes not only the excess reducing agent, but also the reduced technetium itself. Many pitfalls are evident.

Just as there has been limited information on the oxidation state prevalent in various chelates, there have also been few reports on the aqueous chemistry of these reduced species of technetium. Electrophoretic and extraction studies with anionic chelating agents have now provided indirect evidence for the existence of TcO^{2+}, $TcOOH^+$ and TcO_2,[30,31] even though few chelates of ^{99}Tc have been reported.[32] Nevertheless, because of the close chemical relationship between technetium and the element rhenium (Re) which has been much easier to work with in the laboratory, certain assumptions may be applicable with regard to its chemical properties. Reduction of perrhenate (ReO_4^-) with phosphine in methanol-HCl produces a species of $ReOCl_2(OEt)PR_3$ which reacts with carboxylic acids to form binuclear complexes.[33,34] In addition, stable complexes of Re(IV) pyridine in water and rhenium dithiocarbamate also exist.[35] If both elements actually do behave in a similar chemical manner, this would imply the capability for technetium to form compounds in the same manner and demonstrate similar chelate oxidation states.

The oxides, sulfides, and halides of technetium have received considerable attention. Concerning the oxides, oxo compounds are known to predominate in the higher oxidation states, the most common forms being TcO_2 and Tc_2O_7. The hydrated dioxide (TcO_2) is made by the addition of base to Tc(IV) solutions such as $TcCl_6^{2-}$ or $TcBr_6^{2-}$; alternatively, reduction of TcO_4^- in HCl with Zn may also be used. Tc_2O_7 is produced by evaporating acid solutions of pertechnetate. Technetium dioxide's relevance is that it represents a competitive pathway in the production of the desired chelate. To prevent this adverse side reactions from occurring, an adequate amount of chelating agent, e.g. DTPA, must be present. Of importance also are the sequence and timing for the addition of the pertechnetate, the chelating agent, and finally, the reducing agent.[28]

Simple halides of technetium have been reported, although they are unstable and hydrolyze in the presence of base. The complex fluoride of Tc(IV) is stable to hydrolysis and has been prepared as a potential skeletal imaging agent,[36,37] but demonstrates no advantages when compared to the phosphate labelled compounds. A hexachloro salt has also been prepared by pertechnetate reduction with concentrated HCl; this salt hydrolyzes slowly in dilute acid and is of interest only in terms of its chemical existence.

Technetium forms sulfur compounds in the presence of H_2S or sodium thiosulfate. Tc_2S_7 can be obtained by saturation of a 2–6 N HCl solution of $^{99}TcO_4^-$ with H_2S; precipitation of elemental sulfur as the colloid is often incomplete with this method. Consequently, sodium thiosulfate in acid, which is easier to prepare[38,39] and consequently more applicable clinically, has become the sulfur colloid source for liver scintigraphy.

In addition to the chemistry of technetium itself, the chelate chemistry of the metallic reducing agents must be considered. Regarding the most popular reducing agent, stannous chloride, little published data are available. Equilibrium constants, however, are known for the stannous chelates of pyrophosphate[40,41] and DTPA;[42] these constants determine the concentration of chelating agent necessary to maintain the solubility of the metal at neutral pH.[43,44] Since metal oxides competitively interfere with and inhibit the desired binding of technetium,[45] the reducing agent must be in chelate form to prevent this side reaction from occurring.

Having described the objectives of this paper, with references to both historical perspectives into the synthesis of derivatives and reference data for the chemical behavior of technetium, we shall turn to what is known and surmised concerning the labelling of different types of agents with technetium.

DIRECT LABELLING OF CHELATING AGENTS

1. *Radiopharmaceuticals*

Technetium chelates have been developed and are used to varying degrees in the imaging of three organs or organ systems. For the first, renal scintigraphy, a recent review documents some 12 different chelating agents which have been combined with ^{99m}Tc.[46] These include: (1) EDTA,[47] (2) DTPA,[48,49] (3) mannitol,[50] (4) mannitol with gelatin,[51] (5) penicillamine-acetazolamide,[52,53] (6) caseidin,[54] (7) citrate,[55] (8) tetracycline,[56] (9) inulin,[50] (10) dimercaptosuccinic acid (DMSA),[57] (11) glucoheptonate,[58] and (12) gluconate.[59]

Although all of these chelates may be used, those most commonly utilized for diagnostic imaging in nuclear medicine laboratories are DTPA, glucoheptonate and DMSA. The remainder have not gained the same widespread acceptance.

Hepatic and hepatobiliary ^{99m}Tc chelates have

also been proposed as possible replacements for ^{131}I Rose Bengal, although at present, no particular advantages have been gained through their use. They are, however, as follows: (1) penicillamine,[60] (2) pyridoxylideneglutamate (PG) a Schiff base,[61] (3) dihydrothioctic acid,[62] (4) tetracycline,[56] (5) mercaptoisobutyric acid,[63] and (6) 6-mercaptopurine.[64] Most imaging departments continue to rely upon iodinated rose bengal.

The final group of technetium chelates revolutionized nuclear medicine in 1972, for it was at this time that phosphate derivatives were introduced for skeletal scintigraphy. Prior to this, one was forced to deal with the undesirable nuclear characteristics of either fluorine or strontium.

Tripolyphosphate was the first of these chelating agents which resulted in ^{99m}Tc bone localization.[65] Shortly thereafter, ^{99m}Tc polyphosphate was introduced. The rather slow blood clearance and varying chain length [66] led to further investigation, and subsequent synthesis of today's conventional skeletal imaging agents, pyrophosphate[67] and disphosphonate (HEDP).[68] Recent reviews[69–71] compare these compounds to other di- and imidodiphosphonate derivatives in terms of bone uptake and blood clearance, the two most crucial biological properties.

Because ^{99m}Tc phosphates are weak chelates, it is important to define the particular formulation used in the preparation of these phosphate derivatives. The effects of the concentration of, for example, pyrophosphate, the ratio of the amount of the stannous chloride reducing agent to the pyrophosphate, and the absolute quantity of stannous chloride[72] used, all may have secondary ramifications regarding the chemical product and the subsequent scan image quality. This analysis can best be discussed in terms of radiochemical purity.

2. In vitro *radiochemical purity*

In general, technetium chelates are analyzed not for radiochemical purity, but to determine the presence of pertechnetate. This latter approach has been performed to determine the integrity of the product, however, misleading conclusions may be drawn from the chromatographic results.

Of the chelating agents studied, ^{99m}Tc-Sn-DTPA is one which appears to be radiochemically pure.[73] By definition, this would imply that only the chemical form as stated is present in the compound. In contrast, most of the other chelating agents mentioned seem to have a low affinity (are weak chelates) for Tc; in these instances, the risk of radiochemical impurity (species other than those desired) is raised.

The various methods for preparation of one of these weak chelates, ^{99m}Tc-Fe-ascorbate, has been reviewed by Perrson *et al.*,[74] with the kinetics of chelation by ascorbate studied on Sephadex G25 using the gel chromatographic column scanning method (GSC). Likewise, Halpern *et al.*[52] demonstrated that $^{99m}TcO_4^-$ is reduced by penicillamine, and following chelation by the addition of acetazolamide, produces a rapidly moving radioactive species on Whatman No. 1 paper in a butanol–acetic acid–H_2O solvent. Unfortunately, neither of these determinations fulfilled the rigid criteria for radiochemical purity. To do so, only a single band of radioactivity may appear in at least two chromatographic systems, and, the partition coefficient must be such that the compound neither freely migrates nor is strongly adsorbed by the base support. With ^{99m}Tc–Fe-ascorbate, reduced but unbound radiochemical impurities of ^{99m}Tc were still evident under optimal reaction conditions. Similarly, ^{99m}Tc penicillamine acetazolamide did not meet the necessary criteria, as the desired product migrated near the solvent front upon analysis in only one chromatographic system.

Further investigation into the ^{99m}Tc chelates suggested for renal scintigraphy reveals that due to the vast majority being weak chelates, confusing results are obtained when they are studied on certain chromatographic systems. One source of error can result from the interaction of the solid phase of the system with the radiopharmaceutical, as the chelating property of the solid support competes for the ^{99m}Tc. As an example of artefacts induced by this phenomenon, the polysaccharide, Sephadex, actively competes with certain weak chelates such as ^{99m}Tc gluconate and Tc mannitol.[75,76] This shortcoming with Sephadex column chromatography has been demonstrated for a number of weak technetium chelates, and will suggest reduced unbound forms of Tc which, in fact, reflect the effect of this chromatographic system on the chelate product. To avoid the Sephadex artifact, some have used the inert solid support polyacrylamide, Bio-Gel P-10, with which there is no competitive adsorption of technetium by the base support.[77] Alternatively, the Sephadex column can be eluted with the same concentration of chelating agent used in the preparation of the radiopharmaceutical.[75,78]

The phenomenon of the introduction of artefacts by the chromatographic system has also been observed with paper techniques.[79] If the radiochemically pure bone agent, ^{99m}Tc pyrophosphate, is chromatographed with saline, most of the reduced ^{99m}Tc appears to be unbound; however, if the paper is used with pyrophosphate solution, the chelate is found to be radiochemically pure. In this instance, the chelate releases the ^{99m}Tc to another chelating compound because of the dilutional effect of the saline solvent and rapid dissociation constant of technetium pyrophosphate. Consideration must be given this property when evaluating for radiochemical purity with these systems.

Among the agents suggested for hepatobiliary imaging, only ^{99m}Tc-PG and ^{99m}Tc-penicillamine have been shown to be radiochemically pure. It is unclear however, as to the exact structure of the localizing species. One can demonstrate the absence of TcO_4^- or ^{99m}Tc pyridoxal by electrophoresis, although with the desired product remaining at the origin, radiochemical purity cannot be determined.[61] Nonetheless, this electrophoretic study in conjunction with two separately published paper chromatography systems, does seem to indicate radiochemical purity.[80]

Radiochemical purity tests were also performed on Tc penicillamine[52] in a single system: the R_f of ^{99m}Tc penicillamine, TcO_4^-, and unlabelled penicillamine were respectively, 0.7, 0.25 and 0.6. In combination with another ^{99m}Tc penicillamine study[81] using Sephadex G25 and saline, there are indications that a radiochemically pure labelled penicillamine can be produced. Regarding other technetium hepatobiliary agents, only the analysis for TcO_4^- has been performed.

Of the bone imaging agents, ^{99m}Tc-HEDP appears to be a strong chelate and has been chromatographed on Sephadex G25 eluted with saline.[82] Technetium pyrophosphate, on the other hand, dissociates on Sephadex G25 and on Whatman paper when saline is used. This artefact was described earlier and is remedied with a pyrophosphate solution; the change produces a single peak in both systems for radiochemically pure products. Technetium polyphosphate will also elute from Sephadex G25 with saline, but presents an additional radiochemical purity analysis problem. Although one radioactive peak has been obtained in the chromatography of Tc polyphosphate, the variations in phosphate chain lengths[83] raise uncertainties concerning the identity of the localizing species. Hopefully, the use of high pressure liquid chromatography will clarify many of these and other pending questions concerning the radiochemical purity of chelates labelled with technetium.

3. In vivo *radiochemical purity of chelates*

Reports on the analysis of chemical forms of ^{99m}Tc used as renal imaging agents have been sparse. A chromatographic study of plasma and urine samples after injection of Tc–Sn-DTPA indicated at one hour post administration, that the Tc–Sn-DTPA in the plasma was 90% radiochemically pure, while the labelled DTPA in the urine demonstrated 98% radiochemical purity. Comparing these results to the analysis of ^{99m}Tc–Fe-ascorbate, only 38% of this agent was found as the ^{99m}Tc chelate in the plasma and 18% as the original radiolabelled chelate in the urine.[84] These data emphasize the superior stability of the Sn-DTPA compound.

Some information concerning the *in vivo* fate of the skeletal imaging agents is also available. Krishnamurthy *et al.*[85–87], in a series of ^{99m}Tc phosphate and diphosphonate comparisons, established the early binding properties of these compounds. At one hour post-administration, approximately 80% of the ^{99m}Tc activity was bound to serum proteins, most of which was associated with the globulin fraction. In addition, red blood cell binding was also found to be evident. In another investigation, Bowen *et al.*[88] determined that following the intravenous administration of ^{99m}Tc polyphosphate, greater than 50% of the Tc plasma activity was due to ^{99m}Tc polyphosphate; this contrasts the plasma clearance data obtained with ^{99m}Tc pyrophosphate, where less than 33% of the total plasma activity after ^{99m}Tc pyrophosphate injection resulted from ^{99m}Tc pyrophosphate. Interestingly enough, the major urinary constituent after the administration of either ^{99m}Tc pyrophosphate or polyphosphate was Tc pyrophosphate. Blood analysis after polyphosphate injection revealed the following binding distribution of ^{99m}Tc activity: pyrophosphate (3–25%), polyphosphate (53–80%), TcO_4^- (10–15%), protein (10–26%), and pertechnetate (18–58%); here, the amount of reported TcO_4^- generates ambiguity regarding the conclusions to be drawn from this work.

Concerning the *in vivo* behavior of ^{99m}Tc hepatobiliary chelates, no noteworthy data are available as of this review.

DIRECT LABELLING OF COLLOIDS AND PARTICLES

1. *Radiopharmaceuticals*

A great many formulations for radiolabelled colloids for use in hepatic scintigraphy have been reported. In the preparation of Tc sulfur colloid, a variety of stabilizing agents are known. Among these are gelatin (the most popular), albumin, PVP, and polyhydric alcohol.[89–92] Stabilizer-free preparations have also been proposed,[93] as well as compounds with carrier rhenium[94] and antimony.[95] Other types of colloids, presently of interest from an academic viewpoint, include stannous oxide,[96] technetium oxide,[97] and microaggregated albumin.[98]

For lung scanning, a number of ^{99m}Tc labelled particles have been described. Methods and types of preparations include: macroaggregation of ^{99m}Tc labelled albumin,[99] the conversion of Tc sulfur colloids into HSA macroaggregates,[100] coprecipitation of ^{99m}Tc with iron hydroxide,[101] macroaggregation of albumin in the presence of colloid,[102] and incorporation of ^{99m}Tc sulfur colloid or reduced technetium into microspheres.[103] Those most commonly used are the HSA macroaggregates, microspheres, and iron hydroxide particles.

2. In vitro *radiochemical purity*

In this category, only the colloid work of Cifka *et al.*[104] will be discussed, for it was their investigation that discovered pre-mixing of pertechnetate with sodium thiosulfate could produce a non-pertechnetate radiochemical impurity. In routine silica gel TLC systems, this impurity could not be identified because it remained at the origin and was therefore indistinguishable from the ^{99m}Tc sulfur colloid. On Whatman No. 3 paper with 0.3 N HCl however, this impurity did manifest itself. Differences in chromatographic behavior of technetium sulfur colloid preparations are known,[105] but this has not been related to biological behavior.

3. In vivo *radiochemical purity*

There are no reports concerning the chemical form of ^{99m}Tc following the intravenous administration of either radiolabelled particles or colloids. The implications of the previously mentioned *in vitro* work[104] and the qualitative differences in other *in vitro* impurities for sulfur colloid preparations,[105] have not been elucidated relative to their biological behavior.

DIRECT LABELLING OF CELLS AND BLOOD ELEMENTS

1. *Radiopharmaceuticals*

Eythrocytes and albumin (HSA) are the two blood components which have received the most attention in radiolabelling with ^{99m}Tc.[106] Studies have also been published on attempts to label leukocytes,[107] lymphocytes,[108] and platelets,[109] although little success has been achieved with these blood cells. A host of tumor cells have been labelled with technetium including murine fibrosarcoma, human carcinomas of the breast, lung and colon, as well as malignant melanoma.[110] In addition to the above, thymocyte labelling with ^{99m}Tc has been reported.[111]

Because of the frequency of thrombo-embolic disorders, a great deal of interest has been stimulated in the field of labelled fibrinogen[112] and urokinase[113] for blood clot localization. Streptokinase, although not found in humans, has also received significant attention for the same reasons.[114] For these, and all other blood products and cells mentioned, the labelling procedure is based on the stannous chloride method developed for RBCs and HSA.

2. In vitro *radiochemical purity*

For any type of cell, the best method to determine labelling yield is by centrifugation; this technique separates the unbound radioactivity from the labelled cells. It does not separate ^{99m}Tc labelled colloids or particles since they would be collected in the pellet formed at the bottom of the centrifuge tube.

For the protein components of blood, the most widely employed means of separation is the gel filtration method with Sephadex G25 using saline solution.[73] Referring to the early work on HSA analysis, it was shown that radiochemical impurities could exist. The conclusions of this investigation implicated pertechnetate as the only possible impurity, leading to much confusion concerning the *in vivo* stability of this technetium labelled protein.

The complicated method used in the production of ^{99m}Tc HSA by the iron ascorbate reduction process[106] has been clarified to some extent. Recent work by Yokayama *et al.* has shown that the initial increase in pH from 1 to 6 is needed to form a reduced technetium ascorbate complex.[115] The necessity to then reach a pH level of 9–10, and return to an acid pH is thought to be related to the specific binding site configuration of the HSA molecule. Apparently, changes in the pH level are required to expose the sulfhydryl groups involved in the binding of the technetium.[116]

With the development of the tin reduction method for the preparation of ^{99m}Tc HSA, not only is the production of this compound facilitated, but the probability of a pertechnetate impurity is quite remote. This process is extremely simple to perform, requiring only a single low pH reduction which produces very sufficient binding.[117] It should be pointed out however, that those labelling procedures advocating the use of stannous chelate and HSA at neutral pH, may cause the formation of small colloids known to have slow blood clearance.[118] One should be aware of the potential difficulties in the preparation of technetium labelled human serum albumin.

Turning to the *in vitro* behavior of agents investigated for their potential in visualizing or localizing thrombo-embolic disease, Perrson *et al.* [119] demonstrated that even under optimal reaction conditions, ^{99m}Tc streptokinase was only 70–80% radiochemically pure. Regarding fibrinogen, labelling with technetium has occurred following electrolysis using zirconium electrodes. In the chromatographic examination of filtered plasma-buffered ^{99m}Tc fibrinogen, 85% of the activity appeared to be fibrinogen-bound. This finding was supported by electrophoretic work showing a similar binding capability.[112] Finally, it appears that animal fibrinogen binds technetium with a similar affinity. As reported by Harwig *et al.*[120] both canine and rabbit fibrinogen had ^{99m}Tc labelling efficiencies of 70–80%, with clottable protein binding 55–65% of the radioactivity.

3. In vivo *radiochemical purity*

The technical labelling of blood products with technetium is a minor task when compared with the problem of proving that the radiolabelled product truly represents its natural blood counterpart.

For RBCs, red cell volume determinations have indirectly confirmed 95% radiochemical purity as much as one hour after injection.[121] Similar studies with ^{99m}Tc HSA have likewise been carried out, although their true meaning is in question in view of the lack of accompanying *in vitro* radiochemical purity data.

Concerning the labelling of cells other than RBCs, two problems are noteworthy. First, as in many other situations, no conclusive *in vivo* determinations of the chemical form of technetium have been obtained. The second difficulty, which is far more imposing, pertains to the complications due to differences in cell type, for labelling fragile cells such as leukocytes, platelets and thymocytes necessitates an evaluation of cell viability following the labelling procedure. Although viability tests such as trypan blue staining or determinations of the cellular ability to incorporate thymidine and amino acids may indicate that the function tested is intact, the *in vivo* distribution does not represent that of a native cell. A possible explanation is that the cell membrane is compromised during the labelling process, changing the nature, and consequently the characteristics and behavior of the cell.[122] This problem illustrates the difficulties faced by those involved in blood elements.

Recently, a thorough study of the fate of

technetium labelled fibrinogen has shed some light on the potential applicability of this agent in the evaluation for thrombotic disease.[120] It was found that ^{99m}Tc activity is rapidly cleared from the blood, with approximately 25% of the administered dose remaining in the circulation 10 min post-injection. Electrophoresis of blood samples taken at this time revealed that the largest fraction of the radioactivity migrated with the alpha-2 globulin protein, and that only 18% of the tracer was associated with the clotting components for thrombus formation. In spite of this finding however, an experimentally induced thrombus was visualised in a dog with this preparation of ^{99m}Tc fibrinogen.

CHELATE CONTAINING DERIVATIVES

1. *Radiopharmaceuticals*

In the reduced valence states, technetium requires an octahedral coordination structure, i.e. up to six coordination sites in a target molecule would be directly bound to the radionuclide. This can prevent the expected interaction of the biological molecule with the active site responsible for localization. There is also little evidence that ^{99m}Tc is bound directly to many molecules with such an affinity to withstand *in vivo* dilution or ligand substitution. As a result, the concept that tracer studies for the parent molecule are being carried out is probably incorrect in many cases. Finally, the ease of oxidation of certain reduced states when weakly chelated has prevented studies of radiochemical purity and *in vivo* metabolism.

To avoid the problems encountered with direct labelling of a biologically active molecule, derivatives have now been formed by covalent bonding of a chelating agent to a molecule which is known to act on a specific organ, or follow a specific pathway. This type of approach was considered a logical step to combat the difficulties encountered in synthesis by the direct labelling.

The application of the concept of a derivative of a biologically active compound to radiopharmaceuticals is a recent event. Although this approach has been used to alter the properties of pharmaceuticals, the mechanics of its use in the field of diagnostic tracer compounds has obvious differences. Of the radioactive metals, technetium, for reasons already elucidated, appears to be the most logical for the development of agents with widespread applicability.

To bind these positively charged metal ions like ^{99m}Tc requires a relatively large chelating agent. Although the chelate will usually have less charge than the metal ion, the chelate will add a polar group to the biologically active molecule. Because of the non-polarity of most of these molecules, there will be an alteration in the biologically active molecule with the addition of this large polar group. In order to illustrate this chelate derivative approach, pertinent research by various investigators will be presented. SUNDBERG *et al.*[123,124] synthesized an EDTA derivative containing a diazonium group (1-(*p*-benzenediazonium)-ethylenediamine-N, N, N′, N′, tetraacetic acid). This chelating agent could be reacted with a molecule containing an activated benzene ring, i.e. phenol or aniline, and has in fact formed chelate derivatives of fibrinogen, albumin, and bleomycin have been constructed. Although these compounds were subsequently labelled only with ^{111}In, the potential applicability to technetium is evident.

In other work, HEINDEL *et al.*,[125] prepared a series of structural analogs of the pancreatic hypoglycemic, tolbutamide. These derivatives contained either an N,N dimethylaminoethylaminoethyl, an aminoethylaminoethyl, or a 3-carbethoxymethyl - 1 - toluenesulfonyl urea derivative. No practical applications for imaging resulted from this research.

To determine whether biological fatty acids could be used to transport technetium for myocardial scintigraphy,[126] fatty acid and long chain hydrocarbon analogs containing a strong chelating group were evaluated. The chelating agents involved in this investigation included DTPA, EDTA, and diethylenetriamine (DTA). DTPA was chosen as one of these agents since it is known to form stable technetium chelates both *in vivo* and *in vitro*. Furthermore, DTPA offers the advantages of allowing one to incorporate a well-defined structural molecule to the chosen biologic. This chelate also occupies the six coordinate sites of technetium, lessening the possibility of binuclear complexes found with chelating agents possessing fewer sites.[127] Finally, its use eliminates the possibility of bis compounds which may exist with a large excess of tridentate ligand.

An attempt to trace amino acid metabolism with a derivative which could firmly bind technetium has been reported by CASTRONOVA;[128] this amino acid synthetic contained a phosphonic acid group. No successful clinical application has been obtained however.

More recently, LOBERG *et al.*[129,130] has produced an analog of the anti-arrythmic drug, lidocaine. Their chelating agent was iminodiacetic acid (IDA) which characteristically binds metals strongly and easily reacts with functional groups on the biologically active molecule. To produce this lidocaine analog, IDA was reacted with ω-chloro-ω 2,6-dimethylacetanilide.

Presently, the optimal type of chelating agent is unresolved. Interaction of the chelating group with the active site responsible for the biological effect cannot be predicted for DTPA, IDA or DTA with the preliminary data available. However, the most successful derivative suggested by affinity chromatography studies[131] will be that with the chelating agent located the greatest distance from the site-interacting functional groups of the biological molecule. Another more subtle consideration is the effect of the chelating agent on the resultant charge of the molecule. If a particular chelating agent was found to stabilize a certain oxidation state of technetium, one chelate derivative may be more suited to reduce the charge of the chelate. This would then allow minimal disruption in polarity to the lipophilic character of the biologically active molecule.

2. In vitro *radiopharmaceutical purity*

SUNDBERG *et al.*[123] have analyzed the azobenzene EDTA by the standard chemical means; however, there was much difficulty in evaluating this as a labelled compound due to limitations in separating the large molecular weight biologically active molecule from the biologic containing the chelating agent (azobenzene EDTA). For this reason, no *in vitro* determination of the nature of the radiopharmaceutical was obtained.

HEINDEL *et al.*,[125] after determining that the tolbutamide derivatives under analysis were indeed the desired pharmaceutical, evaluated them for their hypoglycemic qualities. This enabled them to chose one which retained its biological effectiveness. The dimethylaminoethylaminoethyl- derivative was selected, labelled with technetium and tested for radiochemical purity. Sephadex G25 was used to purify the radiopharmaceutical prior to injection, and paper chromatography with saline further confirmed radiochemical purity. The structure of the radiolabelled tolbutamide derivative remained unknown however. As possible formulations, bis compounds, binuclear formation, or chelation with the biologically active *p*-toluenesulfonyl urea, have all been suggested as potential forms for technetium binding.

The structure of the fatty acid derivatives was analyzed and verified by physical measurements and elemental analysis. Following technetium labelling, both gel chromatography and TLC demonstrated the radiochemical purity of these radiopharmaceuticals. No study of biologic activity was reported, and as with the tolbutamide derivatives, no structural information is available regarding the radiolabelled molecule. A reasonable conclusion concerning the structure of the DTPA- and EDTA-like derivatives is that they should resemble the transition metal EDTA chelates.[132] These demonstrate no binuclear formation, bis compounds, or chelation with fatty acid groups.

In the CASTRONOVO work on phosphonic acid, the derivative was shown to be radiochemically pure as analyzed by paper chromatography with 85% methanol and with Sephadex G25. The functional groups involved in chelation were determined by observing the ability of the components and the derivative to dissolve stannous oxide. This study demonstrated that the amino acid alone was insufficient to bind stannous oxide, but did not eliminate a joint venture by the phosphoric acid group and the amino acid for chelation with the derivative.

Both methyliminodiacetic acid (MIDA) and a lidocaine analog, N(2,6-dimethylphenylcarbamoylmethyl) iminodiacetic acid (HIDA) labelled with ^{99m}Tc were studied by LOBERG *et al.*[129,130] in regard to their radiochemical purity. Determinations were obtained by a number of paper chromatography systems, the most discriminating of which was acetonitrile:water in a 3:1 ratio. Gel chromatography was also employed for the same purpose, and to evaluate relative binding affinity. For ^{99m}Tc HIDA, qualitative competitive binding demonstrated complete elution of the compound from Sephadex G25, just as is noted with ^{99m}Tc DTPA. In the examination of ^{99m}Tc MIDA, approximately 6% was retained on the Sephadex column which is in marked discordance to a 90% retention of ^{99m}Tc pyrophosphate in the same system. As mentioned previously, chelate stability on Sephadex does not imply *in vivo* stability, however, one would tend to be more optimistic with a compound that has a similar *in vitro* behavior to DTPA. It is always judicious to note that these indices actually reflect competition (at the specific concentrations used) between the polysaccharide groups of Sephadex, and the chelating agent.

3. In vivo *radiochemical purity*

To evaluate his HSA derivative, SUNDBERG injected ^{111}In azobenzene EDTA-HSA into tumor bearing mice; the distribution of this synthetic was then studied 24 hours post-administration and differed markedly from ^{131}I labelled albumin.[124] It is not clear as to whether impurities were responsible for this disparity, or, if the chemical reaction to derivatize the albumin damaged the molecule, changing its biological behavior. Electrophoresis of blood samples taken six days after injection did give some indication as to the integrity of the ^{111}In molecule, for the indium did remain with the albumin derivative, as opposed to being transferrin bound.

The same indium chelate derivative of fibrinogen was also evaluated and was found to retain 56% of its clottability with the radiolabelled indium. It is not certain that a sensitive protein like fibrinogen can truly withstand the reaction conditions necessary to produce the derivative,[120] and there has been little date allowing comparison with ^{125}I fibrinogen.

In a similar fashion, bleomycin was labelled after derivatization with the azobenzene EDTA and injected into tumor bearing mice. Twenty-four hours later, a tumor to blood ratio of 4.7 was achieved, however there was minimal investigation concerning the *in vivo* radiochemical purity of the compound.

As a point of interest which relates to all of these indium labelled derivatives, a component with a short biological half-life was found following dialysis of HSA, fibrinogen, and bleomycin derivatives. Most consider this to be the ^{111}In azobenzene EDTA itself.

When the technetium labelled tolbutamide derivative (1 - *p* - toluenesulfonyl - 3 - (N, N - dimethylaminoethylaminoethyl) urea was evaluated in rats, the distribution pattern did not indicate any preferential pancreatic uptake.[125] Control experiments indicated that the distribution obtained could not be attributed to either pertechnetate or unbound reduced technetium. It has been suggested that the concentration of the parent tolbutamide compound may be insufficient itself for adequate pancreatic concentration due to its own poor pancreatic localization.

In the evaluation of fatty acids which localize in the myocardium, a heart–blood ratio of 30 was obtained following the ^{3}H palmitic acid administration into rabbits. After radiolabelling of the fatty acid derivative with ^{57}Co, the same ratio was 3, sharply contrasting baseline values. It is clear from these data that addition of the ^{57}Co chelating group must radically change the biological transport mechanism of palmitic acid. In addition, it was shown that the tritium labelled derivative also had a poor differential uptake without the cobalt label.[133] Recent iodination studies has demonstrated a similar distribution with oleic acid, i.e. heart–blood ratios of 2–3.[134]

Amino acid derivatives have been labelled in the hope that the distribution would be determined by the amino acid moiety causing concentration of the compound in areas of increased amino acid turnover. To do this, the (3-amino-3-carboxypropyl), phosphonic acid, was labelled with technetium and injected into rabbits.[128] It was found that the distribution of the compound was not determined by the phosphonate group alone because at four hours post-injection, skeletal concentration was low, 18% of the administered dose was in the liver, and 27% remained in the blood. As can be seen, this preliminary evaluation did not suggest amino acid derivatives as valid amino acid tracers.

The bio-distribution of technetium HIDA is especially interesting. Without ^{99m}Tc, ^{14}C HIDA localized to a minimal extent in the heart, while the major fraction is rapidly excreted by the kidneys. However, with ^{99m}Tc HIDA, the

^{99m}Tc is excreted via the hepatobiliary system with little myocardial accumulation. The rationale for the use of ^{99m}Tc HIDA in hepatobiliary scintigraphy is based on the above mentioned property. When urinary and gallbladder contents were re-injected into experimental animals, similar distribution patterns were observed thereby supporting an intact radiopharmaceutical. Bile and urinary samples also exhibited similar R_f patterns to the injected material when tested on paper chromatographs with saline solvent; this finding would support the *in vivo* stability of the compound even further.[129] The nature of technetium binding and the structure of ^{99m}Tc HIDA await further investigation.

DISCUSSION

The nuclear properties of ^{99m}Tc has carried it to many successes which would not have been attained by a nuclide with a less than ideal decay scheme. For example mercury chlormerodrin[46] has better biological properties (in light of the ideal qualities outlined earlier for site directed derivatives) than the ^{99m}Tc agents suggested for kidney imaging. However, because of the poor nuclear properties of the mercury nuclides, ^{99m}Tc chelates are generally used for kidney imaging. The extraordinary success of ^{99m}Tc radiopharmaceuticals in the past 15 years has led to the neglect of the chemical properties of ^{99m}Tc. The first compounds developed were based on the removal of foreign substances from the blood and many factors such as binding affinities, equilibrium constants, and oxidation state were not crucial to the effacacious application of these imaging agents.

However, the binding of technetium to biologically active compounds or drugs necessitates the clear understanding of the effect of the nuclide on the biological properties of the radiopharmaceutical and the affinity for which this radionuclide will be bound. As exemplified by bleomycin, the direct binding of technetium may be quite limited and really only applicable to the labelling of large molecular weight proteins and cells. It has been suggested by analogy with research efforts in pharmaceuticals, chromotherapeutics, and radiolabelled antigens for radioimmunoassay and radioreceptor assay that the synthesis of a chelate containing derivative of a biologically active molecule may be the best approach; however, further investigation will be necessary. The chemistry of ^{99m}Tc is little known and the necessary *in vitro* and *in vivo* data for the present radiopharmaceuticals have not been collected.

Indirect labelling by the preparation of a chelate-containing derivative of a biologically-active molecule seems promising in that it can guarantee a stable chelate of ^{99m}Tc, and the ^{99m}Tc will not interfere with those functional groups which interact with the active site. The problem of air oxidation of reduced technetium may also be overcome to some extent by the formation of a chosen stable ^{99m}Tc chelate. The major obstacle to success will be the synthesis of derivatives of sufficient similarity to the parent molecule to retain the desired biological action. This problem cannot be overemphasized based on the precedent available from other efforts to prepare derivatives which retain the biological activity of the parent.

The first prerequisite to assure further development of ^{99m}Tc radiopharmaceuticals is the resolution of the many uncertainties concerning the chemistry of ^{99m}Tc. This will be achieved by further elucidation of the structure, the oxidation state, the equilibrium constant, and the kinetics associated with ^{99m}Tc chelates. The application of this knowledge in the design of chelate-containing biologically active derivatives will hopefully lead to a second quantum jump in the use of ^{99m}Tc.

Acknowledgement—This work is supported in part by USPHS grant HL 19127 awarded by the Heart and Lung Institute and USPHS grants CA 18661, CA 18674 and CA 18675 awarded by the National Cancer Institute.

REFERENCES

1. RENAULT H., HENRY R., NAPIN J. *et al.* In *Radiopharmaceuticals and Labeled Compounds*, p. 195. IAEA, Vienna (1973).
2. REBA R. C., ECKELMAN W. C., POULOSE K. P. *et al.* In *Radiopharmaceuticals* (Edited by SUBRAMANIAN G.) p. 464. Soc. Nucl. Med., New York (1975).
3. FUJII A., MAEDA K. and UMEZAWA H. *J. Antibiotics* **25**, 755 (1972).

4. Richards P. and Steigman J. In *Radiopharmaceuticals* (Edited by Subramanian G.), p. 23. Soc. Nucl. Med., New York (1975).
5. Asakura H., Hort M. and Umezawa H. *J. Antibiotics* **28,** 537 (1975).
6. Eckelman W. C., Kubota H., Siegel B. *et al. J. nucl. Med.* **17,** 385 (1976).
7. Sinkula A. A. and Yalkowsky S. H. *J. Pharm. Sci.* **64,** 181 (1975).
8. Waddell W. J. *Ann. Rev. Pharmacol.* **13,** 153 (1973).
9. Tsou K. C., Damle S. B. and Crichlow R. W. *J. Pharm. Sci.* **56,** 484 (1967).
10. Larinov L. F., Degteva S. A. and Lesnaia N. A. *Vop Onkol* **8,** 12 (1962).
11. Degteva S. A. *Vop Onkol* **10,** 52 (1964).
12. Wall M. E., Abernethy G. S. and Carroll F. I. *J. Med. Chem.* **12,** 810 (1969).
13. Kirdani R. Y., Mittleman A., Murphy G. P. *et al. J. Clin. Endocrinol. Metab.* **41,** 305 (1975).
14. Shepherd R. E., Huff K. and McGuire W. L. *J. Nat. Canc. Inst.* **53,** 895 (1974).
15. Levy R., Hurwitz E., Maron R. *et al. Cancer Res.* **35,** 1182 (1975).
16. Auerbach G. D., Fedak S. A., Woodward C. J. *et al. Science* **186,** 1223 (1974).
17. Eckelman W. C. and Levenson S. M. In *Quality Control in Nuclear Medicine* (Edited by Rhodes B. A.), C. V. Mosby Inc., St. Louis (1977).
18. Richards P. Rome, *Comitato Nazionale Ricerche Nucleari* **2,** 223 (1960).
19. Cotton F. A. and Wilkerson G. In *Advanced Inorganic Chemistry* p. 972. Interscience, New York (1972).
20. Benjamin P., Rejali A. and Friedell H. *J. nucl. Med.* **11,** 147 (1970).
21. Steigman J., Eckelman W. C., Meinken G. *et al. J. nucl. Med.* **15,** 75 (1974).
22. Dreyer R. and Munze R. Z Karl Marx Univ. Leipzig, *Math Naturwiss R.,* 18, Jg., p. 629 (1969).
23. Harper P. V., Lathrop K. and Gottschalk A. In *Radioactive Pharmaceuticals,* p. 335. U.S. Atomic Energy Commission, (1966).
24. Lin M. S., Winchell H. S. and Shipley B. A. *J. nucl. Med.* **12,** 204 (1971).
25. Johnson A. E. and Gollan F. *J. nucl. Med.* **11,** 564 (1970).
26. Williams J. J. and Deegan T. *Int. J. appl. Radiat. Isotopes* **22,** 767 (1971).
27. Steigman J., Meinken G. and Richards P. *Int. J. appl. Radiat. Isotopes* **26,** 601 (1975).
28. Eckelman W. C., Meinken G. and Richards P. *J. nucl. Med.* **13,** 577 (1972).
29. Hambright P., McRae J., Valk P. E. *et al. J. nucl. Med.* **16,** 478 (1975).
30. Gorski B. and Koch H. *J. inorg. Nucl. Chem* **31,** 3565 (1969).
31. Guennec J. Y. and Guilloumont R. *Radiochem. Radional. Lett.* **13,** 33 (1973).
32. Gorski B. and Koch H. *J. inorg. Nucl. Chem.* **32,** 3832 (1970).
33. Cotton, F. A., Eiss R. and Foxman B. M. *Inorg. Chem* **8,** 950 (1969).
34. Cotton F. A. and Foxman B. M. *Inorg. Chem.* **7,** 1784 (1969).
35. Shandles R., Schlemper E. O., Murinann R. *et al. Inorg. Chem.* **10,** 2785 (1971).
36. Chervu L. R., Novich I. and Blaufox M. D. *Radiology* **107,** 436 (1973).
37. Eckelman W. C., Meinken G., Richards P. *et al.* BNL Report (in press).
38. La Mer V. K. and Dinegar R. H. *J. Am. Chem. Soc.* **72,** 4847 (1950).
39. Zaiser E. M. and Le Mer V. K. *J. Colloid Sci.* **3,** 571 (1948).
40. Davis J. A. (Thesis) Indiana University (1955).
41. Mesner R. E. and Irani R. R. *J. inorg. Nucl. Chem.* **28,** 493 (1966).
42. Smith T. D. *J. Chem. Soc.* Part 2, 2544 (1961).
43. Martell A. E. In *Stability Constants.* Special Publication No. 17, The Chemical Society, Burlington House, London W.1 (1964).
44. Martell A. E. In *Stability Constants.* Special Publication No. 25, Supplement No. 1 to Special Publication No. 17, The Chemical Society, Burlington House, London W1V 0BN (1971).
45. Lin M. S. and Winchell H. S. *J. nucl. Med.* **13,** 58 (1972).
46. Arnold R. W., Subramanian G., McAfee J. G. *et al. J. nucl. Med.* **16,** 357 (1975).
47. Fleay R. F. *Aust. Radiol.* **12,** 265 (1968).
48. Hauser W., Atkins H. L., Nelson K. G. *et al. Radiology* **94,** 679 (1970).
49. Eckelman W. and Richards P. *J. nucl. Med.* **11,** 761 (1970).
50. Subramanian G., McAfee J. G., Bell E. G. *et al. J. nucl. Med.* **12,** 399 (1971).
51. Lebowitz E., Atkins H. L., Hauser W. *et al. Int. J. appl. Radiat. Isotope* **22,** 786 (1971).
52. Halpern S. E., Tubis M., Endow S. J. *et al. J. nucl. Med.* **13,** 45 (1972).
53. Halpern S. E., Tubis M., Golden M. *et al. J. nucl. Med.* **13,** 723 (1972).
54. Winchell H. S., Lin M., Shipley B. *et al. J. nucl. Med.* **12,** 678 (1971).
55. Kountz S. L., Yeh S. H., Wood J. *et al. Nature, Lond.* **215,** 1397 (1967).
56. Fliegel C. P., Dewanjee M. K., Holman L. B. *et al. Radiology* **110,** 407 (1974).
57. Lin T. H., Khentigan A. and Winchell H. S. *J. nucl. Med.* **15,** 34 (1974).

58. New England Nuclear Corp., Boston, Mass.
59. Charamza O. and Budikova M. *Nucl., Stuttg.* **8,** 301 (1969).
60. Tubis M., Krishnamurthy G. T., Endow J. S. *et al. J. nucl. Med.* **13,** 653 (1972).
61. Baker R. J., Bellen J. C. and Ronai P. M. *J. nucl. Med.* **16,** 720 (1975).
62. Dugal P., Eikman E. A., Natarajan T. K. *et al. J. nucl. Med.* **13,** 428 (1972).
63. Lin T. H., Khentigan A. and Winchell H. S. *J. nucl. Med.* **15,** 613 (1974).
64. Hunt F. C., Maddalena D. J. and Yeates M. G. *Proc. 1st World Cong. Nucl. Med.* Tokyo, Japan, p. 869 (1974).
65. Subramanian G. and McAfee J. G. *Radiology* **98,** 192 (1971).
66. Subramanian G., McAfee J. G., Bell E. G. *et al. Radiology* **102,** 701 (1972).
67. Perez R., Cohen Y., Henry R. *et al. J. nucl. Med.* **13,** 788 (1972).
68. Yano Y., McRae J., Van Dyke D. C. *et al. J. nucl. Med.* **14,** 73 (1973).
69. Subramanian G., McAfee J. G., Blair R. J. *et al. J. nucl. Med.* **16,** 1137 (1975).
70. Jones A. G., Francis M. D. and Davis M. A. *Semin. Nucl. Med.* **6,** 3 (1976).
71. Davis M. A. and Jones A. G. *Semin Nucl. Med.* **6,** 19 (1976).
72. Chandler W. M. and Shuck L. D. *J. nucl. Med.* **16,** 690 (1975).
73. Eckelman W. C. and Richards P. *J. nucl. Med.* **13,** 202 (1972).
74. Perrson R. B. R., Strand S. E. and White T. *Int. J. Nucl. Med. Biol.* **2,** 113 (1975).
75. Steigman J., Williams H. P., Richards P. *et al. J. nucl. Med.* **15,** 318 (1974).
76. Valk P. E., Dilts C. A. and McRae J. *J. nucl. Med.* **14,** 325 (1973).
77. Bullinghurst M. W. and Palser R. F. *J. nucl. Med.* **15,** 722 (1974).
78. Schneider P. B. *J. nucl. Med.* **14,** 843 (1973).
79. Eckelman W. C., Reba R. C., Kubota H. *et al. J. nucl. Med.* **15,** 279 (1974).
80. Kubota H., Eckelman W. C., Poulose K. P. *et al. J. nucl. Med.* **17,** 36 (1976).
81. Domek N. S., Custer J. R. and Loken M. K. Personal communication.
82. Castronovo P., Jr. and Callahan R. J. *J. nucl. Med.* **13,** 843 (1972).
83. King A. G., Christy B., Hupf H. B. *et al. J. nucl. Med.* **14,** 695 (1973).
84. Atkins H. L., Cardinale K. G., Eckelman W. C. *et al. Radiology* **98,** 674 (1971).
85. Krishnamurthy G. T., Hoebotter R. J. and Walsh C. F. *J. nucl. Med.* **16,** 109 (1975).
86. Krishnamurthy G. T., Thomas P. B., Tubis M. *et al. J. nucl. Med.* **15,** 832 (1974).
87. Krishnamurthy G. T., Tubis M., Endow J. S. *et al. J. nucl. Med.* **15,** 848 (1974).
88. Bowen B. M. and Garnett E. S. *J. nucl. Med.* **15,** 652 (1974).
89. Harper P. V., Lathrop K. A. and Richards P. *J. nucl. Med.* **5,** 382 (1964).
90. Larson J. M. and Bennett L. R. *J. nucl. Med.* **10,** 294 (1969).
91. Ege G. N. and Richards L. P. *Br. J. Radiol.* **42,** 552 (1969).
92. Hunter J. W. *J. nucl. Med.* **10,** 607 (1969).
93. Webber M. M., Victery W. and Cragin M. D. *Radiology* **92,** 170 (1969).
94. Larson S. M. and Nelp W. B. *J. nucl. Med.* **7,** 817 (1966).
95. Garzon O. L., Palcos M. C. and Radicella R. *Int. J. appl. Radiat. Isotopes* **16,** 613 (1965).
96. Mass R., Alvarez J. and Arriaga C. *Int. J. appl. Radiat. Isotopes* **18,** 653 (1967).
97. Johnson A. E. and Gollan F. *J. nucl. Med.* **11,** 564 (1970).
98. Yamada H., Johnson D. E., Griswold M. L. *et al. J. nucl. Med.* **10,** 453 (1969).
99. Gwyther M. M. and Field E. O. *Int. J. appl. Radiat. Isotopes* **17,** 485 (1966).
100. Cragin M. D., Webber M. M., Victery W. K. *et al. J. nucl. Med.* **10,** 621 (1969).
101. Yano Y., McRae J., Honbo D. S. *et al. J. nucl. Med.* **10,** 683 (1969).
102. Lin M. S. and Winchell H. S. *J. nucl. Med.* **13,** 928 (1972).
103. Rhodes B. A., Stern H. S., Buchanan J. A. *et al. Radiology* **99,** 613 (1971).
104. Cifka J. and Vesely P. In *Radiopharmaceuticals and Labelled Compounds,* Vol. 1, p. 53. IAEA, Vienna (1973).
105. Kristensen K. and Pedersen B. *J. nucl. Med.* **16,** 440 (1975).
106. Eckelman W. C. *Semin. Nucl. Med.* **5,** 5 (1975).
107. Dugan M. A., Kozar J. J., Ganse G. *et al. J. nucl. Med.* **13,** 782 (1972).
108. Barth R. F., Singla O. and Gillespie G. Y. *J. nucl. Med.* **15,** 656 (1974).
109. Uchida T., Yasunaga K., Kariyone S. *et al. J. nucl. Med.* **15,** 801 (1974).
110. Gillespie G. Y., Barth R. F. and Goburty A. *J. nucl. Med.* **14,** 706 (1973).
111. Barth R. F. and Singla O. *J. nucl. Med.* **16,** 633 (1975).
112. Wong D. W. and Mishkin F. S. *J. nucl. Med.* **16,** 344 (1975).
113. Millar W. J. and Smith J. F. *Lancet* **II,** 695 (1974).
114. Dugan M. A., Kozar J. J., Ganse G. *et al. J. nucl. Med.* **14,** 233 (1973).

115. Yokayama A., Kominami G., Harada S. *et. al. Int. J. appl. Radiat. Isotopes* **26,** 291 (1975).
116. Steigman M. A., Williams H. P. and Solomon N. A. *J. nucl. Med.* **16,** 573 (1975).
117. Eckelman W. C., Meinken G. and Richards P. *J. nucl. Med.* **11,** 707 (1971).
118. Kyker G. C. and Rafter J. J. In *Radioactive Pharmaceuticals,* p. 503. USAEC (1966).
119. Perrson R. B. R. and Kempi V. *J. nucl. Med.* **16,** 475 (1975).
120. Harwig S. S. L., Harwig J. F., Coleman R. E. *et al. J. nucl. Med.* **17,** 40 (1976).
121. Korubin V., Maisey M. and MacIntyre P. *J. nucl. Med.* **13,** 760 (1972).
122. Barth R. F. Personal communication.
123. Sundberg M. W., Meares C. F. and Goodwin D. A. *et al. J. Med. Chem.* **17,** 1304 (1974).
124. Goodwin D. A., Sundberg M. W., Diamanti C. I. *et al.* In *Radiopharmaceuticals* (Edited by Subramanian G.), p. 80. Soc. Nucl. Med., New York (1975).
125. Heindel N. D., Risch V. R., Burns H. D. *et al. J. Pharm. Sci.* **64,** 687 (1975).
126. Eckelman W. C., Karesh S. M. and Reba R. C. *J. Pharm. Sci.* **64,** 704 (1975).
127. Cannon R. D. and Stillman J. S. *Inorg. Chem.* **14,** 2207 (1975).
128. Castryonova F. P., McKusick K. A. Potsaid M. S. *et al.* In *Radiopharmaceuticals* (Edited by Subramanian G.), p. 68. Soc. Nucl. Med., New York (1975).
129. Loberg M. D., Cooper M., Harvey E. *et al. J. nucl. Med.* **17,** 633 (1976).
130. Callery P. S., Faith W. C., Loberg M. D. *et al. J. Med. Chem.* **19,** 962 (1976).
131. Smith T. W., Wagner H., Jr., Markis J. E. *et al. J. Clin. Invest.* **51,** 1777 (1972).
132. Seccombe R. C., Lee B. and Henry G. M. *Inorg. Chem.* **14,** 1147 (1975).
133. Karesh S. M., Eckelman W. C. and Reba R. C. *J. Pharm. Sci.* (in press).
134. Schelbert H. R., Ashburn W. L., Chancey D. M. *et al. J. nucl. Med.* **15,** 1092 (1974).

International Journal of Applied Radiation and Isotopes, 1977, Vol. 28, pp. 83–95. Pergamon Press. Printed in Northern Ireland

Problems Associated with Stannous ^{99m}Tc-Radiopharmaceuticals

S. C. SRIVASTAVA*, G. MEINKEN, T. D. SMITH and P. RICHARDS
Department of Applied Science, Brookhaven National Laboratory, Upton, New York 11973, U.S.A.

(*Received* 31 *March* 1976)

Possible reasons for anomalous *in vivo* behavior of ^{99m}Tc radiopharmaceuticals are evaluated. The complicated and poorly understood solution chemistry of technetium as well as tin (II), the most widely used reducing agent, is shown to contribute to problems. Unreliable performance of stannous kits is deduced to be partly due to initial oxidation and hydrolysis of tin (II) as a result of poor formulation, and also due to low stoichiometric ratios of complexing agent to tin. The kits could fail in two ways: (1) only a fraction of the original tin may be available in the desired form at reconstitution; (2) undesirable side reactions of tin and technetium may occur. Evaluation of generator as well as instant technetium has occasionally revealed situations where the carrier content of ^{99m}Tc solutions could exceed the reductive capability of the stannous ion; this becomes critical with kits containing a very small quantity of Sn (II) in the "usable" form. Parameters for effective performance of tin (II) containing kits are examined, and a titration method allowing for *in vitro* evaluation of available kits is presented. A model procedure for preparing stable Sn (II) kits has been developed.

INTRODUCTION

THE USE of ^{99m}Tc radiopharmaceuticals continues to grow. A large number of these employ tin (II) reagents to reduce technetium (VII) (pertechnetate) to a lower valence state, thereby making it more amenable to complex forming reactions. Complexes of technetium with agents like DTPA, serum albumin, red cells, phosphates and polyphosphates, etc. are routinely used in nuclear medicine laboratories. Despite the inherent shortcomings of the procedure, stannous ion has remained the most popular reducing agent for ^{99m}Tc radiopharmaceuticals. Following the development of the first "instant" stannous ion containing kit at Brookhaven,[1] a variety of kits have become commercially available.

Recently, various investigators have reported anomalous results with pertechnetate solutions because of non-pertechnetate impurities. Occasionally, liver images of diagnostic quality have been reported during normal brain scan studies. It appears that the pertechnetate solutions in some generator eluates on occasion are contaminated with reduced hydrolyzed forms of technetium. The reasons are difficult to define since the chemistry of technetium in the molybdenum generators is not well understood. In general, such anomalous results have been encountered only occasionally, and mostly with the larger generators. Further research in the area of Tc-generator chemistry thus becomes desirable.

Reports about unreliable performance of tin (II) containing kits have appeared sporadically in the recent past, particularly with the use of lung and bone scanning agents (pyrophosphate and others). Free pertechnetate (unreduced) and/or colloidal (reduced, hydrolyzed) technetium have occasionally been detected in the preparations sometimes in quantities sufficient to produce faulty scans. At times, even kits from the same batch have given preparations with different *in vivo* distribution. This kind of

*Address all correspondence to this author at: Brookhaven National Laboratory, Medical Radionuclide Development Division, Building 801, Upton, N.Y. 11973, U.S.A.

problem could be serious and warrants further consideration.

Samples of instant technetium as well as eluates from generators with a long period of ingrowth often contain a large quantity of $^{99}TcO_4^-$ which, by consuming Sn (II) itself, can cut down the usable tin available in certain stannous radiopharmaceuticals and produce depressed labeling yields because of incomplete reduction of $^{99m}TcO_4^-$. This does become critical for kits that contain an extremely small amount of "usable" stannous ion. The latter is defined as the Sn (II) present effectively to reduce Tc (VII) to a given oxidation state for desired complex formation without side reactions of either tin or technetium taking place. The carrier effect causing depressed labeling yields was first observed after a simple kit for preparing ^{99m}Tc-labeled red blood cells was developed using trace levels of $^{99m}TcO_4^-$.[2] The kit frequently performed poorly when 20–30 mCi of ^{99m}Tc were used. When gradual increments of technetium atoms were added to a series of RBC kits, labeling yields were good for low technetium levels but became progressively worse with greater technetium content of the samples. Experimentally determined technetium capacity of the kits could be used to predict RBC labeling success with any given generator eluate since the total technetium buildup (^{99}Tc + ^{99m}Tc) prior to a milking is easily determinable. Labeling success with instant technetium was difficult to predict because of unknown sample history. An excellent correlation was observed, however, between the technetium content of the sample (based on ^{99}Tc assay after decay of ^{99m}Tc) and RBC labeling yields.

This research was undertaken in order to: (a) define the nature and the magnitude of the above mentioned problems; (b) be able to predict and evaluate the occurrence of such problems by *in vitro* techniques, and (c) develop possible solutions and the methodology to eliminate the existing problems. Reported results include an investigation of several selected commercial stannous radiopharmaceutical kits. Evidence for a correlation among usable vs non-usable stannous ion content of the kits, carrier content of ^{99m}Tc solutions, non-pertechnetate impurities in Tc eluates, and unreliable performance of Tc radiopharmaceuticals is presented.

EXPERIMENTAL

Materials and instruments

All chemicals used were high purity reagents. ^{99m}Pertechnetate was obtained from commercial generators as a solution in saline and used without further purification unless otherwise noted. Ammonium ^{99}pertechnetate was obtained from the Oak Ridge National Laboratory. A ^{99}Tc standard from Amersham-Searle was used for calibration purposes. ^{113}Sn was obtained from General Electric as a high specific activity solution of $SnCl_4$ in conc. HCl. Radionuclidic purity of samples was checked using a Ge(Li) detector spectrometer. Selected commercial stannous radiopharmaceutical kits were used. Sephadex G-25, medium (Pharmacia Fine Chemicals) was used for gel filtration. The following instruments were used: Nuclear Chicago NaI(Tl) well-type Autogamma spectrometer, liquid scintillation counter, Cary 14 u.v./vis spectrophotometer, Capintec ionization chamber/dose calibrator, and the commonly available constant voltage power supply sources, pH meters and potentiometers, and a Virtis model 10–800 freeze drier assembly.

Preparation of standard stannous chloride solution

A typical procedure for preparing quantitative Sn (II) solutions from electrochemically deposited tin metal, based on a method developed by Brown *et al.*,[3,4] was as follows: An aliquot of a 10 mg/ml tin (IV) solution in 1 M HCl was transferred to a 10 ml beaker and a measured quantity of ^{113}Sn (IV) tracer solution (initially 13.77 mCi/mg Sn) was added. A small platinum wire helix wound from 0.010-in. diameter platinum wire on a 0.118-in. o.d. mandrel was used as the cathode, and a straight 0.010-in. diameter platinum wire as the anode. Tin metal was deposited from the continually stirred electrolyte solution onto the platinum helix cathode over 90 min at a current of 50 μA from a constant current power supply. The extent of tin deposition (>99% at end of run) was monitored by (a) counting the spent electrolyte and the platinum cathode, (b) weighing the helix before and after each run, and (c) by titrating an aliquot of the dissolved tin metal with a standard ceric solution to a potentiometric end point. At the end of electrodeposition, the platinum helix was very carefully rinsed with water, acetone, and dried. After obtaining the actual weight, the helix was transferred into a glove box filled with prepurified nitrogen. The tin metal was dissolved by immersing the helix in a test tube containing 1 ml conc. HCl and heating on a sand bath. Complete dissolution was achieved in a few minutes as evidenced by cessation of bubbling. The solution was quantitatively transferred to a 10 ml volumetric flask and raised to mark with distilled water. This stock solution was used immediately to prepare the desired stannous kits. Standardization was accomplished by titrating aliquots with a 0.0101 N Ce (IV) solution to a potentiometric end point. The stannous/stannic content of the stock solution was also determined by a differential sulfide precipitation procedure.[4] A small aliquot was added to 10 ml cold HCl containing a 1.3 mg/ml synthetic mixture of Sn (II) and Sn (IV) ions. The sample was mixed well, made 0.214 M in HF, saturated with H_2S for 30 sec, and centrifuged 0.5 min to separate SnS. The supernatant was decanted into a centrifuge tube containing 5 ml saturated H_3BO_3, and bubbled with H_2S to precipitate the SnS_2. The sulfide precipitates were dissolved in 2 ml hot conc. HCl and counted 18 hr later for ^{113}Sn after allowing the re-establishment of the ^{113}Sn–^{113m}In parent–daughter equilibrium. For a series of electrochemically prepared $SnCl_2$ stock solutions, stannic ion content was never found to exceed 1–3 %.

Preparation of stannous radiopharmaceutical kits

Both stannous-DTPA and stannous-pyrophosphate kits were prepared.

Two series of stannous-DTPA kits (lyophilized and non-lyophilized) with varying total stannous ion content were formulated. In one series, the molar ratio of DTPA to tin (II) was 890, in the second it was 9. All operations were carried out inside a nitrogen-filled glove box. Typically, 2 ml of a stock solution of $SnCl_2$ (*vide supra*) containing 500 μg Sn (II) per ml were added to 18 ml of a pH 6 solution of 0.4 M DTPA to give a 50 μg per ml concentration of Sn (II) and a DTPA/Sn (II) molar ratio of 890. Further dilutions of this solution afforded varying Sn (II) concentrations with the same DTPA/Sn (II) ratio. One ml aliquots of the appropriate stannous solutions were dispensed into individual multi-injection vials containing 1 ml deaerated normal saline added as an inert salt filler to improve reconstitution properties of the final lyophilized preparation. Samples were frozen in a dry ice bath before removal from the nitrogen glove box, and then lyophilized for 48 hr.[2] Vials were backfilled with nitrogen before final capping. The appearance of lyophilized pellets did not change upon storage for several months at room temperature.

Lyophilized stannous-pyrophosphate kits were prepared in a similar manner. Two series were prepared: in one, the molar ratio of pyrophosphate to Sn (II) was 196 and in the other it was 13.7. To aliquots of electrochemically prepared standard stannous chloride solutions (containing ^{113}Sn), a standard solution of $Na_4P_2O_7 \cdot 10H_2O$ was added to give the desired stoichiometric ratios of pyrophosphate to Sn (II). The final pH of the preparations was ~7. The kits were freeze dried as before.

Titration of formulated and commercial stannous kits for their usable Sn (II) *content*

Aliquots of a standard $^{99}TcO_4^-$ solution in a total 4 ml saline volume were added to both formulated kits, DTPA and pyrophosphate, under a nitrogen atmosphere. The aliquots contained increasing stoichiometric quantities of technetium based on a four-electron change, and the solutions were spiked with trace ^{99m}Tc for counting purposes. The reconstituted kits were stirred at room temperature, either 30 min for the DTPA kit or 15 min for the pyrophosphate kit, and then a 0.5 ml aliquot was applied to a Sephadex G-25 M column (0.9 × 35 cm), precalibrated with appropriate tin and technetium standards. The DTPA aliquot was eluted from the column with saline having a DTPA concentration identical to that of the applied sample. Similarly, saline solutions with pH and pyrophosphate concentration equivalent to those of the sample served as the eluent for the pyrophosphate aliquot. The column was kept under a nitrogen atmosphere and the eluting solution was continuously purged with N_2. Two ml fractions were collected in all cases and counted for total activity (^{113}Sn + ^{99m}Tc) the following day after re-establishment of the ^{113}Sn–^{113m}In parent–daughter equilibrium. Five days later, the samples were recounted for ^{113}Sn, and with appropriate decay corrections, the original ^{113}Sn and ^{99m}Tc activities of the fractions were determined. For each level of Sn (II), per cent yields of the complex (DTPA or pyrophosphate) when plotted vs the number of added technetium atoms provided the technetium saturation point and thereby the usable tin in the formulations.

Similar titrations were conducted on a commercial stannous DTPA kit (molar ratio DTPA/Sn (II) = 9), and a commercial stannous pyrophosphate kit (molar ratio pyrophosphate/Sn = 14.1).

The kit technetium capacity of four commercial radiopharmaceuticals, three lung agents and one bone agent was compared to the technetium content of commercial instant and generator $^{99m}TcO_4^-$ sources. The four kits were titrated as previously described to determine their technetium capacity. Labeling yields were determined by sedimentation and separation for the lung agents and by Sephadex G-25 gel filtration for the bone agent. The technetium content of instant and generator $^{99m}TcO_4^-$ samples was determined by liquid scintillation assay of decayed 1 ml sample aliquots in a modified Brays solution. Saline dilutions* (1 ml aliquots) of an Amersham-Searle ^{99}Tc standard were counted in a similar fashion. To avoid possible errors due to contaminating nuclides present in decayed $^{99m}TcO_4^-$, samples were counted periodically over several months until two successive counts were equivalent. The absence of gamma contaminants was confirmed by NaI (Tl) crystal assays. The liquid scintillation assays of the stock $^{99}TcO_4^-$ solution were confirmed spectrophotometrically.[5] Comparison of the Tc content of clinical $^{99m}TcO_4^-$ sources to these kits indicated that some commercial kits have insufficient usable tin to reduce the TcO_4^- present in some sources.

Assay of ^{99m}Tc generator eluates for non-pertechnetate contaminants

Two different-size fission-product ^{99}Mo generators,* 440 and 2200 mCi at receipt (100 and 500 mCi size in previously used terminology) were milked after varying periods of ingrowth during their life. Five ml saline was used for elutions and undiluted as well as 10- or 25-fold diluted eluates were subjected to analysis by G-25 M Sephadex gel filtration. Elution was carried out with N_2-purged saline after applying 0.5 ml sample on a 0.9 × 35 cm precalibrated column. Fractions from column (2 ml each) were collected and counted as usual. Periodically, the milkings were allowed to stand for various time intervals prior to analysis.

*The standard must be diluted in saline for comparison with clinical technetium samples since sodium chloride causes quenching.

*The 2200 mCi generator used in this study was kindly supplied by E. R. Squibb & Sons.

Irradiation experiments with $^{99m+99}TcO_4^-$ samples

Saline solutions (2 ml) containing trace $^{99m}TcO_4^-$ and varying concentrations of $^{99}TcO_4^-$ were irradiated in 10 ml pharmaceutical vials with a ^{60}Co source for varying time periods. The total radiation dose in different experiments corresponded to between 2.5 and 115 × 10^6 rads. After irradiation, 0.5 ml aliquots of the samples were analyzed as usual by Sephadex G-25 M filtration.

RESULTS

Evaluation of generator eluates for non-pertechnetate contaminants

Gel filtration analyses of periodic eluates from several 440 mCi generators and one 2200 mCi generator were carried out. A non-pertechnetate fraction ranging between 5 and 80% of the total was present in the initial milkings of the 440 mCi generators. Subsequent elutions showed markedly reduced or insignificant quantities of this fraction. The 2200 mCi generator (one experiment only) showed only pertechnetate in various periodic elutions. The non-pertechnetate fraction when present eluted immediately following the column void volume. When the initial eluate from the 440 mCi generators was diluted 10- or 25-fold, the non-pertechnetate species was reduced drastically to between 0.5 and 28%. In different experiments, the initial generator eluates were also autoclaved for 30 min with or without added H_2O_2, and bubbled with oxygen as well as air. Results of these experiments were

TABLE 1. Titration data using Sn (II)–DTPA kits*

Sn (II) in kit μg	Ratio ^{99}Tc/Sn (II) ($4e^-$ change)	% yield Tc–DTPA Theor.	Expt.	% Tc reduced and hydrolyzed†	% Tc unreduced (TcO_4^-)	"Usable" tin (II)‡ μg	% of Total
0.96	0.011	100	96.7	3.3	—	0.78	81.2
	1.09	91.5	71.5	5.9	22.6		
	3.28	30.5	21.4	11.0	67.6		
1.92	0.011	100	98.9	1.1	—	1.55	80.7
	0.098	100	98.7	1.3	—		
	0.49	100	98.2	1.0	0.8		
	1.09	91.5	59.9	2.3	37.8		
	1.96	50.9	36.1	3.6	60.3		
	3.28	30.5	20.1	8.8	71.1		
4.8	0.011	100	99.0	1.0	—	4.0	83.3
	1.09	91.5	67.8	5.1	27.1		
	3.28	30.5	24.1	4.9	71.0		
9.6	0.011	100	98.8	1.2	—	9.4	98.0
	0.098	100	99.7	0.3	—		
	0.49	100	99.7	0.3	—		
	1.09	91.5	84.2	4.3	11.5		
	1.96	50.9	38.5	12.1	49.4		
	3.28	30.5	25.9	7.9	66.2		
48.0	0.011	100	100	—	—	47.5	99.0
	0.098	100	99.9	0.1	—		
	0.49	100	99.3	0.7	—		
	1.09	91.5	86.5	8.5	5.0		
	1.96	50.9	46.0	3.8	50.2		
	3.28	30.5	25.1	4.9	70.0		
131.0§	1.07×10^{-4}	100	95.8	4.2	—	15–28	11.5–21.4
	1.07×10^{-3}	100	97.5	2.5	—		
	1.07×10^{-2}	100	94.4	5.6	—		
	0.054	100	93.7	6.3	—		
	0.107	100	94.5	5.5	—		
	0.535	100	85.7	14.3	—		
	0.80	100	85.9	13.5	0.6		
	1.01	99.0	83.3	10.6	6.1		
	2.03	49.3	35.6	13.5	50.9		
134.3‖	0.0104	100	99.1	0.9	—	16–35	11.9–26.1
	0.1043	100	96.7	2.2	1.1		
	0.783	100	91.0	9.0	—		
	1.052	95.1	87.1	9.5	3.4		
	2.097	47.7	37.8	16.2	46.0		
	3.142	31.8	21.0	14.2	64.8		

*Molar ratio of DTPA to Sn (II) = 890.

†Values include technetium absorbed on column (not removable with H_2O_2).

‡From theoretical curve in Fig. 1, after determining the Tc saturation point (≥95% yield of complex) from appropriate curves in Fig. 2.

§Commercial kit (lyophilized), molar ratio of DTPA to Sn (II) = 9.

‖Present work, non-lyophilized preparation, molar ratio of DTPA to Sn (II) = 9.

uncertain; however, it appeared that the non-pertechnetate fraction remained essentially unchanged in the eluates so treated. Eluates containing as much as 80% of this fraction when injected in mice, rabbits or dogs showed normal pertechnetate behavior.

Irradiation of $^{99}TcO_4^-$ solutions

A series of solutions of $^{99}TcO_4^-$ (spiked with trace ^{99m}Tc) when analyzed by Sephadex G-25 M filtration following irradiation with varying doses of ^{60}Co gamma radiation failed to show evidence for formation of reduced technetium.

Titration of stannous-DTPA kits with pertechnetate

Table 1 shows results obtained with stannous-DTPA kits containing varying total amounts of Sn (II) ion. The molar ratio of DTPA to Sn (II) was 890 in kits containing 0.96–48 μg stannous ion. The ratio was 9 in the commercial lyophilized kit (131 μg tin) and the kits containing 134.3 μg tin (II) (this work, non-lyophilized). The ratio of ^{99}Tc to Sn (II) was calculated on the basis of a four electron change in the tin–technetium–DTPA system.[6] After titration of each level of tin (II) with different stoichiometric quantities of TcO_4^-, resulting mixtures were analyzed for technetium present in the DTPA complex, as unreduced pertechnetate and in the reduced hydrolyzed form. Occasionally, a fraction of the technetium absorbed on the column was not removable with H_2O_2; this is included in the values in column 4 of the table. Figure 1 is a graphical representation of the titration data obtained with kits containing a 890-fold excess of DTPA. A straight line is obtained upon plotting log (Tc^{3+}–DTPA) vs the amount of tin used in accordance with the following relationships:

$$2\,Sn^{2+} + Tc^{7+} \rightleftharpoons 2\,Sn^{4+} + Tc^{3+} \quad (1)$$

$$[Tc^{3+}] = \frac{K[Sn^{2+}]^2\,[Tc^{7+}]}{[Sn^{4+}]^2} \quad (2)$$

$$\log\,[Tc^{3+}] = 2\log\,[Sn^{2+}] + \log\,[Tc^{7+}] + \log K - 2\log\,[Sn^{4+}]. \quad (3)$$

Parentheses denote equilibrium concentrations and K is the formation constant for the reaction. The theoretical curve shows a slope of 2 and the intercept provides the value of the remaining terms on the right hand side of equation (3). The slope would deviate from 2 in keeping with the adherence of the system to the above relationships. The plotted data were obtained under loading conditions; i.e. the molar ratio Sn (II)/TcO_4^- was ≤2. Figure 2 shows a log–log plot of the total technetium added vs the percent yield of the DTPA complex for each level of tin in different kits. From this curve, the experimental technetium saturation point for each tin concentration is determinable—i.e. the

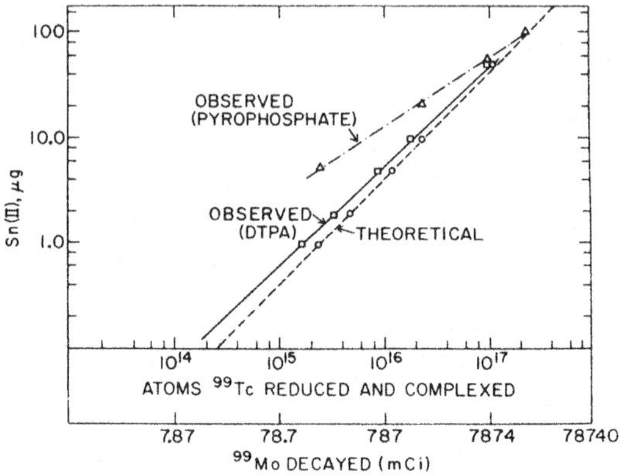

FIG. 1. Observed and theoretical plots for Sn–Tc-DTPA and Sn–Tc-Pyrophosphate systems. Titration of formulated stannous kits (molar ratios DTPA/Sn (II) = 890 and pyrophosphate/Sn (II) = 196) with $^{99}TcO_4^-$. (1 μg $^{99}Tc = 6.08 \times 10^{15}$ atoms, producible from decay of 479 mCi ^{99}Mo).

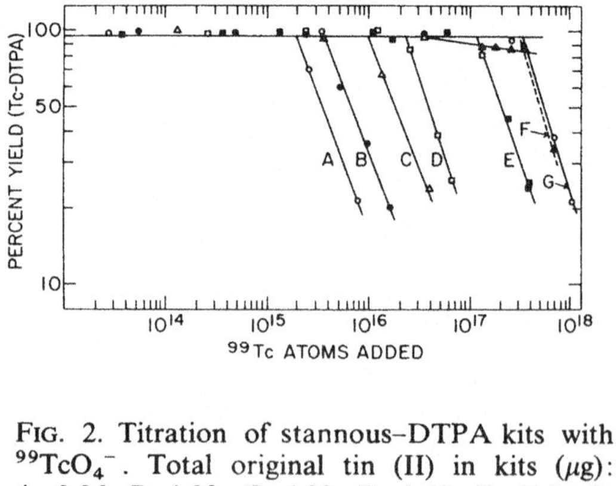

FIG. 2. Titration of stannous–DTPA kits with $^{99}TcO_4^-$. Total original tin (II) in kits (μg): A, 0.96; B, 1.92; C, 4.80; D, 9.60; E, 48.0; F, 131.0 (commercial); G, 134.3. Molar ratio DTPA/Sn (II) = 890 (A–E) and 9.0 (F, G).

maximum number of Tc atoms that can be reduced and complexed with DTPA using a given tin (II) level. (A 95% yield is taken as the breaking point). The usable tin in the kit then is determined as the quantity required to reduce and complex this amount of technetium from the theoretical curve of Fig. 1. The last two columns of Table 1 show the values so obtained. The observed curve in Fig. 1 reflects these values for usable tin in the kits prepared using a 890-fold molar excess of DTPA. Fig. 2 also includes data obtained with the commercial kit, as well as kits prepared in this work, both with a ratio DTPA/Sn = 9.

Titration of stannous-pyrophosphate kits with pertechnetate

Results of these titrations are shown in Table 2 and represented graphically in Figs. 1 and 3. Molar ratios of pyrophosphate to tin were 14.1 (commercial kit), 13.7 and 196 in various formulations. A four electron change was presumed in calculating the ^{99}Tc/Sn (II) ratio.[7] Kits with different levels of tin (II) were titrated with increasing stoichiometric amounts of pertechnetate. Analysis by G25 M gel filtration gave per cent technetium in various forms—as pyrophosphate complex, unreduced pertechnetate, and reduced hydrolyzed technetium.

TABLE 2. Titration data using lyophilized Sn (II)–pyrophosphate kits

Sn (II) in kit, μg*	Ratio ^{99}Tc/Sn (II) ($4e^-$ change)	% Tc as pyrophosphate complex Theor.	% Tc as pyrophosphate complex Expt.	% Tc reduced and hydrolyzed	% Tc unreduced (TcO_4^-)	Usable tin (II)‡ μg	Usable tin (II)‡ % of Total
	0.10	100	44.5	44.8	10.7		
5.11[a]	0.51	100	39.2	6.1	54.8	1.1§	21.5
	2.04	49.0	10.7	16.1	73.2		
20.44[a]	0.51	100	92.7	3.0	4.3	11	53.8
	2.04	49.0	27.7	6.0	66.2		
51.1[a]	2.04	49.0	38.1	1.3	60.6	39.7§	77.8
	0.51	100	97.4	1.4	1.2	102.2	100
102.2[a]	1.02	98.0	97.9	1.3	0.8		
	1.50	66.7	72.0	0.9	27.1		
	2.50	40.0	42.3	1.6	56.1		
	0.0102	100	95.7	3.4	0.90	102.2	100
	0.102	100	97.2	2.1	—		
	0.509	100	97.3	1.8	—		
102.2[b]	1.02	98.0	97.6	1.4	0.8		
	2.04	49.0	55.0	1.1	43.9		
	3.06	32.7	38.6	0.9	61.0		
	0.01	100	94.6	4.6	—	60	55.5
	0.10	100	94.3	4.8	0.7		
	0.75	100	72.1	0.9	26.6		
108.2†	1.00	100	54.5	0.5	44.1		
	2.00	50.0	28.0	1.3	70.5		
	3.00	33.0	19.0	0.6	80.2		

*Molar ratio of pyrophosphate to Sn (II): a, 196; b, 13.7.

†Commercial kit; total tin per vial from label, 758 μg. The kits were reconstituted with 7 ml saline and a 1 ml aliquot was used for titrations. Molar ratio pyrophosphate to tin = 14.1. Information supplied by manufacturer upon request: at assay time, this batch of kits contained 770 μg total tin per vial and 520 μg (67.5% of total) as Sn (II).

‡From theoretical curve in Fig. 1, after determining the technetium saturation point ($\geq$95% yield of complex) from appropriate curve in Fig. 3.

§Values obtained by extrapolation in Fig. 3, and may be in error.

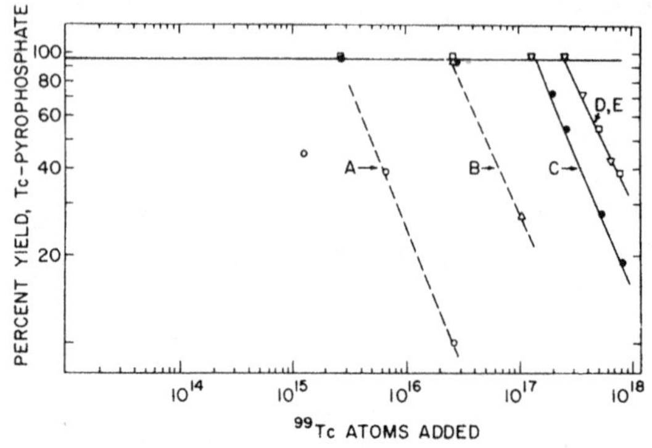

FIG. 3. Titration of stannous-pyrophosphate kits with $^{99}TcO_4^-$. Total original tin (II) in kits (μg): A, 5.11; B, 20.44; C, 108.2 (commercial); D and E, 102.2. Molar ratio pyrophosphate/Sn (II) = 196 (A, B, E), 14.1 (C), 13.7 (D).

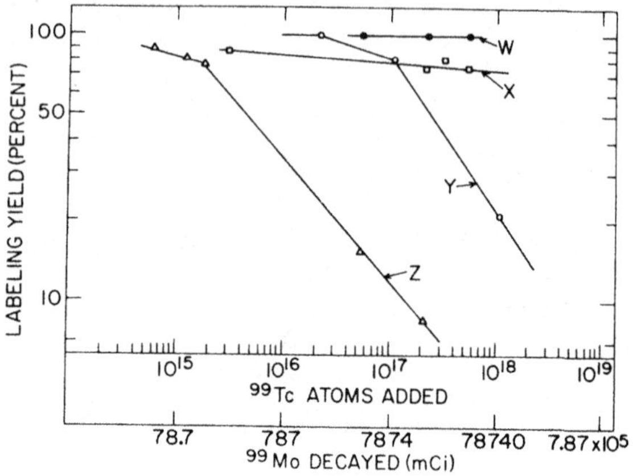

FIG. 4. Technetium capacity for typical commercial stannous kits. Total original tin (II), μg, is as follows: *W*, lung agent, 195–395; *X*, bone agent, 84; *Y*, lung agent, 100–250; *Z*, lung agent, 80.

The observed curve in Fig. 1 is a graphical representation of the titration data using kits with the following molar ratios: pyrophosphate/tin = 196; TcO_4^-/Sn (II) = 1. Similar to the DTPA system, a log–log plot of Tc-pyrophosphate complex vs the amount of tin used gives a straight line, when the molar ratio of Sn (II) to TcO_4^- is ≤ 2. The theoretical curve, based on a four electron change is identical for both the pyrophosphate and the DTPA systems. Figure 3 shows the titration data obtained for determining the Tc saturation point for each tin level in different kits. From the determined technetium capacity, the usable tin in kits is calculated as the quantity required to reduce and complex this amount of technetium from the theoretical curve in Fig. 1. These values are depicted in the last two columns of Table 2. The kinetics of reduction and complexation in the Sn-Tc-pyrophosphate system, particularly at levels of tin (II) < 20 μg, is much slower than in the DTPA system. The observed curves in Fig. 1 demonstrate this situation. When kits containing 20 μg Sn (II) were titrated with an equimolar quantity of $^{99}TcO_4^-$ for various time periods, no significant improvement in yield of the complex was noticeable after up to 4 hr.

Effect of carrier content of ^{99m}Tc *samples on labeling yields using commercial kits*

Four commercial kits representing a range of Sn (II) content were selected for titration with $^{99}TcO_4^-$ solution. Kits were first reconstituted with trace $^{99m}TcO_4^-$ (no added ^{99}Tc) in order to determine the highest obtainable labeling yield. The results are depicted graphically in Fig. 4.

Company *W* lung agent and Company *X* bone agent (% yield with trace ^{99m}Tc: > 99 and 86.5 respectively) showed no significant labeling yield depression with the addition of $> 10^{17}$ atoms of technetium. Depressed yields resulted, however, for lung agents *Y* and *Z* (% yields with trace ^{99m}Tc: 99 and 93.5 respectively). The yield was ~90% for agent *Z* at 6×10^{14} atoms of technetium; thereafter, it fell off more sharply. The break-off point for agent *Y* was around 2×10^{16} Tc atoms. Figure 4 also includes a conversion scale for correlating the ^{99}Mo decay with the total Tc atoms produced. For example, 78.7 mCi of ^{99}Mo decay generates 10^{15} atoms of technetium (equivalent to 6.22×10^3 dpm of ^{99}Tc).

A number of instant $^{99m}TcO_4^-$ samples were assayed for their ^{99}Tc content and the data are presented in Table 3. Company *A* samples contained more ^{99}Tc per mCi of ^{99m}Tc than others. Samples were progressively better during the week, Monday samples frequently being the poorest. Company *B* was non-uniform, and Thursday samples in general, had a higher carrier content. Companies *C* and *D* were more nearly ideal, both had uniformly low ^{99}Tc level

TABLE 3. Carrier content of instant ^{99m}Tc solutions

Source (Samples)	Period	(liquid scintillation counting) ^{99}Tc assay dpm/ml	^{99}Tc/mCi ^{99m}Tc dpm/mCi
A (4)	1974	23,700–57,400	950–2300
A (8)	1975	14,700–67,000	587–2680
B (14)	1974	1500–22,300	130–1635
C (4)	1974	3400– 5420	192– 270
D (12)	1974	760– 9150	76– 305

in samples throughout the week. It was not uncommon to find instant ^{99m}Tc samples containing enough ^{99}Tc to exceed the reducing capacity of usable stannous ion present in several commercial pharmaceutical kits.

DISCUSSION

Despite the recently described empirical to semi-quantitative studies on the chemistry of technetium-pharmaceutical systems,[6–13] the reduction chemistry of technetium relevant to nuclear medicine (at the carrier-free level) remains intricate and poorly understood. The present work provides an insight into the anomalous behavior of stannous kits and the peculiarities of the tin–technetium radiopharmaceuticals. The following discussion also includes hypotheses concerning the non-pertechnetate impurities occasionally found present in generator eluates.

Non-pertechnetate contaminants

Characterization of the nature of these impurities so far has been unsuccessful. Reported incidents strongly suggest the presence of reduced hydrolyzed technetium in the eluates. Since the occurrence of this phenomenon is random, elucidation of the nature of the technetium species in question becomes quite difficult. A thorough understanding of the chemistry of ^{99}Mo–^{99m}Tc generator is imperative in order to solve this problem. It has generally been presumed that pertechnetate gets reduced by ionizing radiation and that lower valence states of Tc are strongly absorbed by the alumina column.[14] It is conceivable that the common radicals produced upon radiolysis of aqueous solutions (H_2, H, OH, HO_2, etc.)[15] are responsible for the reduction. In the absence of complexing agents, hydrolysis or precipitation of reduced technetium becomes a real possibility. In the case of big generators in particular, curie amounts of ^{99}Mo are loaded on the alumina columns. The initial pH of molybdate solutions is generally ≥ 9; at this pH, even trace amounts of ^{99}Tc if present in the reduced form, will hydrolyze. Behavior of hydrolyzed technetium (soluble or precipitated) on the column or during elution cannot be predicted because of an insufficient understanding of the system. The pH of molybdate solutions is usually adjusted to 5–6 before loading on the column. Reduced technetium if formed could initially be present as a soluble hydroxo cation, or as a hydrated oxide in solution (in a concentration dictated at a particular p(OH) by the K_{sp}). Depending, then, on the pH and the extent of the buildup of technetium atoms, precipitation may ensue upon exceeding the solubility product concentration of the technetium oxide, initially present in solution. Many of the reduced hydrolyzed species of technetium (the soluble ones in particular) could be quite labile, and may produce pertechnetate upon slow reoxidation. The solubility product constant, $[TcO^{2+}][OH^-]^2$, is calculated to be 6.29×10^{-25}, based on the reported constants for the following hydrolysis reactions:[16]

$$TcO^{2+} + H_2O \rightleftharpoons TcO(OH)^+ + H^+ \quad (K_1 = 4.3 \times 10^{-2}) \qquad (4)$$

$$TcO(OH)^+ + H_2O \rightleftharpoons TcO(OH)_2 + H^+ \quad (K_2 = 3.7 \times 10^{-3}) \qquad (5)$$

$$\downarrow$$

$$TcO_2 \cdot H_2O.$$

Thus if Tc (IV) is present as the product of reduction, at pH 6, precipitation will occur upon the total reduced technetium exceeding 6.29×10^{-9} M or 3.25×10^{12} atoms/ml. In other words, if the technetium atoms per ml generated from the decay of 299 μCi of ^{99}Mo got reduced to the +4 state, precipitation would occur (1 mCi ^{99}Mo = 1.269×10^{13} atoms). At pH of 5 and 4, total Tc (IV) atoms per ml corresponding to the decay of 29.85 mCi and 2.99 Ci of ^{99}Mo, respectively, would have to be exceeded before the onset of precipitation. Apparently the hydrogen ion concentrations of the column load, column, and the eluting solution are quite important, and using pH <5 at each step might lead to an improved situation. Of course, if the technetium is reduced to either a + 6 or a + 5 state, more of it will remain soluble before precipitation. It does appear that soluble species of reduced technetium are relatively labile and prone to re-oxidation to Tc (VII). Insoluble reduced technetium may or may not elute from the column, depending upon its interaction with the alumina bed and the particle size of the precipitate. The non-pertechnetate fraction present in the initial milkings of 440 mCi generators is perhaps an intermediate soluble species of reduced technetium that is quite labile and gets readily oxidized to pertechnetate upon dilution or *in vivo*.

Irradiation of $^{99}TcO_4^-$ solutions with a ^{60}Co source did not show evidence of reduction. Further experiments under various controlled conditions would be necessary in order to define the problem of radiation induced reduction of technetium in solution or on the generator columns.

Tin–technetium-DTPA system

Stannous–DTPA kits were examined for optimal performance since the chemistry of technetium reduction in this system is relatively better understood.[6] Titrations of model as well as commercial kits with increasing stoichiometric amounts of pertechnetate are described in Table 1. A four electron change has been presumed in calculating the TcO_4^-/Sn (II) ratios, the implication being that the Sn (II)–DTPA redox couple drives the reduction of Tc (VII) all the way to Tc (III). In earlier studies,[6] it was empirically established (for mM concentrations) that the predominant species present was Tc (III)–DTPA. At micro or nanomolar levels, it appears that the kinetics of the reduction in the last two steps is relatively slow. A two electron change (complementary reaction) takes place rather rapidly followed by slow reduction to Tc (IV) or Tc (III). Tc (V)–DTPA complex would be relatively less stable and thus partial hydrolysis of Tc could result. Also, the Tc (V) can disproportionate to a mixture of either Tc (IV) and Tc (VII) or Tc (III) and Tc (VII), prior to complexation with DTPA. Results in Table 1 indicate the above possibilities. At ratios (4 e^- change) of Tc/Sn approaching ≥1 the reducing characteristics of the Sn–DTPA system seem to change and the resulting complex appears to involve mixed oxidation states of technetium, perhaps III and IV. An average electron change of 3.5 was earlier found in this system under such conditions.[6] In common radiopharmaceutical usage, such a situation where technetium is in excess over the Sn (II) would generally not be encountered and the predominant species would be Tc (III)-DTPA. In kits with very low tin (II) levels (<10 μg) however, slow kinetics may produce either incomplete reduction of Tc to the desired state or mixed oxidation states even at high Sn/Tc ratios.

Initial levels of tin and the complexing agent in the kit, both, are important. Very little tin could be used for reducing clinical levels of technetium providing the tin is kept available in the useful chemical form. The non-useful portion (present as oxidized or hydrolyzed tin in the formulation) not only diminishes the total reductive capacity of the system, but by engaging in further side reactions can cause depressed yields of the desired Tc complex; e.g. by trapping part of the reduced Tc as a Sn–Tc colloid. Support for this comes from the observation that some commercial kits with high tin levels did reduce nearly theoretical quantities of technetium but gave low complex yields at much before the theoretical Tc saturation point. A molar ratio of DTPA to tin (II) = 9 failed to give good yields under certain conditions (Table 1). A relatively small quantity of tin in the formulations (e.g. 10μg, which under kinetically favorable situations, can reduce far more technetium than would normally be required in radiopharmaceutical use) would allow use of a large excess of the complexing agent. This would keep the tin complexed, and cut down on the competition between Sn and Tc for the ligand. Improved yields of the technetium complex, therefore, would result. A molar ratio of DTPA to tin = 890 which was used arbitrarily, produced close to theoretical labeling yields with all levels of tin (II) examined (0.96–48 μg) (Figs. 1 and 2). The usable tin in these kits (Table 1) ranged between 80 and 100%. At a moderate level of tin (9.6 μg), 98% of the initial tin was available in the usable form. This will reduce 2.4×10^{16} atoms of technetium, or the total Tc (99 + 99*m*) produced as a result of the decay of 1890 mCi of ^{99}Mo. Good labeling yields would result for up to 200 mCi Tc from the worst instant Tc situation listed in Table 3. Many commercial kits have too much original tin and a low DTPA to tin ratio (the commercial kit tested had a molar ratio of 9), resulting in unreliable performance. A high ligand ratio would be even more desirable in the case of weaker complexing agents. Factors such as kinetics of reduction and/or complex formation could become important in certain systems. Each kit system, therefore, should be investigated individually in order to optimize the levels of tin (II) and the complexing agent for effective performance.

Tin–technetium-pyrophosphate system

This system was investigated as a bone agent model, again since some useful chemical information on stannous-pyrophosphate[17,18] and Sn–Tc-pyrophosphate[7] systems is available. Studies carried out at millimolar concentrations with an excess of tin (II) indicate a four electron change in this reduction system over a wide range of conditions.[7] At the radiopharmaceutical level and generally in the presence of excess reducing agent, quantitative reduction of technetium would be expected to result in Tc (III), which in turn should be stabilized as the pyrophosphate complex. Significant quantities of unreduced technetium prevail, however, when low levels of tin (II) are used. Results in Table 2 demonstrate this kinetic effect, *viz.* low complex yield at <100 μg/4 ml Sn (II). It seems that the initial reduction to Tc (V) is fast, but the subsequent reduction steps are not kinetically favored. Disproportionation of the initially formed unstabilized Tc (V) then could produce a mixture of Tc (VII) and Tc (III) or Tc (IV). The pyrophosphate complex involves either Tc (III) or Tc (IV), or a mixture of both. The observed curve in Fig. 1 merges with the theoretical curve at a tin concentration of ~25μg/ml. Below this level, the slope of the plot indicates a net electron change of about 2.6; however, this would be meaningful (in terms of the resultant valence state of Tc) only if there was no unreduced pertechnetate. Definite conclusions are not possible due to lack of more quantitative information. The theoretical curve in Fig. 1 allows the determination of usable tin in a formulation following the determination of the latter's technetium capacity from actual titration (Fig. 3).

In the final radiopharmaceutical preparation, a concentration of usable tin (II) ~25 μg/ml, and a moderate excess of pyrophosphate over tin should provide optimum reduction of clinical levels of $^{99m}TcO_4^-$ and a satisfactory yield of the ^{99m}Tc–pyrophosphate complex.

Commercial bone and lung agents

The technetium capacity of a given stannous kit could be correlated with the total Tc content of generator eluates because of the known generator history. The number of mCi's of ^{99}Mo which must decay to generate enough Tc atoms to saturate a particular kit, and thereby the permissible ingrowth period for the generator, can be easily derived. This method assumes, of course, 100% generator elution, no generator overloading and use of the total eluate with the selected kit. Complete generator elution can be approached by an extra saline wash following each milking; this would also eliminate presence of unwanted ^{99}Tc in the next clinical elution.

The most sensitive kit, lung agent *Z* began to exhibit sharp decrease in yields with $>6 \times 10^{14}$ atoms (3732 dpm) of ^{99}Tc. Eluates obtained from the worst generator case would far exceed the capacity of this kit, but it is unlikely that the total eluate would be used. Using a tenth or smaller fractions of the total eluate will provide satisfactory yields. It must be emphasized that, though there is enough tin in the kit to begin with, a large portion is not available to reduce the technetium either due to physical entrapment or slow kinetics or both. Labeling yields could certainly be improved at progressively higher ^{99}Tc levels with the same kit by choosing conditions to make more of the tin available for reduction.

The data in Fig. 4 can be used to calculate permissible generator ingrowth times which will assure good yields with selected kits. For example since lung agent Z could tolerate $\sim 6 \times 10^{14}$ Tc atoms (from decay of 47 mCi ^{99}Mo), and taking the case where one-tenth of the eluate will be added to the kit, a total of 470 mCi of ^{99}Mo can decay on the generator column. In the worst case, i.e. a large generator that contains 2918 mCi of ^{99}Mo at loading time, 470 mCi of ^{99}Mo would decay in 16.7 hr ($t(\text{hr}) = -\ln A/A_0 \times 66/0.693$; $A_0 = 2918$ mCi, $A = 2448$ mCi). The generator, therefore, must not grow-in more than 16.7 hr and at that time, no more than one-tenth of the eluate could be used with the kit in order to obtain labeling yields of $\sim 90\%$.

Instant technetium performance cannot be predicted since the sample history is unknown. However, upon determining the ^{99}Tc content of samples after decay of ^{99m}Tc, the specific activity (mCi ^{99m}Tc/atoms of Tc) of the originally assayed sample (mCi ^{99m}Tc/ml) can be calculated (Table 3). With a knowledge of the Tc capacity of a kit from actual titration, a calculation will provide the maximum mCi's of ^{99m}Tc that could have been added to the kit at the ^{99m}Tc sample assay time for a satisfactory labeling yield. This method was found reliable upon testing instant technetium samples of known specific activity with stannous kits of predetermined technetium capacity. If the worst instant Tc sample from Table 3 is used, the lung agent Z would tolerate only ~ 1.4 mCi of ^{99m}Tc since these samples contain ^{99}Tc equivalent to ~ 2680 dpm/mCi of ^{99m}Tc. Other kits in Fig. 4 can tolerate much more ^{99}Tc and would not be likely to exhibit problems with the use of instant technetium.

Kits containing identical amounts of Sn (II) do not always perform similarly, contrary to what one might have expected. Although the stannous ion content of lung agent Z and bone agent X is essentially similar, the bone agent has a considerably greater Tc capacity. In the lung kit, apparently, a smaller fraction of the total tin is available for reducing technetium. The chemistry of both tin (II) and technetium in the lung agents is different in many respects from the soluble majority of Tc-radiopharmaceutical preparations. Any hydrolysis of tin (II) might result in coprecipitation of reduced Tc, and this Sn–Tc colloid (depending on its particle size) may or may not become a part of the Tc–Sn-lung agent particles system. Labeling yield determinations based on centrifugation of the Tc-labeled particles do not differentiate between these two forms of technetium. Also, here is a situation where the availability of tin may be important in the physical sense. It is likely that the lung agent particles trap Sn (II), and only the superficially adsorbed or bound tin is available for reduction. After this "layer" of tin is exhausted, a subsequently very slow availability of the remaining tin results in unreduced or unbound Tc in the final preparation. Indeed, when a longer mixing period was used, e.g. 60 min at R.T., or 15 min at 55° (the usual manufacturer's directions are for ~3 min mixing at reconstitution), greatly improved labeling yields resulted (as determined by centrifugation*). Whether longer incubations or heating result in a changed biodistribution of the pharmaceutical needs to be determined by actual experiment. It would appear that, in this system, an initially higher amount of tin (II) would provide a better yield with short mixing periods. This is actually what seems to be the case with the lung agents W and Y (Fig. 4). The use of more tin (II), of course, is not to be recommended. For clinical levels of ^{99m}Tc, the tin (II) actually required for reduction is so small that the best solution would be to keep the level of original tin to a minimum and, by using proper conditions, to make most of it available when needed, in the desired useful form.

CONCLUSION

In order to eliminate the existing problems and obtain optimal performance, each stannous kit system would have to be examined individually, e.g. according to procedures developed in this study. Factors such as the total stannous ion content, ligand to tin ratio, "usable" tin (II), and the kinetics and chemistry of reduction and

*It is possible that longer incubation or heating resulted in coagulation of the Sn–Tc colloid initially present in the supernatant solution and this came down in the centrifugate.

complexation would have to be considered. It is recommended that: (1) stannous solution used for formulations should be prepared with great caution to avoid oxidation and hydrolysis; (2) a minimum quantity of Sn (II) and an excess of the complexing agent (as optimized by exploratory experiments for a particular kit system) should be used; (3) if the kit contains very little usable tin (II), the carrier content of $^{99m}TcO_4^-$ solutions should be evaluated. The total technetium sometimes may exceed the reductive capacity of tin. An understanding of the nature of various tin species in the final radiopharmaceutical preparation and their fate as determined by actual *in vitro* and *in vivo* studies is highly desirable. Also, further studies are needed to characterize the occasional non-pertechnetate impurities in ^{99}Mo generator eluates. Work in these areas is in progress and will be reported elsewhere upon completion.

Acknowledgements—The authors would like to acknowledge the assistance of H. L. Atkins, A. Ansari, J. Klopper and P. Som of the Medical Department, Brookhaven National Laboratory, during stages of this work. Thanks are due to J. Steimers of the Health Physics Division for assaying ^{99}Tc samples by liquid scintillation counting. The use of a tin plating procedure developed by L. C. Brown of Abbott Laboratories, Chicago, during a Guest Scientist appointment at Brookhaven in 1974, is gratefully acknowledged. This research was performed under the auspices of the U.S. Energy Research and Development Administration.

REFERENCES

1. Eckelman W. and Richards P. *J. nucl. Med.* **11**, 761 (1970).
2. Smith T. D. and Richards P. *J. nucl. Med.* **17**, 126 (1976).
3. Brown L. C. Unpublished work.
4. Brown L. C. and Wahl A. C. *J. inorg. nucl. Chem.* **29**, 2133 (1967).
5. Boyd G. E. *J. chem. Educ.* **36**, 3 (1959).
6. Steigman J., Meinken G. and Richards P. *Int. J. appl. Radiat. Isotopes* **26**, 601 (1975).
7. Steigman J., Meinken G. and Richards P. Manuscript in preparation.
8. Bratu C., Bratu Gh., Galateanu I. and Roman M. *J. radioanalyt. Chem.* **26**, 5 (1975).
9. Hambright P., McRae J., Valk P. E., Bearden A. J. and Shipley B. A. *J. nucl. Med.* **16**, 478 (1975).
10. McRae J., Hambright P., Valk P. and Bearden A. J. *J. nucl. Med.* **17**, 208 (1976).
11. Steigman J. and Richards P. *Semin. nucl. Med.* **4**, 269 (1974).
12. Eckelman W., Meinken G. and Richards P. *J. nucl. Med.* **12**, 596 (1971).
13. Eckelman W., Meinken G. and Richards P. *J. nucl. Med.* **13**, 577 (1972).
14. Veselý P. and Ćifka J. Some chemical and analytical problems connected with technetium-99m generators. *Radiopharmaceuticals from Generator-produced Radionuclides*, p. 71, (Proc. Panel Vienna, 1970). IAEA, Vienna (1970).
15. Allen A. O. *The Radiation Chemistry of Water and Aqueous Solutions.* Van Nostrand, Princeton, New Jersey (1961).
16. Gorski B. and Koch H. *J. inorg. nucl. Chem.* **31**, 3565 (1969).
17. Mesmer R. E. and Irani R. R. *J. inorg. nucl. Chem.* **28**, 493 (1966).
18. Van Wazer J. R. and Callis C. F. *Chem. Rev.* **58**, 1011 (1958).

International Journal of Applied Radiation and Isotopes, 1977, Vol. 28, pp. 97–104. Pergamon Press. Printed in Northern Ireland

Labeling Plasmin with Technetium-99*m* for Scintigraphic Localization of Thrombi

BERTIL R. R. PERSSON and LENNART DARTE
Department of Radiation Physics, University of Lund,
Lasarettet, S-22185, Lund, Sweden

(*Received* 17 *February* 1976)

A detailed study has been made of the method for labeling plasmin (NOVO Industri A/S Denmark) with ^{99m}Tc in order to prepare a radioactive indicator for early scintigraphic visualization of thrombi and tumors. The best method found for preparing ^{99m}Tc-plasmin involved the reduction of 2.5 ml ^{99m}Tc-pertechnetate solution with 0.5 ml of 4 mM $SnCl_2$, 2 M NaCl and 70 mM HCl. This mixture was then added to 5 mg of plasmin to give a final pH of about 2. After 60 min of reaction the labeling efficiency was 80–90% as determined by gel chromatography column-scanning. The labeling kinetics and influence of pH, concentration of $SnCl_2$, NaCl, plasmin and lysine were studied. The enzymatic activity of plasmin was reduced by less than 15% by the labeling process. Preliminary experiments in rabbits with artificially induced thrombi indicate accumulation of ^{99m}Tc activity in the same area as ^{125}I-fibrinogen after the administration of ^{99m}Tc-plasmin.

INTRODUCTION

THE DETECTION of intravascular thrombi by means of labeled plasmin was demonstrated by OUCHI and WARREN.(1) They used plasmin labeled with ^{131}I and external counting to detect the presence of deep venous thrombi. Later, however, GOMEZ *et al.* reported difficulty in the localization of ^{131}I-labeled plasmin in the clot and related this to inadequate labeling and probable denaturation of the protein.(2) This may have been due to the fact that the stability of plasmin in neutral and alkaline solution is very poor and therefore plasmin is not suited to standard iodination procedures. In acid solution (pH 1–3) however, plasmin is very stable.(3) We therefore suggested that it would be well suited to labeling with ^{99m}Tc by using the stannous method at low pH.(4) This method has been previously used for labeling streptokinase with ^{99m}Tc for clinical detection of thrombi.(4, 5)

The aim of the present work was to study in detail the labeling of plasmin with ^{99m}Tc to find a labeling procedure that resulted in a reproducible and high labeling yield without denaturation of the protein. Plasmin was supplied by NOVO Industry A/S Denmark, who prepare it for fibrinolytic therapy and for treatment of cancer.(3) Plasmin NOVO labeled with ^{131}I has been reported to accumulate in the tumor area of a patient with osteosarcoma.(6) Thus ^{99m}Tc-labeled plasmin might be useful for scintigraphic tumor localization as well as clot detection.

The use and testing of ^{99m}Tc-plasmin as a radioactive indicator for thrombi and tumors is under investigation and will be reported elsewhere.

MATERIAL AND METHODS

The plasmin NOVO used in this work was produced from porcine blood, dialyzed at pH 2.5 and lyophilized but not stabilized with lysine. The enzymatic activity of the preparation was approximately 3 NOVO units/mg.(3) Lysofibrin NOVO is a plasmin preparation with the stabilizing amino acid lysine added.(3)

The labeling technique was based on reduction of pertechnetate with stannous chloride (Matheson, Coleman & Bell, USA), which was then added to a solution of plasmin.

Gel chromatography column scanning (GCS) was used to analyze the fraction of ^{99m}Tc-plasmin, ^{99m}Tc-complex (of lysine and other constituents of low molecular weight), ^{99m}Tc-pertechnetate and reduced, hydrolyzed ^{99m}Tc under various conditions of labeling (7–9). The influence of incubation time, pH, concentrations of $SnCl_2$, NaCl, plasmin and lysine was studied.

A 0.1 ml aliquot of ^{99m}Tc-plasmin preparation was applied to the top of a gel column eluted with 10.0 ml of 0.9% NaCl/HCl solution with the same pH as the sample. The columns, which had an inner diameter of 15 mm, were filled to a height of about 300 mm with Sephadex G-25 Fine (Pharmacia Fine Chemicals AB, Uppsala, Sweden) and saturated with the same NaCl/HCl solution used for elution. No radioactive components are eluted from the column under these conditions. The columns were sealed and scanned with a collimated (1 mm) NaI(Tl) detector.

An example of the use of the GCS method is shown in Fig. 1, which is the radioactivity profile of a ^{99m}Tc-plasmin sample. The peak at 17 cm indicates the presence of ^{99m}Tc-plasmin. The peak at 4 cm corresponds to pertechnetate and at 7 cm to ^{99m}Tc-complex of lysine and chloride. The zone at the top of the column is reduced, hydrolyzed ^{99m}Tc which is adsorbed by the Sephadex gel.

The enzymatic activity after labeling was determined by the casein method according to NOVO.[10] A solution of casein is decomposed by the enzyme for 20 min, pH 7.5, 35.5°C. The reaction is stopped by precipitating the protein with perchloric acid and the amount of substrate decomposed is determined by measuring the optical density at 275 nm. If the optical density increase during these conditions is 1 unit, the plasmin activity is defined as 1 NOVO plasmin unit.[3,10]

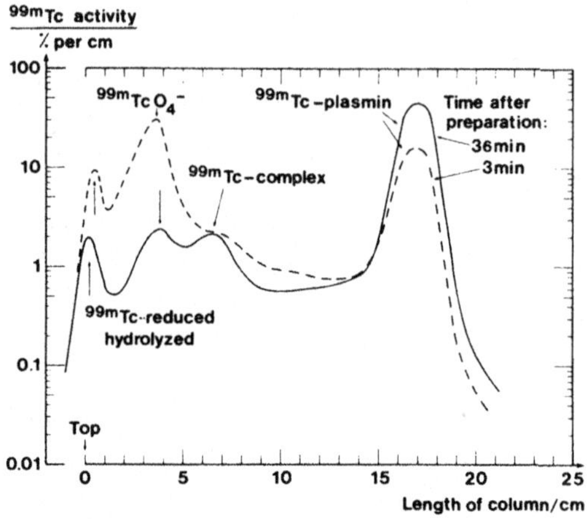

FIG. 1. Gelchromatography column scanning profiles recorded at two different times after adding a mixture of 2.5 ml ^{99m}Tc-pertechnetate and 0.5 ml $SnCl_2$ solution (4 mM $SnCl_2$, 2.0 M NaCl, 0.1 M HCl) to 5 mg of plasmin dissolved in 0.5 ml saline. The final pH-value was about 2.0.

EXPERIMENTAL RESULTS AND DISCUSSION

Plasmin concentration

The fraction of ^{99m}Tc-activity at 17 ± 3 cm in the GCS profile, corresponding to ^{99m}Tc-plasmin, was studied under various labeling conditions. We started with the labeling conditions that have been found to give the best labeling yield for ^{99m}Tc-streptokinase.[4,5] Pertechnetate (2.5 ml) was reduced with 0.5 ml 4 mM $SnCl_2$, 1 M NaCl, 0.1 M HCl. This mixture (3 ml) was added to various amounts of lyophilized plasmin-NOVO previously dissolved in 0.5 ml saline. The final pH was about 2. After 30 min of reaction, samples were analyzed by GCS. The fraction of ^{99m}Tc-activity in the plasmin peaks is given in Table 1. Five mg plasmin gives a promising labeling yield and this amount has been used in the following preparations.

TABLE 1. Per cent of the ^{99m}Tc-activity in different zones of the GCS-profile for samples taken at 30 min after adding 3.0 ml ^{99m}Tc reduced with $SnCl_2$ to various amounts of plasmin at pH 2 in 0.5 ml saline

	Average of the radioactivity-zone distance below the top of the GCS-column	Per cent of the ^{99m}Tc-activity in the different zones at various amount of plasmin (mg)				
		0.2 mg	0.5 mg	1.0 mg	2.0 mg	5.0 mg
^{99m}Tc-reduced-hydrolyzed	3 mm	76	70	70	72	30
^{99m}Tc-pertechnetate	41 mm	7	7	4	5	7
^{99m}Tc-complex	70 mm	15	9	9	7	17
^{99m}Tc-plasmin	171 mm	2	14	17	16	46

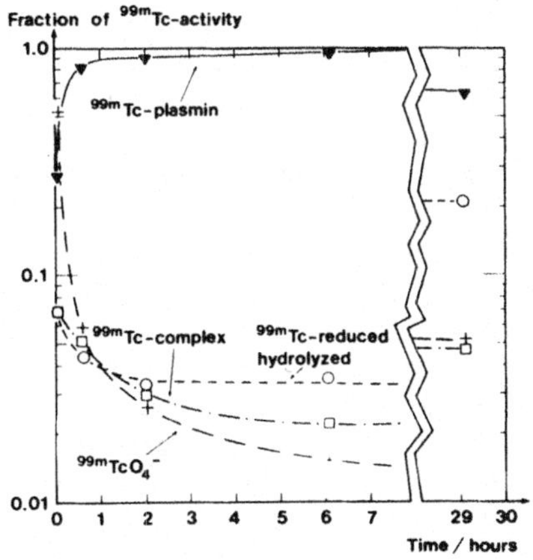

FIG. 2. Fractions of ^{99m}Tc-activity in different zones of gelchromatography-column-scanning-profiles that represent reduced hydrolyzed ^{99m}Tc (top—20 mm), ^{99m}Tc-pertechnetate (20–50 mm), ^{99m}Tc-complex (50–90 mm), and ^{99m}Tc-plasmin (140–210 mm) recorded at different times after adding a mixture of 2.5 ml ^{99m}Tc-pertechnetate and 0.5 ml $SnCl_2$-solution (4 mM $SnCl_2$, 2.0 M NaCl, 0.1 M HCl) to 5 mg of plasmin dissolved in 0.5 ml saline. The final pH-value was about 2.0.

Labeling kinetics

The kinetics of ^{99m}Tc-plasmin labeling was studied at various NaCl concentrations of the reducing solution. A mixture of 2.5 ml ^{99m}Tc-pertechnetate and 0.5 ml $SnCl_2$ solution (4 mM $SnCl_2$, 0.1 M HCl and various NaCl concentrations) was added to 5 mg plasmin in 0.5 ml saline.

The final pH of the preparation was 1.8–1.9. Samples were taken at different times and analyzed by GCS. The GCS profiles obtained after 3 and 36 min of reaction using 2 M NaCl are shown in Fig. 1. Fig. 2 shows the corresponding fractions of ^{99m}Tc-activity representing ^{99m}Tc-plasmin, ^{99m}Tc-complex, ^{99m}Tc-pertechnetate and reduced, hydrolyzed ^{99m}Tc at different times after addition of the ^{99m}Tc-$SnCl_2$ mixture to plasmin. The time course of the yields obtained with various NaCl concentrations were very similar, but the best labeling yield was obtained with 2 M NaCl. The results for other NaCl concentrations obtained after 4 hr of equilibration are given in Table 2.

The labeling of plasmin with ^{99m}Tc was rather slow, with equilibrium obtained after about 1 hr. These results are similar to those obtained when labeling streptokinase with ^{99m}Tc.[4,5]

Influence of pH *on labeling*

The labeling of plasmin with ^{99m}Tc was studied at different pH values by the GCS method. The ^{99m}Tc-pertechnetate (2.5 ml) was reduced with 0.5 ml of 4 mM $SnCl_2$, 2 M NaCl at various pH values. Adjustment of pH was with HCl or NaOH. The reduced ^{99m}Tc was added to 5 mg of plasmin dissolved in 0.5 ml saline. Preparations were allowed to stand at room temperature for 4–6 hr and samples were analyzed by GCS. The results obtained for ^{99m}Tc-plasmin and reduced, hydrolyzed ^{99m}Tc are displayed in Fig. 3. The best labeling yield is obtained in the pH interval 1.5–2.7 for the final solution. In the pH interval 3–8 no labeling is obtained but at pH above 10, fair labeling is indicated. Thus the best labeling was obtained at the pH where the plasmin is most soluble and most stable. For pH below 3, the amount of free pertechnetate increases with decreasing pH value.

TABLE 2. Percent of the ^{99m}Tc-activity in different zones of the GCS-profile for samples taken at 4 hr after adding 3.0 ml ^{99m}Tc reduced with $SnCl_2$ solutions of various NaCl concentrations to 5 mg plasmin at final pH 2

	Average of the radioactivity-zone distance below the top of the GCS-column	Per cent of the ^{99m}Tc-activity in the different zones at various NaCl concentrations (M)		
		0.154 M	1.0 M	2.0 M
^{99m}Tc-reduced-hydrolyzed	3 mm	2	3	3
^{99m}Tc-pertechnetate	41 mm	3	3	2
^{99m}Tc-complex	70 mm	7	4	2
^{99m}Tc-plasmin	171 mm	87	90	91

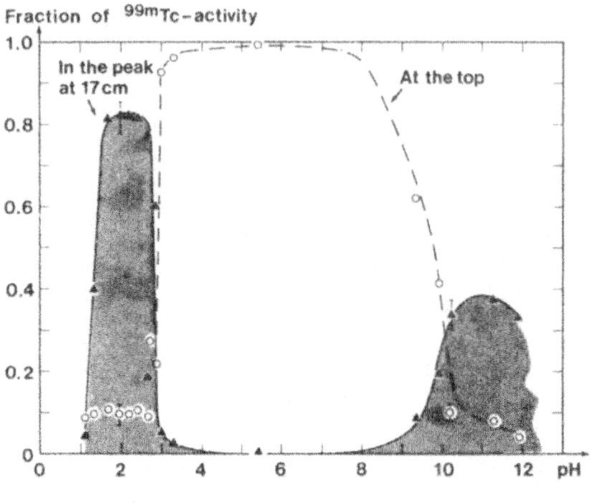

FIG. 3. Fractions of ^{99m}Tc-activity in different zones of gelchromatography-column-scanning-profiles that represent reduced hydrolyzed ^{99m}Tc (top—20 mm), ^{99m}Tc-pertechnetate (20–50 mm), ^{99m}Tc-complex (50–90 mm), and ^{99m}Tc-plasmin (140–210 mm) recorded at 4–6 hr after adding mixtures of 2.5 ml ^{99m}Tc-pertechnetate and 0.5 ml $SnCl_2$-solution (4 mM $SnCl_2$, 2.0 M NaCl) of various pH-values to 0.5 ml of plasmin. The pH-values given in the figure are final values.

At present it is not possible to determine if the fraction of ^{99m}Tc at the top of the GCS column is due to insoluble ^{99m}Tc-hydroxide or precipitated ^{99m}Tc-plasmin. By the experiment described above it is probably due to ^{99m}Tc-hydroxide because the pH was adjusted before adding plasmin.

Other experiments where the pH was 3, 5 or 7 after the ^{99m}Tc-plasmin preparation was first equilibrated at pH 2 for about 2–4 hr gave results similar to those shown in Fig. 3, after 2 hr at the higher pH.

$SnCl_2$ *concentration*

The amount of stannous chloride found optimal for streptokinase labeling (2 μmol) also gave a good labeling yield for plasmin.[5] All the first experiments were carried out with this amount of tin, but we have studied the variation of the labeling yield with various concentrations of $SnCl_2$. Mixtures of 2.5 ml ^{99m}Tc-pertechnetate and 0.5 ml 2 M NaCl solutions of various $SnCl_2$ concentrations were added to 5 mg plasmin in 0.5 ml saline. After 4 hr of reaction, samples were analyzed by GCS. The results obtained are displayed in Fig. 4 where the $SnCl_2$ concentration is given for the final preparation. The $SnCl_2$ concentration which gives the best labeling yield and the lowest fraction of competing labeled products was 0.57 mM. This concentration corresponds to 2 μmol $SnCl_2$, the same amount which was originally assumed to be optimal.

In one experiment the labeling procedure was carried out excluding the $SnCl_2$ solution. Instead a piece of tin metal was added to the mixture of ^{99m}Tc-pertechnetate and plasmin at pH 2. After 5 min of reaction the yield of ^{99m}Tc-plasmin was 20% and after 4 hr 52%, about the same as obtained at a final $SnCl_2$ concentration of 0.015 mM.

Stability and animal studies

High concentrations of the amino acid lysine have been shown to increase the solubility and stability of plasmin.[3] The influence of lysine on the labeling process was tested by labeling lysine without the presence of plasmin. To lysine of various concentrations was added ^{99m}Tc reduced with 4 mM $SnCl_2$ in 1 M NaCl at a final pH of 2.

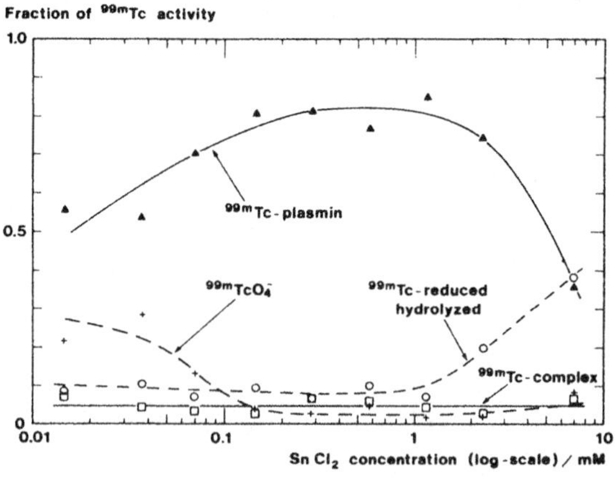

FIG. 4. Fractions of ^{99m}Tc-activity in different zones of gelchromatography-column-scanning-profiles that represent reduced hydrolyzed ^{99m}Tc (top—20 mm), ^{99m}Tc-pertechnetate (20–50 mm), ^{99m}Tc-complex (50–90 mm) and ^{99m}Tc-plasmin (140–210 mm) recorded at 4 hr after adding mixtures of 2.5 ml ^{99m}Tc-pertechnetate and 0.5 ml $SnCl_2$ solutions of various concentrations, to 5 mg of plasmin in 0.5 ml saline. The $SnCl_2$ concentrations in the diagram are the final concentrations and the final pH value was about 2.0.

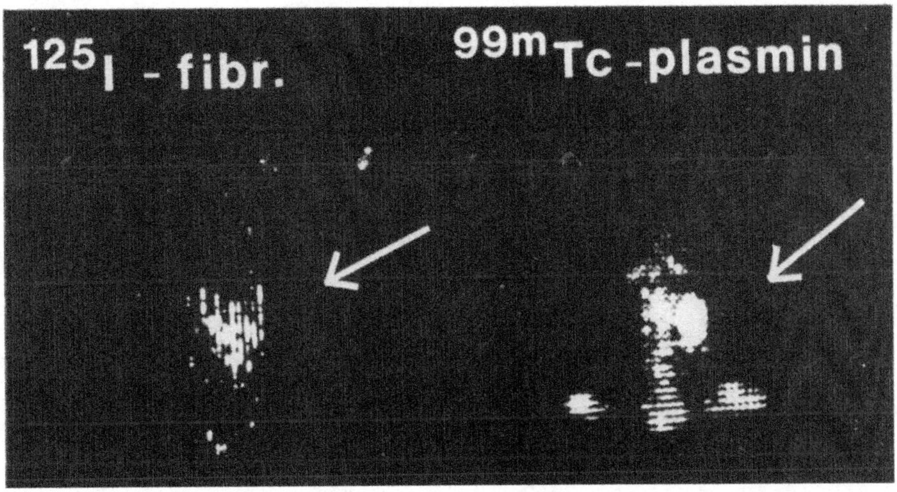

FIG. 5. Scintillation camera views of the neck and head of a rabbit with an artifically induced thrombus in the left jugular vein which was induced after ^{125}I-fibrinogen administration (100 μCi, Amersham, England) and thyroid blocking with KI. The ^{125}I-fibrinogen uptake was recorded at 2 hr after the formation of the clot. The arrow in the picture indicates the localization of the clot. The ^{99m}Tc labeled plasmin was administered in the right ear vein and the ^{99m}Tc uptake was recorded 1 hr later.

TABLE 3. Per cent of the ^{99m}Tc-activity in different zones of the GCS-profile for samples taken at 30 min after adding 3.0 ml ^{99m}Tc reduced with $SnCl_2$ to 0.5 ml lysine solution at pH 2

	Average of the radioactivity-zone distance below the top of the GCS-column	Per cent of ^{99m}Tc-activity in the different zones at various lysine concentrations (mM) 0 mM	10 mM	20 mM	50 mM	100 mM	200 mM	*
^{99m}Tc-reduced-hydrolyzed	3 mm	84	78	78	81	88	92	3
^{99m}Tc-pertechnetate	41 mm	7	8	8	8	3	4	6
^{99m}Tc-complex	70 mm	9	14	14	11	9	4	12
^{99m}Tc-plasmin	171 mm	0	0	0	0	0	0	73

*For comparison the results obtained when labeling plasmin with no lysine added are given in this column.

The results shown in Table 3 indicate that lysine is not labeled with ^{99m}Tc to any high degree.

The influence of lysine on the stability of ^{99m}Tc-plasmin prepared according to the previously obtained optimal conditions was studied by adding 0.2 M lysine to an equal volume of the preparation at about 2–3 hr after starting the labeling procedure. The final pH was adjusted to 2 and 7.4 respectively and the ^{99m}Tc activity in the plasmin zone of the GCS profile was about 50% after 1.5 hr at pH 2 and 40% after 2.0 hr at pH 7.4. With no lysine present the ^{99m}Tc activity under equivalent conditions was about 10% after 2 hr at pH 7.0. The stability of the ^{99m}Tc-labeled plasmin under various conditions is still under investigation and will be reported in more detail elsewhere.[11]

^{99m}Tc-labeled plasmin at pH 2 has been administered to rabbits with artificially induced blood clots in one of the jugular veins. The clot was first localized with ^{125}I-fibrinogen and 5 min after the administration of ^{99m}Tc-plasmin a high uptake was clearly seen in the same area using a scintillation camera. In Fig. 5 is shown the scintillation camera view of ^{125}I-fibrinogen at 2 hr after the formation of the blood clot and of ^{99m}Tc at 1 hr after administration of ^{99m}Tc-plasmin and 3 hr after the formation of the blood clot.

The animal testing of ^{99m}Tc-plasmin is still in progress and more details about its use for thrombus detection and tumor localization will be reported elsewhere.[11]

Acknowledgement—This research has been supported by the JOHN and AUGUSTA PERSSON Foundation for Scientific Medical Research. Thanks are due to NOVO Industri A/S for supplying us with plasmin and for the analysis of enzymatic activity. We also wish to extend our thanks to Dr. C. G. OLSSON at the Department of Clinical Physiology and Medicine at the University Hospital in Lund, for valuable comments and suggestions.

The cooperation of S. E. STRAND in the animal testing Department of Radiation Physics, is also gratefully acknowledged.

REFERENCES

1. OUCHI H. and WARREN R. *Surgery* **51,** 42 (1962).
2. GOMEZ R. L., WHEELER H. B., BELKO J. S. and WARREN R. *Ann. Surg.* **158,** 905 (1963).
3. Lysofibrin, Data-sheet, NOVO Industri A/S, Copenhagen, Denmark (1974).
4. DARTE L., OLSSON C. G. and PERSSON R. B. R. ^{99}Tc^{m}-labelling of streptokinase and its use for detection of deep vein thrombosis using scintillation camera or hand detector. In *Proc. XIII Int. Ann. Meeting Soc. Nucl. Med.* Copenhagen, 10–13 Sept. (1975).
5. PERSSON R. B. R. and KEMPI V. *J. nucl. Med.* **16,** 474 (1975).
6. AMRISE C., LARSEN V., MORGENSEN B. and STORM O. *Dan. med. Bull.* **11,** 146 (1964).
7. PERSSON R. B. R. and STRAND S. E. Labelling processes and short term biodynamical behaviour of different types of ^{99m}Tc-labelled complexes. In *Radiopharmaceuticals and Labelled Compounds,* Vol. I, p. 169. IAEA, Vienna (1973).

8. PERSSON R. B. R. Gelchromatography column scanning (GCS) a method for identification and quality control of technetium-99*m* radiopharmaceuticals. In *Proc. Int. Symp. Radiopharmaceuticals, Atlanta Ga. USA* (1974), p. 228. Soc. Nucl. Med. NY, USA (1975).
9. PERSSON R. B. R. and DARTE L. *J. Chrom.* **101,** 315 (1974).
10. TANG P. NOVO casein method for the determination of plasmin *In NOVO-enzyme information*, 32-263-2, NOVO Industri A/S, Copenhagen, Denmark, March (1971).
11. DARTE L., OLSSON C. G., PERSSON R. B. R. and STRAND S. E. Preparation and testing of ^{99m}Tc-plasmin for thrombus detection. (to be published).

International Journal of Applied Radiation and Isotopes, 1977, Vol. 28, pp. 105–112. Pergamon Press, Printed in Northern Ireland

A Kit for the Preparation of Basic ^{99m}Tc Penicillamine for Renal Scanning

RALPH G. ROBINSON, DIANA BRADSHAW, BUCK A. RHODES,*
JAY A. SPICER, ROSANNE JUDITH VISENTIN
and ALLAN H. GOBUTY

Division of Nuclear Medicine, Department of Diagnostic Radiology, Kansas University Medical Center, Kansas City, Kansas 66103, U.S.A.

(*Received* 5 *February* 1976)

A kit for preparing basic (pH 8.4) ^{99m}Tc-penicillamine complex for renal studies is described. The radiopharmaceutical prepared from the kit localizes in the kidneys, primarily in the renal cortex. Data is presented which demonstrates that the kit method for ^{99m}Tc-penicillamine results in biological distribution of the ^{99m}Tc-penicillamine complex equivalent to that observed for the older extemporaneous method of preparation. The per cent of injected dose localized in the kidneys of rabbits at one hour is $18.3 \pm 3.1\%$, which compares favorably with other ^{99m}Tc-complexes used for renal imaging.

INTRODUCTION

RENAL imaging with a basic complex of ^{99m}Tc-penicillamine was introduced in 1973.[1] This complex differs from ^{99m}Tc-penicillamine acetazolamide complex (TPAC) for renal imaging[2,3] in that acetazolamide was eliminated from our formulation because studies at this institution revealed that pH instead of acetazolamide was the key to renal localization of the tracer.

Clinical studies in over 200 patients have demonstrated the value of basic ^{99m}Tc-penicillamine for renal imaging procedures. The continued use of this agent, however, requires that it compare favorably with other renal imaging agents, and that its preparation be reduced to a simple kit procedure. ARNOLD *et al.* recently completed an exhaustive intercomparison of 17 different renal imaging agents in rabbits,[4] including data on several new ^{99m}Tc complexes. We believe that organ distribution studies with basic ^{99m}Tc-penicillamine compare favorably with data obtained by Arnold for the better ^{99m}Tc-complexes for renal scanning, and report the development and evaluation of a kit for the routine preparation of this renal imaging agent.

MATERIALS AND METHODS

Extemporaneous and kit methods

Table 1 outlines the extemporaneous method as originally developed. Prior to converting this procedure to a kit, the effects of pH on labeling yields and biodistribution were investigated in combination with procedures to permit more precise control of pH. Better control was accomplished by (1) reduction of the concentration of the acid and base solutions and (2) the introduction of a buffer. Table 2 outlines the kit method. Three lots of kits (Solution A & B) of 100 units each were prepared with a mean final pH of 9.35, 7.73 and 8.44.

Preparation of reagents

Solution A consists of 3.100 g D-(—)-penicillamine, 99% pure, (Aldrich Chemical

* Present address: College of Pharmacy, University of New Mexico, Albuquerque, New Mexico 87106, U.S.A.

For reprints contact: Ralph G. Robinson, M.D., Department of Diagnostic Radiology, Kansas University Medical Center, Kansas City, Kansas 66103, U.S.A.

TABLE 1. Extemporaneous preparation of basic ^{99m}Tc-penicillamine complex

1. Transfer 1 ml of D-(−)-penicillamine stock solution (60 mg/ml) to a sealed, 10 ml vial.
2. Add $^{99m}TcO_4^-$ in 3 ml saline and shake.
3. Add 1 ml of concentrated HCl and mix gently.
4. 0.05 ml of phenolphthalein solution (1 mg/ml in 50% ethanol in water).
5. Add enough 12 N NaOH solution to bring solution to a slightly pink color.
6. Autoclave at 15 psi for 15 minutes.
7. Cool and pass solution through a sterile 0.22 micron Millipore filter into an evacuated 10 ml vial.
8. Assay radioactivity and label.

TABLE 2. Kit preparation of basic ^{99m}Tc-penicillamine complex

1. Add $^{99m}TcO_4^-$ in 3 ml saline to vial containing Solution A and shake.
2. Add contents of syringe (Solution B) to vial, flushing syringe 3 times with the solution to assure complete mixing and neutralization.
3. Autoclave at 15 psi for 15 min.
4. Cool, assay radioactivity and label.

Co.), 1.237 g Boric Acid, and 8.5 ml concentrated hydrochloric acid dissolved in sterile pyrogen-free distilled water (U.S.P) brought up to 100 ml in a depyrogenated volumetric flask. This solution was transferred through a 0.22 μm millipore filter to a sterile evacuated bottle as shown in Fig. 1, and 1 ml aliquots were transferred to 10 ml sterile, pyrogen-free glass vials as shown in Fig. 2.

Solution B consists of 1.451 g KCl and 8.936 g NaOH (reagent grade) dissolved in sterile, pyrogen-free water (U.S.P) brought up to 200 ml in a depyrogenated volumetric flask. The solution was then filtered into a sterile evacuated bottle as shown in Fig. 1. After titration of triplicate, 1 ml aliquots of Solution A with Solution B it was established that 1.1 ml of Solution B was required to adjust the final pH to 8.44. One point one ml aliquots of Solution B were then transferred to 3 ml plastic disposable syringes as shown in Fig. 3. The syringes were closed with Luer tip caps (Beckton, Dickinson and Company).

Preparation and aliquoting of Solutions A and B were done aseptically in a laminar flow hood.

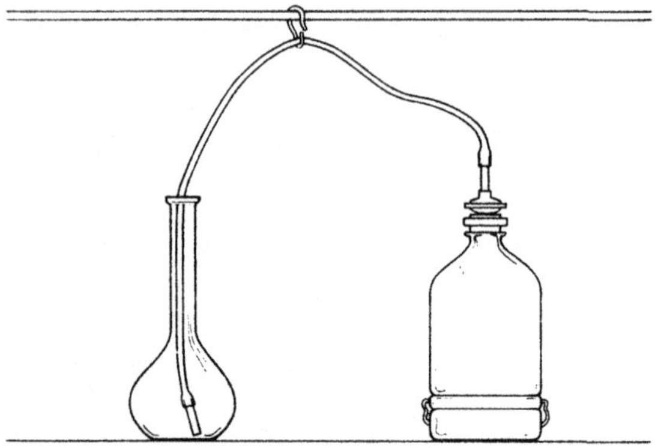

FIG. 1. Both solutions are prepared in depyrogenated volumetric flasks and then transferred to sterile evacuated vials. This is done aseptically in a laminar flow hood. The transfer set is Sterile Pac Extension Set No. SM-03-84, the filter is Millipore (Millex) 0.22 micron, Travenol Laboratories, Inc., Deerfield, Illinois 60015 and the receiving bottle is 1000 ml, Evacuated Container, No. NDC 0074-1614-05, Abbott Laboratories, North Chicago, Illinois 60064.

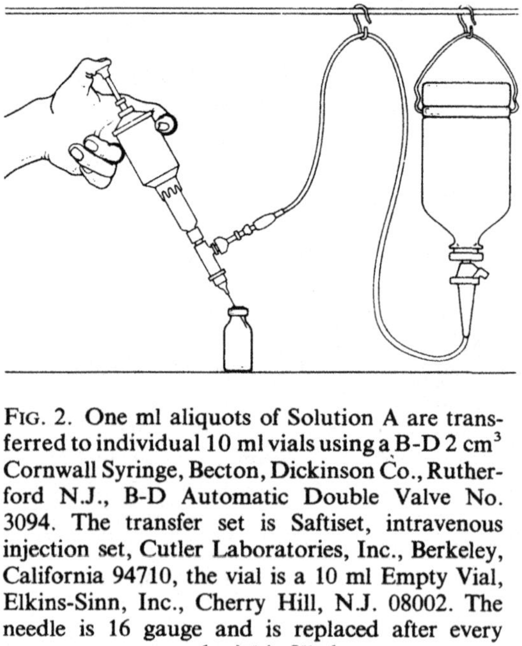

FIG. 2. One ml aliquots of Solution A are transferred to individual 10 ml vials using a B-D 2 cm^3 Cornwall Syringe, Becton, Dickinson Co., Rutherford N.J., B-D Automatic Double Valve No. 3094. The transfer set is Saftiset, intravenous injection set, Cutler Laboratories, Inc., Berkeley, California 94710, the vial is a 10 ml Empty Vial, Elkins-Sinn, Inc., Cherry Hill, N.J. 08002. The needle is 16 gauge and is replaced after every tenth vial is filled.

Titration of the boric acid buffered penicillamine: HCl *solution*

In order to evaluate the effect of adding borate buffer to Solution A, aliquots of Solution A and

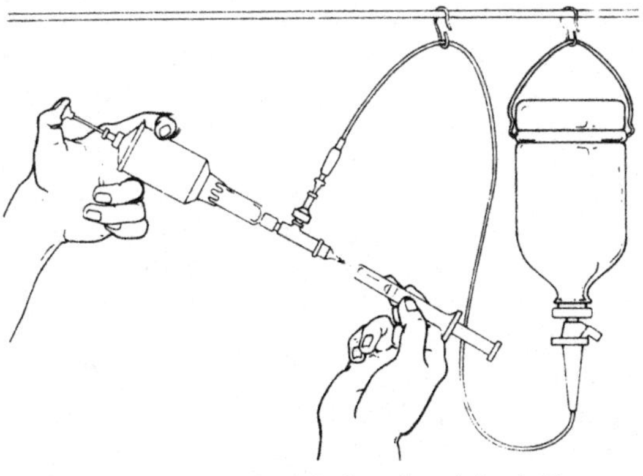

FIG. 3. Solution B is transferred as 1.1 ml aliquots in plastic syringes using the same apparatus as before. Once loaded the syringes are capped with Luer tip caps (see text).

of penicillamine: HCl solution without the boric acid were titrated with Solution B. The pH was measured with a Fisher Model 320 expanded scale pH meter equipped with a Microprobe combination electrode. A commercial buffer solution at pH 7.00 was used for calibration of the pH meter.

pH variance for each lot of kits was determined by taking 10 or more random units of Solution A and Solution B and mixing them as per the procedure outlined in Table 2, except the ^{99m}Tc-radioactivity was omitted. The final pH values were recorded and means and standard deviations were derived for each lot.

Tissue distribution studies

The organ distribution of basic ^{99m}Tc-penicillamine was determined in female New Zealand white rabbits weighing 2–3 kg. Biodistribution was determined ($n = 3$) within 2 days of preparation of each lot of kits. Organ distribution was also determined ($n = 3$) at intervals of 1 and 2 weeks after the kits were prepared.

The rabbits were injected intravenously with 1 ml of the radiopharmaceutical containing 300–500 μCi of ^{99m}Tc-penicillamine. Duplicate syringes were prepared for each rabbit to allow preparation of standards for each animal, and to determine the net activity injected. They were sacrificed one hour post-injection by intravenous administration of sodium pentobarbital. Blood was obtained by cardiac puncture and the liver, right and left kidneys, thigh muscle and bladder (if containing urine) were removed. The organs and carcass were weighed and the radioactivity assayed using a gamma scintillation spectrometer and probe. The results were expressed as percent injected dose per organ and per gram of tissue.

Safety testing

White laboratory mice (9–15 g) were injected intraperitoneally with 1 ml of penicillamine made from each of the 3 kits. Six mice were used per kit, each mouse receiving a dose prepared from separate units of each lot. Controls received 1 ml normal saline. All mice were weighed 3 times weekly for at least one week before and two weeks after administration of penicillamine. The mice were kept 6 per cage and fed a standard laboratory ration. The mean weight gain for each group was compared to the control groups using Student's t test of the hypothesis that two populations have the same mean when the population variance is identical but unknown.[5]

RESULTS

The addition of the boric acid to the penicillamine: HCl solution decreased the slope of the titration curve as the pH rose above 7.5, while the extemporaneous procedure the curve was extremely steep up to pH 10. By varying the amount of Solution B added to Solution A, preparations with pH values of 5–11 were obtained. The biodistribution studies in rabbits revealed that the 1 hr uptake of radioactivity/g renal tissue was highest if the pH was between 7 and 9.5 (Fig. 4). The percent of injected dose

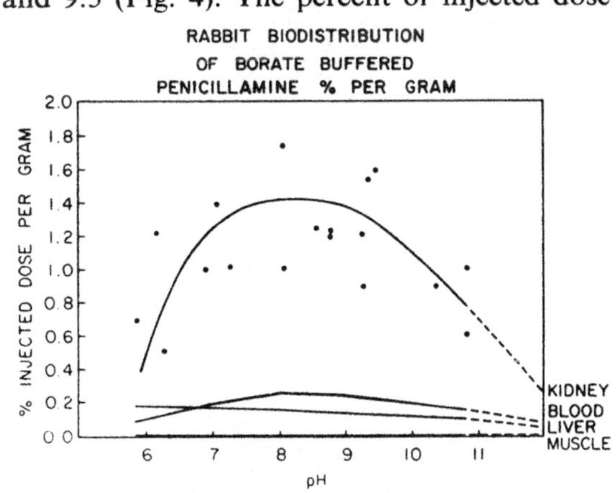

FIG. 4. The effect of pH (or ml of Solution B) on the 1 hour distribution of ^{99m}Tc-penicillamine complex in rabbits.

TABLE 3.

Method	% Dose in whole organs at 1 hr		
	Kidney	Blood	Liver
Extemporaneous, previous (1)	18.3	34.7	12.8
Extemporaneous, current	15.8	N.D.	14.1
Kit, Lot III	18.4	34.4	6.9

remaining in the blood at 1 hr increased from 15% at pH 6 to 31% at pH 9 (Fig. 3). The percent dose in the liver showed only a slight drop between pH 6 and 11 (Fig. 3). Our kits gave results comparable to the original extemporaneous method (Table 3). The addition of the concentrated HCl in the extemporaneous method was shown not to be an essential step, however some HCl is needed in order to achieve maximum kidney uptake.

Biodistribution studies using the extemporaneously prepared complex showed that the concentration of basic ^{99m}Tc-penicillamine in renal tissue continues to increase up to three hours after administration. At 3 hr the renal cortex concentration is approximately twice the 1 hr value. The radioactivity in the renal medulla increases between 1 and 2 hr then begins to fall so that at 3 hr it is only 1/20th. of the concentration in the renal cortex.

The mean organ distributions with pH data are reported in Table 4. Lot III, with a mean pH of 8.44 gave the highest 1 hr kidney uptake. The results for lot III in this experiment were in close agreement with the earlier results of a previous experiment shown in Fig. 3. The precision of final pH control increased with each successive lot prepared, so that all samples of lot III tested fell within the narrow pH range of 8.14–8.68.

The tissue distribution of lot I (mean pH 9.35) was measured as a function of age of the reagents stored at room temperature. As early as one week after preparation a significant loss of 1 hr renal deposition of the tracer was observed (Table 5). Until further stability data is available, it will be necessary to store Solution A in a freezer or add the HCl separately.

No significant (p 0.9) differences in weight gain were observed between control and test animals (Table 6).

TABLE 4. pH and biodistribution data of different lots of penicillamine kits

Lot #	pH		Mean % Dose per organ ± S. D.*		
	Mean ± S: D.	Range	Kidney	Blood	Liver
I $n = 14$	9.35 ± 1.27	7.45 – 11.98	11.55 ± 1.51	34.49 ± 5.10	7.46 ± 1.11
II $n = 10$	7.73 ± 0.65	6.28 – 8.21	9.35 ± 3.00	16.85 ± 6.97	13.52 ± 3.30
III $n = 16$	8.44 ± 0.18	8.14 – 8.68	18.34 ± 3.06	34.37 ± 3.31	6.90 ± 1.81

* $n = 3$ for the biodistribution studies, values for rabbits, 1 hr after i.v. administration

TABLE 5. Effect of storage of kits at room temperature on biodistribution of ^{99m}Tc-penicillamine

Age	Mean % Dose/g ± S. D.*			Relative % Dose/g	
Wks	Kidney	Blood	Liver	Kidney/Blood	Kidney/Liver
0	0.76 ± 0.10	0.14 ± 0.02	0.11 ± 0.06	5.2	6.7
1	0.36 ± 0.22	0.14 ± 0.05	0.11 ± 0.06	2.6	3.2
2	0.22 ± 0.12	0.09 ± 0.02	0.06 ± 0.01	2.4	3.5

* $n = 3$, values are in rabbits 1 hr after i.v. administration.

TABLE 6. Summary of safety test data

Group	Number of animals	Mean ± S. D.*
Penicillamine	12	5.42 ± 2.19‡
Controls†	6	6.33 ± 2.42

*Weight gain in grams during first two weeks post i.p. injection of 1 ml of penicillamine solution.

†Controlled animals received 1 ml of isotonic saline i.p.

‡Student's statistic is 1.036; $p < 0.9$.

DISCUSSION

When prepared at acid pH, ^{99m}Tc-D-penicillamine is largely excreted in the bile, and it has been reported to be a promising agent for cholescintigraphy.[6–8] However, when prepared at basic pH, a different complex of ^{99m}Tc-penicillamine is formed that avidly localizes in the kidneys, primarily in renal cortex. Although the original basic ^{99m}Tc-penicillamine complex reported by HALPERN included acetazolamide,[2,3] he subsequently has dropped the acetazolamide from his ^{99m}Tc-penicillamine complex for renal imaging (Halpern, cited in ref. 7). An example of a patient studied with basic ^{99m}Tc-penicillamine complex was shown in a recent report from Halpern's institution.[9]

The exact ^{99m}Tc-penicillamine complex produced at pH 8.4 is unknown, but its excretion and biodistribution is clearly different from the ^{99m}Tc-penicillamine complex prepared at acid pH. We have observed that when the penicillamine: HCl solution is made basic with NaOH, u.v. absorbance at 232 and 210 mu occurs. After the autoclave step, new u.v. absorbance peaks occur at 217 and 263 mu and the solution also becomes yellow in color giving off an odor of hydrogen sulfide similar to that noted during the preparation of ^{99m}Tc-sulfur colloid. The tracer could be further improved by identification and isolation of the molecular complex which gives the highest renal uptake. Isolation of the active renal complex would allow investigation of the only potential drawback for the widespread clinical use of this ^{99m}Tc-complex, i.e. the remote possibility of a reaction to penicillamine in a penicillin sensitive individual. To date we have not observed such a reaction; however, the radiopharmaceutical has not been used in patients with a history of penicillin sensitivity.

The biodistribution of the ^{99m}Tc-penicillamine complex formed at basic pH compares favorably with other complexes of reduced ^{99m}Tc for renal imaging. ARNOLD *et al.*[4] compared the renal uptake (percent of injected dose in the kidneys one hour following injection) of several ^{99m}Tc complexes in rabbits and found that ^{99m}Tc-lactobionate (18.0 ± 3.8%) ^{99m}Tc-dimercaptosuccinate (20.0 ± 4.0%) and ^{99m}Tc-caseidin (24.0 ± 4.0) had the highest renal uptakes. We observed that basic ^{99m}Tc-penicillamine gave 18.4 ± 3.1% renal uptake at one hour in rabbits and that two hours after the i.v. administration the uptake in the kidneys increased further to 27% These values compare favorably with the results of ARNOLD. IKEDA *et al.* have also studied several ^{99m}Tc compounds for renal imaging.[10] They obtained the highest renal uptake in rats with a ^{99m}Tc-cysteine-acetazolamide complex. This complex showed renal uptake of 17.1% at 1 hr, increasing to 24.9% at 3 hr. A slow exchange from plasma protein probably accounts for the continued rise in renal activity after the first hour.

Incomplete ^{99m}Tc-reduction has not been observed for either the extemporaneous or kit procedures, provided the solutions are autoclaved for the full 15 min. The paper chromatography system of HALPERN *et al.*[2] was used to determine the amounts of TcO_4^- and ^{99m}Tc-penicillamine complex in the preparation. Complete reduction of TcO_4^- was observed for all pH ranges tested (pH 5–11). The addition of the boric acid and the reduction of the acid and base concentration increased the tolerance for variations in the amount of Solution B required for the pH adjustment. When the reaction conditions were adhered to, the final product had a brilliant yellow color. Faulty preparations were visually apparent, as a yellow color did not develop fully and/or a black precipitate formed. The failure to develop a yellow color apparently resulted from uncontrolled pH, but was often associated with satisfactory renal uptakes. The second condition apparently resulted from a combination of high pH and overheating, and was usually associated with decreased renal uptakes. The Sephadex column chromatography

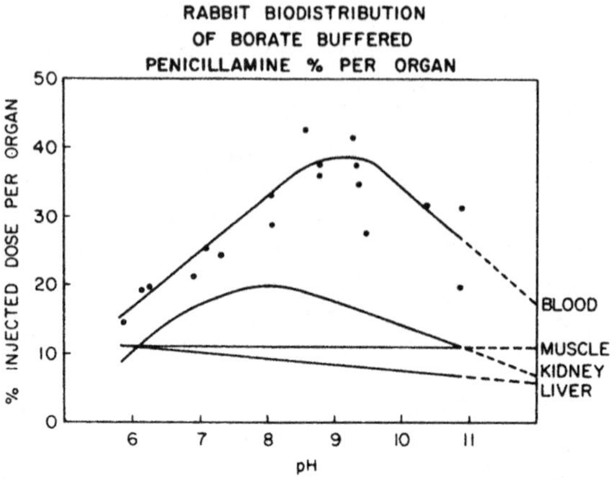

FIG. 5. The effect of pH (or ml of Solution B) on the 1 hour concentration of ^{99m}Tc-penicillamine complex in rabbits.

experiments of YOKOGAMA *et al.*[11] could probably be extended to include the basic as well as the acidic ^{99m}Tc-penicillamine reaction products, and will hopefully lead to a quality control test which will correlate with renal uptake measurements in test animals.

The pH during autoclaving is important. Although the percent of injected dose in the liver showed only a slight drop between pH 6 and 11 in rabbits (Fig. 5), we have observed significant liver radioactivity in man when the final pH was as low as 6.6 (Fig. 6). A final pH of about 8.4 is optimum for renal uptake.

The basic ^{99m}Tc-penicillamine complex prepared by the kit method described gives biodistribution data identical to that obtained by the earlier extemporaneous method. Furthermore, by use of the kit, preparation is simplified and the precision of pH control is increased. The basic ^{99m}Tc-penicillamine kit results in a radiopharmaceutical which compares favorably with the other complexes of ^{99m}Tc for renal imaging.

Acknowledgements—The authors are grateful to MAURICE SMITH, R.Ph. and RALPH MERRICK for help in the development of the aseptic techniques of aliquoting the reagents into the individual units for the kit procedure.

REFERENCES

1. GOBUTY A. H. and ROBINSON R. G. *Invest. Rad.* **8,** 279 (1973) Abstract.
2. HALPERN S., TUBIS M., ENDOW J. S. *et al. J. nucl. Med.* **13,** 45 (1972).
3. HALPERN S. E., TUBIS M., GOLDEN M. *et al. J. nucl. Med.* **13,** 723 (1972).
4. ARNOLD R. W., SUBRAMANIAN G., MCAFEE J. G. *et al. J. nucl. Med.* **16,** 357 (1975).
5. DIXON W. J. and MASSEY F. J., JR. *Introduction to Statistical Analysis,* 2nd Edition, p. 121. McGraw-Hill, New York (1975).
6. TUBIS M., KRISHNAMURTHY G. T., ENDOW J. S. *et al. J. nucl. Med.* **13,** 652 (1972).
7. TUBIS M., KRISHNAMURTHY G. T., ENDOW J. S. *et al.* New radiopharmaceuticals for organ scintigraphy. In *Proc. 1st World Cong. Nucl. Med.* 30 Sept.–4 Oct. Tokyo 1974, World Federation of Nuclear Medicine and Biology pp. 919–921 (1974).
8. KRISHNAMURTHY G. T., TUBIS M., BLAHD W. H. and ENDOW J. S. *Radiology* **115,** 201 (1975).
9. IDEDA I., INOUE O., UCHIDA S. *et al.* New renal scanning agents of ^{99m}Tc-compounds. In *Proc. 1st World Cong. Nucl. Med.* 30 Sept.–4 Oct. 1974, pp. 892–894. World Federation of Nuclear Medicine and Biology, Tokyo (1974).
10. YOKOYAMA A., OKUMURA, SAZI H. *et al.* Studies on the labelling Products of penicillamine with ^{99m}Tc. In *Proc. 1st World Cong. Nucl. Med.* 30 Sept.–4 Oct., 1974, pp. 872–874. World Federation of Nuclear Medicine and Biology, Tokyo (1974).

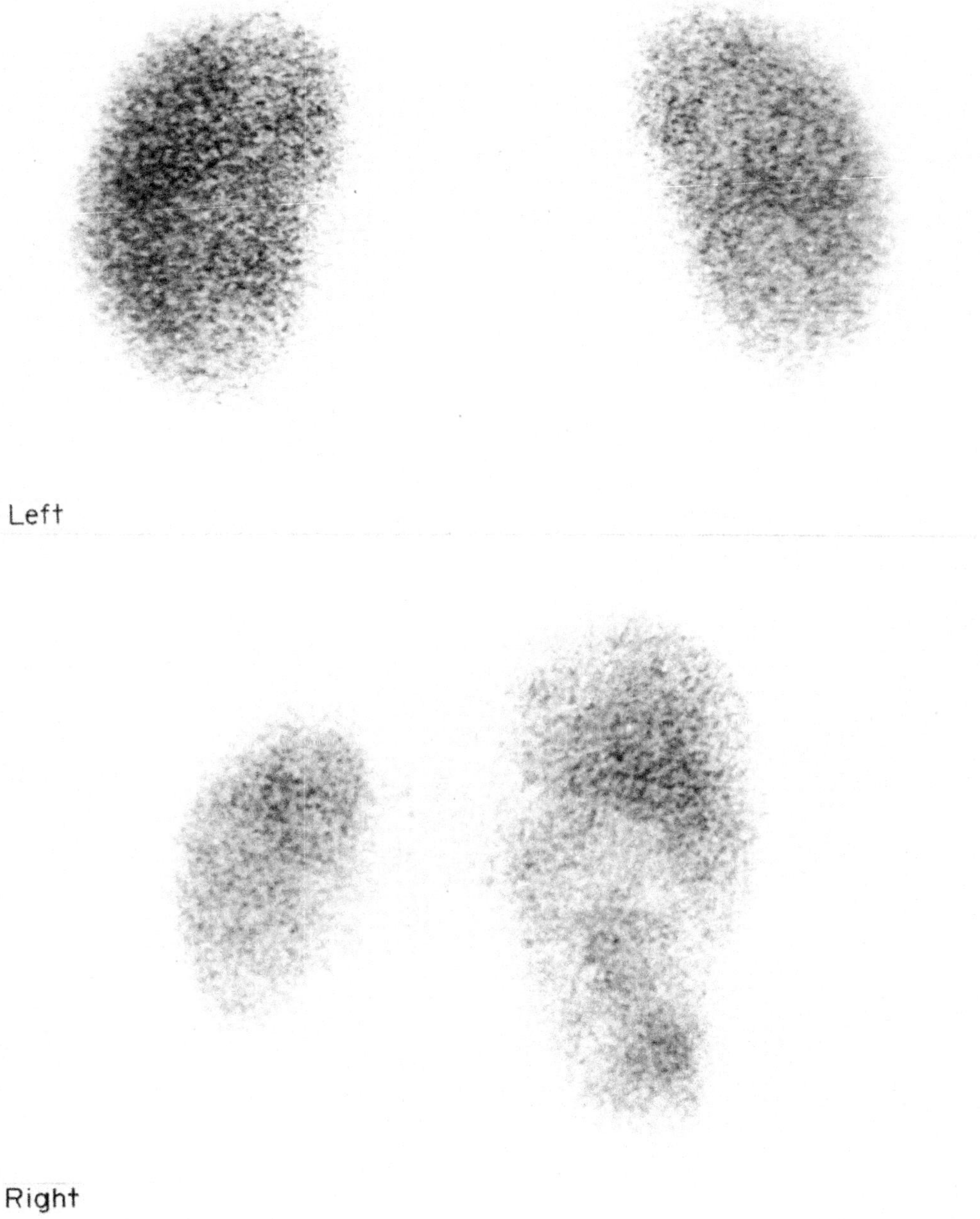

FIG. 6. Left: A gamma camera image of the kidneys of a normal subject given basic ^{99m}Tc-penicillamine complex at pH 8.4. Right: A gamma camera image of the kidney after administration of ^{99m}Tc-penicillamine complex with a final pH of 6.6. Note decreased renal uptake and increased hepatic uptake.

International Journal of Applied Radiation and Isotopes, 1977, Vol. 28, pp. 113–121, Pergamon Press. Printed in Northern Ireland

Development and Evaluation of a Rapid and Efficient Electrolytic Preparation of ^{99m}Tc-Labeled Red Blood Cells*

JOHN F. HARWIG,† PHILIP O. ALDERSON,‡ JOAN L. PRIMEAU, SUPOT BOONVISUT and MICHAEL J. WELCH

Edward Mallinckrodt Institute of Radiology, Washington University School of Medicine, 510 South Kingshighway Blvd., St. Louis, MO 63110, U.S.A.

(*Received* 22 *March* 1976)

This paper reports a new, simple and reproducible method for routine preparation of ^{99m}Tc-labeled red blood cells (RBC). This method employs electrolytic generation of minute quantities of stannous ion to serve as the reducing agent for ^{99m}Tc-pertechnetate. The RBC are incubated with Sn(II), washed with saline, incubated with $Na^{99m}TcO_4$, and washed again with saline. The entire procedure requires about 45 min and consistently provides labeling efficiencies above 90%. The half-clearance time in the blood in humans is approximately 24 hr. Activity released from the labeled RBC *in vivo* is largely in chelate form which is rapidly excreted in the urine. High quality blood pool images can be obtained for many hours after injection. In a damaged form, ^{99m}Tc-RBC provide spleen images of diagnostic quality. This ^{99m}Tc-RBC preparation appears to have good potential for routine use.

INTRODUCTION

RADIOLABELED red blood cells (RBC) have been investigated for many years for use in a number of procedures in nuclear medicine. Applications of labeled undamaged RBC include cardiovascular imaging,[1,2] placental imaging,[3,4] cardiac dynamic studies,[1,5] and red cell volume determinations.[6,7] When damaged in a suitable manner, the labeled RBC are useful for splenic imaging.[8–11] Until recently, RBC have generally been labeled with ^{51}Cr.[12,13] However, the low abundance (9%) and rather high energy (320 keV) of the ^{51}Cr gamma emission are not favorable for use with a scintillation camera, and the 28-day half-life is undesirable for short-term imaging procedures. Technetium-99*m*, with a 140 keV photon and 6-hr half-life, is a more suitable label for RBC.[14] Although a large number of methods have been reported for labeling RBC with ^{99m}Tc[3,6,7,11,14–18] none has gained widespread acceptance. This paper reports the development and evaluation of a new, rapid, efficient and reproducible electrolytic method for preparing ^{99m}Tc-RBC.

MATERIALS AND METHODS

The labeling technique

To develop the most suitable method for preparing ^{99m}Tc-RBC a number of reaction conditions were investigated (Table 1). The procedure outlined below represents the optimum conditions found in terms of labeling efficiency, time and ease of preparation, and reproducibility.

Electrolytic stannous ion generation vials are prepared from readily available components. Two 3-cm pieces of high-purity 0.5 mm-dia. tin wire (Alfa-Ventron Corporation, Danvers, MA)

*Supported by NIH SCOR in Thrombosis 1 P17 HL-14147 and NIH SCOR in Ischemic Heart Disease HL-17646.

†Address all reprint requests to this author at address above.

‡Present address: Armed Forces Radiobiological Research Institute, National Naval Medical Center, Bethesda, MD 20014, U.S.A.

TABLE 1. Summary of various reaction conditions investigated to optimize the ^{99m}Tc-RBC labeling procedure

Factor investigated	Effect on labeling efficiency*
Nature of Sn electrolysis medium	
Buffer solution	−
Saline only	+
pH of Sn electrolysis medium	
Low (≤3)	−
High (≥8)	−
Slightly below neutral (6)	+
Quantity of Sn (II) added to RBC	
Large (5–50 μg)	−
Small (≤1 μg)	+
EDTA in saline washes to complex and remove excess Sn (II)	−
Order of incubation of RBC with Sn (II) and $^{99m}TcO_4^-$	
$^{99m}TcO_4^-$ first	−
Sn (II) first	+
Incubation time of RBC with Sn (II) and $^{99m}TcO_4^-$	
Short (≤5 min)	−
Long (≥10 min)	+
Volume of RBC	No effect
Volume of ^{99m}Tc generator eluate	No effect
Ingrowth time of ^{99m}Tc generator	No effect
Amount of $^{99m}TcO_4^-$ activity	No effect

*Factors which improved labeling are denoted with a (+); those which had an adverse effect are denoted with a (−).

are inserted with the aid of a hypodermic needle through a standard rubber serum stopper into a 6 ml vial containing 5 ml normal saline, pH 6. The saline is prepared by adjusting the pH of preservative-free commercial physiologic saline to 6 with dilute hydrochloric acid or sodium hydroxide, as appropriate. A 3 mm length of wire is allowed to protrude through the top of the stopper and is bent over, after which a standard metal retaining cap is fixed in place. The wires should not make contact with the saline when the vial is in the upright position. After assembly the vials can be easily sterilized by autoclave. Since they contain only saline the vials should have a long shelf life and can be prepared in advance. At the time of use the center of the metal cap is removed and the exposed ends of the wires are attached to the leads of a simple variable DC power supply. The vial is inverted and swirled while a 1 mA current is applied for 10 sec (Fig. 1). The power supply is then disconnected and the inverted vial is swirled for an additional 60 sec, after which the stannous solution is ready for use.

The desired volume of blood, typically 10 ml or less, is obtained by venipuncture and placed immediately in a sterile blood collection tube containing 1 ml ACD. The blood is centrifuged to pack the RBC and the plasma is removed. The remaining steps can be summarized as follows:

(1) Add 1 ml stannous solution to the packed RBC. Incubate 10 min on a rotator.

(2) Wash cells with 4 ml saline. At this stage, commercial physiologic saline can be used with no precautions as to pH or preservatives. Centrifuge and remove supernate.

(3) Add $Na^{99m}TcO_4$ from a generator. Incubate 10 min on a rotator.

(4) Wash cells with saline. Centrifuge and remove supernate.

(5) Draw labeled RBC into a syringe and assay radioactivity. Labeling efficiency is determined by expressing the activity associated with the RBC, corrected for decay, as a percent of the activity added in step 3. In some cases the ^{99m}Tc-RBC preparation is tested by standard U.S.P. methods to confirm apyrogenicity.

(6) If damaged ^{99m}Tc-RBC are desired, the labeled cells are incubated in a water bath at 49°C for 15 min with gentle mixing.[11]

In vitro tests

The supernates from the washing steps were visually checked as an indication of hemolysis of the RBC. The *in vitro* stability of the labeled RBC was determined by resuspension in an equal volume of physiologic saline. Aliquots of this sample were removed at various times, and the activity in the RBC and saline fractions was determined with an automatic gamma scintillation well counter. From these data the percent elution of activity into the saline was calculated.

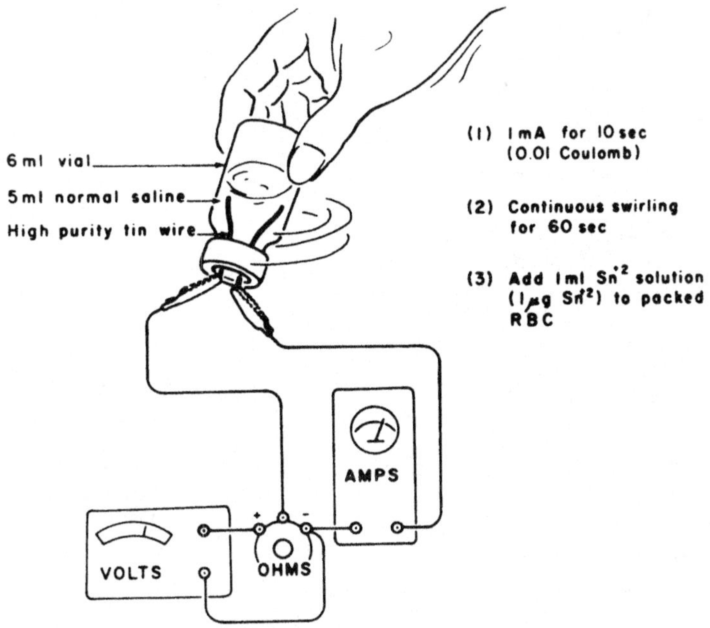

FIG. 1. The electrolytic stannous ion preparation system, showing the steps involved in its operation.

In vivo behavior

The rate of clearance of the ^{99m}Tc-RBC from the blood and the *in vivo* stability of binding of the ^{99m}Tc to the RBC was determined by intravenous reinjection in 8 human subjects. Serial blood samples were obtained from each patient over a 48 hr period following reinjection, and these samples were analyzed as follows:

(1) The activity in 1 ml whole blood was determined and expressed as a percent of the initial (5 min) activity.

(2) The activity in the RBC and plasma fractions of 1 ml whole blood was determined, and the activity in the RBC fraction was expressed as a percent of the total activity in the blood sample.

In five of these patients, all urine was collected during the first 24 hr following reinjection of the ^{99m}Tc-RBC and a blood sample was obtained at 24 hr. The activity in the blood and urine samples and in a standard of the injected dose was determined, and the percent injected dose cleared from the blood and excreted in the urine was calculated. In three of the patients chromatographic evidence of the chemical form of the non-RBC bound ^{99m}Tc in the plasma was also obtained. A 1 ml aliquot of plasma from a blood sample obtained 4–6 hr after injection was analyzed on a 0.9 × 45 cm Biogel P-100 column eluted with saline at 4°C.

Imaging procedures

Scintillation camera images of various portions of the blood pool were obtained in 20 patients at various time intervals up to 24 hr after administration of 30 mCi of ^{99m}Tc-RBC. The regions examined included major arteries, the cardiac blood pool, and the spleen. The principal reason for studying these patients was the determination of the left ventricular ejection fraction by the area-length (gated imaging) method.[19,20]

The use of this ^{99m}Tc-RBC preparation for spleen imaging was assessed in rabbits. Multiple spleen images were obtained in the same animal to permit comparison. First each rabbit had a spleen image with non-damaged ^{99m}Tc-RBC, then with heat-damaged ^{99m}Tc-RBC, and finally with commercial ^{99m}Tc-sulfur colloid. Sufficient time for decay of the previous injected ^{99m}Tc (48–72 hr) was allowed before reinjection.

RESULTS

Labeling

The ^{99m}Tc-RBC labeling efficiency is strongly dependent on several reaction conditions, although a number of other factors have no effect (Table 1). When the procedure is performed using the optimum conditions reported in this paper simple, rapid and efficient labeling is reproducibly attained. The electrolytic vials serve to generate stannous ion according to Faraday's Law:

$$(\text{Sn (II)}\ \mu g = \text{coulombs} \div 96{,}500 \text{ coulombs equiv}^{-1} \div 2 \text{ equiv Sn (II) mole}^{-1} \times \text{mol wt Sn} \times 10^6.$$

If a non-Faradaic current of 200 uA is assumed, the charge of 0.01 coulomb applied to the vials results in the generation of approximately 5 μg Sn (II). Although the method was originally based on the use of a separate vial for each labeling, the Sn (II) produced in this way was found to be usable for multiple labeling with no loss of efficiency over at least a 12-hr period.

This technique can be applied to human, dog, rabbit and rat RBC as a research tool with labeling efficiencies of 90–95%. When the procedure was performed clinically by nuclear medicine technologists for patient studies, labeling efficiency was still consistently greater than 80%. Complete preparation required about 45 min. As originally developed, the procedure included 3 saline washes following pertechnetate incubation. However, with the high labeling efficiency attainable, only 1 final wash is necessary to remove the small amount of non-RBC-bound technetium remaining.

Stability

Little or no hemolysis occurs during the labeling procedure, except for a slight amount occasionally observed following pertechnetate incubation. This is most notable with rat RBC and has no effect on labeling efficiency or subsequent behavior of the RBC. No clumping of the labeled RBC or difficulty in resuspension in saline was observed. The stability of the preparation following resuspension was excellent with 94 ± 4% ($N = 6$) of the original ^{99m}Tc remaining RBC-bound over a 20-hr period. The ^{99m}Tc-RBC cleared exponentially from the blood with a biologic half-time of 25 ± 4 hr (Fig. 2). Over a period of 48 hr 96 ± 3% of the activity remaining in the blood was RBC-bound. Figure 3 shows a typical gel column chromatogram of the non-RBC bound ^{99m}Tc in the blood. A small amount of activity is seen at the void volume, representing ^{99m}Tc bound to plasma proteins. The two peaks at longer retention time suggest small chelates formed between reduced ^{99m}Tc and blood constituents. A ^{99m}Tc-phosphate complex prepared as described previously[21] gives single peak at fraction 35, superimposable with one of the two peaks in question. The peak at fraction 26 is probably also a ^{99m}Tc complex, perhaps also involving phosphate or perhaps a species such as carbonate. There is no peak for pertechnetate. In the comparison of activity cleared from the blood and excreted in the urine during the first 24 hr, the activity excreted (30 ± 4%) accounted for much, but not all, of the activity cleared (54 ± 9%).

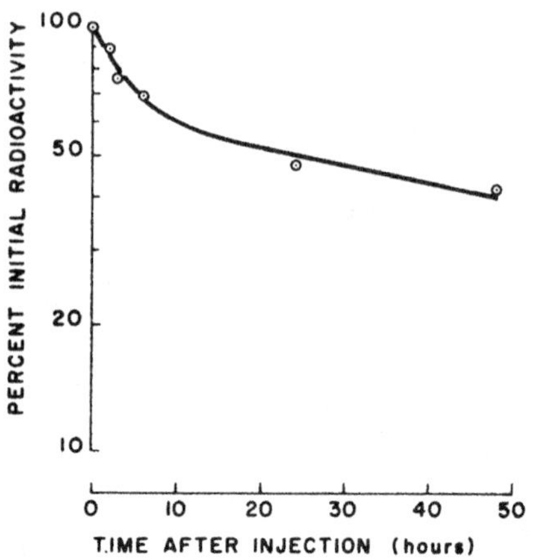

FIG. 2. Typical whole-blood clearance curve following intravenous injection of ^{99m}Tc-RBC in human subject. The half-clearance time is approximately 25 hr.

Diagnostic utility

Imaging studies in human subjects performed up to 24 hr after injection of ^{99m}Tc-RBC demonstrated visualization of blood pool structures without interference from activity in other tissues. The utility of the preparation for spleen imaging is demonstrated in Fig. 4, which

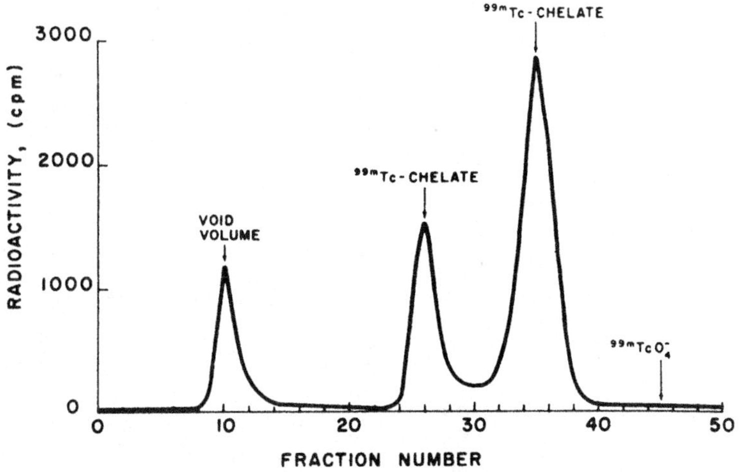

FIG. 3. Typical Biogel P-100 column chromatogram of a plasma sample obtained 4 hr after injection of ^{99m}Tc-RBC in a human subject. There is a protein-bound fraction at the void volume and 2 lower molecular weight chelate fractions at longer retention volumes. There is no free pertechnetate.

shows that splenic visualization with heat-damaged ^{99m}Tc-RBC is superior to that obtained with undamaged ^{99m}Tc-RBC or with ^{99m}Tc-sulfur colloid.

DISCUSSION

Labeling characteristics

The literature pertaining to RBC labeling is extensive.[22] The proliferation of techniques reported in recent years for preparing ^{99m}Tc-RBC suggests that no single reliable method has previously been found. In these reported methods, both the amount and the form of tin is questionable. This is particularly apparent when stannous chloride solutions are prepared by dissolving solid $SnCl_2$ in water. The difficulty lies in the necessity of preparing extremely dilute solutions by repetitive dilutions of a concentrated stock solution. Even if the stock solution is freshly prepared daily the various pipeting and transfer procedures, performed under an ambient air atmosphere, will likely lead to air oxidation of Sn (II). The entire process could be performed under nitrogen, but this would be impractical for routine preparation. The use of lyophilized preparations of $SnCl_2$ or Sn (II) chelates[11,18,23] constitutes a potential approach to the problem of manipulating small quantities of Sn (II). The inability, however, to vary the amount of Sn (II) other than by serial dilutions of the reconstituted material[23] can be a disadvantage. The approach described in the present paper is based on the concept of preparation of an accurate, dilute Sn (II) solution at the time of need by passing a controlled current between 2 tin wires for a controlled time. The amount of Sn (II) generated can readily be varied by changing the electrolysis time and/or current.

Previously reported techniques suffer from other disadvantages besides the tin problem. The requirement that the pertechnetate be contained in a very small volume of solution for incubation with the RBC is a severe limitation.[8,10] It restricts these labeling techniques to use with a fresh generator, which gives an eluate with a high specific concentration, or necessitates a methyl ethyl ketone extraction of the eluate from an older generator.[10] The present method is insensitive toward the volume of pertechnetate (Table 1). As the specific concentration of the generator eluate drops during the week, the volume can be increased as needed, while maintaining a consistently high labeling efficiency. The level of carrier ^{99}Tc present in generator eluate has also been reported to be a problem in RBC labeling.[18] The insensitivity of the present technique toward the generator ingrowth time[24] and the amount of $^{99m}TcO_4^-$ activity added to the RBC (Table 1) indicates that the level of carrier ^{99m}Tc is not a critical concern.

Potential as a radiopharmaceutical

The ^{99m}Tc-RBC preparation exhibits a negligible loss of activity *in vitro*. The half-clearance time of 25 hr *in vivo* is satisfactory for nuclear medicine imaging procedures. The circulating activity remains largely RBC-bound at all times, suggesting either that the labeled RBC are removed from circulation intact or that the RBC-bound ^{99m}Tc is converted to a form which is itself very rapidly cleared. Chromatographic data (Fig. 3) indicate that some of the non-RBC bound activity is associated with plasma proteins, while most of the activity is in small chelate form. Clearance of these chelates through the kidneys and excretion in the urine can account for much of the activity which leaves the vascular compartment. Similar ^{99m}Tc binding properties, leading to relatively rapid blood clearance and considerable urinary excretion, have recently been observed with ^{99m}Tc-fibrinogen.[25] Studies of simultaneous determination of red cell volume with ^{99m}Tc-RBC and ^{51}Cr-RBC have demonstrated that elution of activity from the labeled cells is more of a problem with the ^{99m}Tc agent than with the ^{51}Cr agent,[26] which has a biologic half-time of 28 days. However, if the determination of red cell volume is made within 30 min after injection, good correlation is obtained between the two radiopharmaceuticals, and ^{99m}Tc-RBC has the advantage of a lower radiation dose.[26]

In the undamaged form, the present ^{99m}Tc-RBC preparation is ideal for blood pool imaging and cardiac function studies. The counting rate remains suitably high for many hours after *in vivo* administration. No activity accumulates in tissues or organs which might interfere with visualization of the cardiovascular system. In a damaged form, this ^{99m}Tc-RBC preparation is useful for spleen imaging. Localization of damaged cells in the spleen without accumulation in the liver is an advantage over conventional particulate agents used for spleen scanning.

Feasibility of routine preparation

A critical test of a ^{99m}Tc-RBC labeling procedure is its simplicity and reproducibility in routine daily use. The present method has been employed successfully by nuclear medicine technologists for numerous patient studies. Labeling efficiencies have been consistently high, and *in vivo* stability and behavior have been consistently good. The stannous generation vials are simple and inexpensive to prepare and the electrolysis procedure is fast and easy. The additional 60 sec swirling time following electrolysis is apparently necessary to ensure that the total amount of Sn (II), which seems to initially adsorb on the electrode, is solubilized. Although the method was originally based on the use of a new vial for each labeling, the Sn (II) from a single vial was found to be satisfactory for multiple preparations of ^{99m}Tc-RBC during an entire day. Apparently, the Sn (II) is quite stable in the sealed vial and undergoes little oxidation to Sn (IV). Of course, if more than 5 labeling reactions are to be performed from one vial, a larger volume of saline and a proportionately higher electrolytic charge will be necessary. The overall ^{99m}Tc-RBC preparation time of about 45 min is well within the realm of routine procedures in nuclear medicine.

The preparative ease, excellent properties, and consistent reliability of electrolytically-prepared ^{99m}Tc-RBC should allow the full potential of this radiopharmaceutical to be realized through widespread routine use.

Acknowledgements—Portions of this work were presented at the 22nd Annual meeting of the Society of Nuclear Medicine, Philadelphia, PA, June, 1975.

The authors wish to express their appreciation to Mr. Julius Hecht for preparation of the illustrations and to the several volunteers who donated blood.

REFERENCES

1. Atkins H. L., Eckelman W. C., Klopper J. F. and Richards P. *Radiology* **106**, 357 (1973).
2. Ryo U. Y., Lee J. I., Zarnow H., Schwartz M. P. and Pinsky S. *J. nucl. Med.* **15**, 1014 (1974).
3. Haubold U., Pabst H. W. and Hör G. In *Symp. Medical Radioiosotope Scintigraphy*, Vol. 2, pp. 665–674. IAEA, Vienna (1969).
4. Mahon D. F., Subramanian G. and McAfee J. G. *J. nucl. Med.* **14**, 651 (1973).
5. Berman D. S., Salel A. F., DeNardo G. L., Bogren H. G. and Mason D. T. *J. nucl. Med.* **16**, 865 (1975).

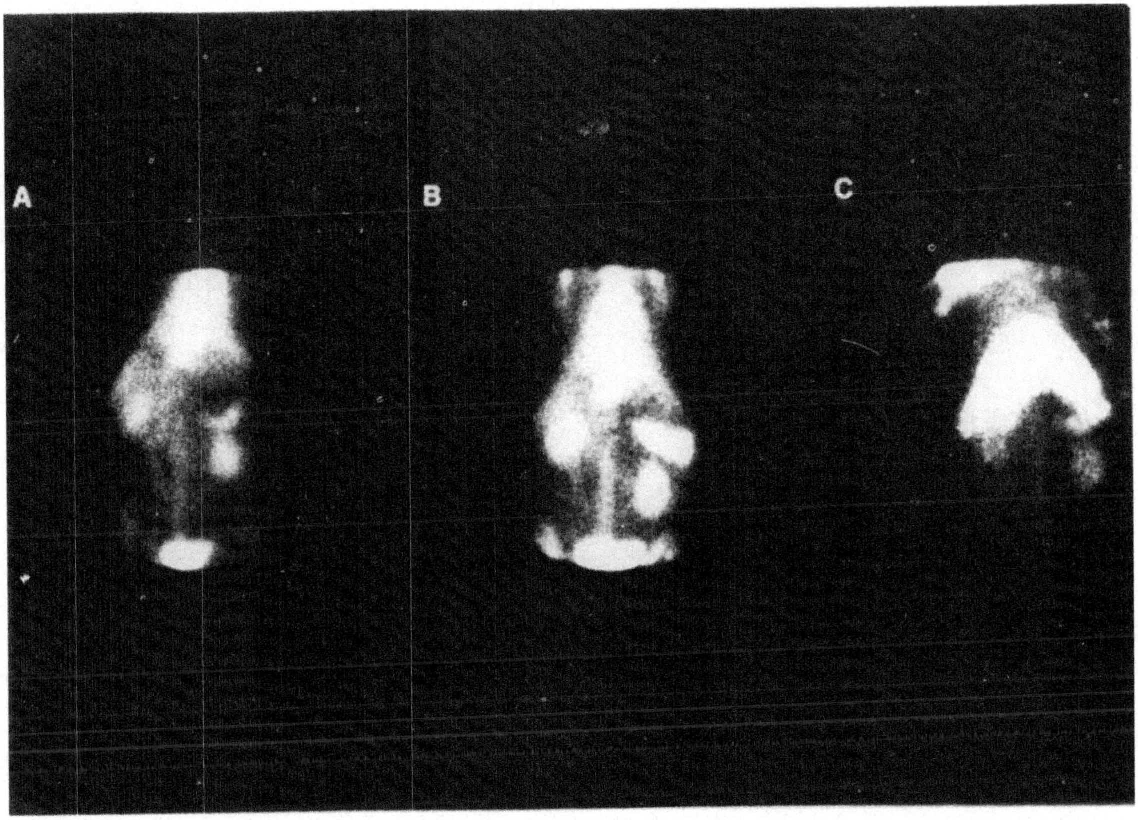

FIG. 4. A series of scintigrams obtained on different days in the same rabbit. (A) 1 hr after injection of undamaged ^{99m}Tc-RBC (300 K counts). (B) 1 hr after injection of heat-damaged ^{99m}Tc-RBC (300 K counts). (C) 20 min after injection of ^{99m}Tc-sulfur colloid. (150 K counts). The spleen is visualized best in (B), with no interference from the large transverse liver.

6. Korubin V., Maisey M. N. and McIntyre P. A. *J. nucl. Med.* **13**, 760 (1972).
7. Eckelman W. C., Reba R. C. and Albert S. N. *Am. J. Roentgenol. Radium Ther. Nucl. Med.* **118**, 861 (1973).
8. Eckelman W. C., Richards P., Atkins H. L., Hauser W. and Klopper J. F. *J. nucl. Med.* **12**, 310 (1971).
9. Atkins H. L., Eckelman W. C., Hauser W., Klopper J. F. and Richards P. *J. nucl. Med.* **13**, 811 (1972).
10. Smith P. H. S. *Int. J. appl. Radiat. Isotopes* **25**, 137 (1974).
11. Gutowski R. F. and Dworkin H. J. *J. nucl. Med.* **15**, 1187 (1975).
12. Berlin N. I. In *Radioactive Pharmaceuticals* (Edited by Andrews G. A.) pp. 337–354. United States Atomic Energy Commission, Conf 651111 (1966).
13. Gray S. J. and Sterling K. *J. clin. Invest.* **29**, 1604 (1950).
14. Fischer J., Wolf R. and Leon A. *J. nucl Med.* **8**, 229 (1967).
15. Weinstein M. B. and Smoak W. M. *J. nucl. Med.* **11**, 41 (1970).
16. Eckelman W., Richards P., Hauser W. and Atkins H. *J. nucl. Med.* **12**, 22 (1971).
17. Schwartz K. D. and Kruger M. *J. nucl. Med.* **12**, 323 (1971).
18. Smith T. D. and Richards P. *J. nucl. Med.* **17**, 126 (1976).
19. Strauss H. V., Zaret B. E., Hurley P. J., Nataranjan T. I. C. and Pitt B. *Am. J. Cardiol.* **28**, 575 (1971).
20. Alderson P. O., Bernier D. R., Ludbrook P. A., Harwig J. F., Roberts R. and Sobel B. E. *Radiology*, **119,** 729 (1976).
21. Harwig J. F., Harwig S. S. L., Wells L. D. and Welch M. J. *Int. J. appl. Radiat. Isotopes* **27**, 5 (1976).
22. Eckelman W. C. *Sem. nucl. Med.* **5**, 3 (1975).
23. Bardy A., Fouye H., Gobin R., Beydon J., de Tovar G., Panneciere C. and Hegesippe M. *J. nucl. Med.* **16**, 435 (1975).
24. Lamson M. L., Kirschner A. S., Hotte C. F., Lipsitz E. L. and Ice R. D., *J. nucl. Med.* **16**, 639 (1975).
25. Harwig S. S. L., Harwig J. F., Coleman R. E. and Welch, M. J. *J. nucl. Med.* **17**, 40 (1976).
26. Ferrant A. and Szur L. *J. clin. Path.* **27**, 983 (1974).

International Journal of Applied Radiation and Isotopes, 1977, Vol. 28, pp. 123–130. Pergamon Press, Printed in Northern Ireland

A Modified Tc-99*m*-Phytate Colloid for Liver–Spleen Imaging

MICHAEL A. DAVIS, MICHAEL L. KAPLAN,
DONALD S. AHNBERG and CLYDE N. COLE
Department of Radiology, Harvard Medical School, and Division of Nuclear Medicine,
Children's Hospital Medical Center, Boston, MA 02115, U.S.A.

(*Received* 4 *April* 1976)

In contrast to the excellent images of liver and spleen obtained with Tc-99*m*-sulfur colloid, the poor splenic visualization following intravenous administration of Tc-99*m*-stannous phytate to patients is considered a serious drawback. Attempts to measure the biologic distribution and target to nontarget ratios in mice and rats for sulfur colloid and phytate revealed a discrepancy between the animal data and human scan findings. The rodent data indicated a nearly identical organ distribution for both radiocolloids 30 or 60 min after injection with 80 ± 5% of the Tc-99*m* activity in the liver and 3 ± 1% in the spleen suggesting both agents to be equally effective in imaging the spleen. As this was clearly not the clinical finding, the rodent models were considered poor indicators for predicting liver–spleen ratios in human subjects. The two agents were administered to 3 rhesus monkeys and 2 baboons in an effort to find an animal model simulating the clinical findings. In all primates tested the spleen/liver ratio for Tc-99*m*-sulfur colloid was 3–3.5 times greater than that obtained for Tc-99*m*-stannous phytate (0.70 vs 0.22). The spleen/liver ratios were calculated with the aid of a computer using an interactive light pen and a region of interest program calculating the average number of counts/cell for each organ. On a total organ basis these values are equivalent to a spleen uptake of 3–4% for the sulfur colloid and 1% for the stannous phytate. Modification of the radiopharmaceutical was accomplished by adding various amounts of calcium ion and stabilizer (albumin or gelatin) to the reconstituted solution of Tc-99*m*-stannous phytate. One to two mg of Ca and 50 mg of albumin were found to give an *in vitro* colloid with biologic properties nearly identical to sulfur colloid. Less than 1 mg of Ca/kit failed to give adequate spleen visualization and more than 2 mg/kit caused variable lung uptake and macroaggregate formation.

INTRODUCTION

INOSITOL hexaphosphoric acid (phytic acid) is known to be directly involved in the adsorption and metabolism of dietary calcium and phosphorus.[1–6] Phytic acid forms complexes with many metallic cations, i.e. calcium, iron and zinc and many of these complexes have a markedly reduced solubility compared to sodium phytate in biological systems.[7–11] The strong complexation of calcium ions by phytic acid and the lack of appreciable solubility of the calcium phytate in aqueous systems makes this material a potent inhibitor of calcification.[12–17]

Technetium-99*m* as the pertechnetate ion (TcO_4^-) can be reduced to a lower oxidation state by numerous methods of which the most commonly used are the ferric ion–ascorbic acid system, electrolysis using zirconium or tin electrodes, sodium borohydride, and stannous chloride. The reduced species of ^{99m}Tc behaves similarly to calcium ion when in the presence of chemical agents known to possess high chelation or complexation strengths toward calcium. Thus, calcium complexing agents such as tetracycline, pyrophosphate and diphosphonate, glucoheptonate, dimercaptosuccinic acid, EDTA and DTPA have been successfully labeled with ^{99m}Tc and used clinically.[18–22]

In 1973, SUBRAMANIAN *et al.* first proposed and carried out the labeling of stannous phytate with ^{99m}Tc, and subsequently used this complex as a liver scintigraphic agent by allowing the administered ^{99m}Tc-Sn-phytate to form the insoluble calcium salt *in vivo*.[23] The colloidal precipitate formed *in vivo* is actively sequestered by the reticulo–endothelial cells of the liver, spleen and bone marrow, particularly the kupffer cells of the liver. Although this agent has achieved moderate clinical utility throughout the world, very little has been reported on its use as a liver imaging agent since the initial article by SUBRAMANIAN.

SEWATKAR *et al.* have reported on the biologic distribution of ^{99m}Tc-phytate in mice and rats, and comparative scintigrams in patients[24] as well as on the stability of the complex.[25] Their primary finding was the relatively decreased splenic uptake of the agent in normal and diseased human subjects compared to ^{99m}Tc-sulfur colloid. The bulk of the clinical studies published to date have been carried out in Japan.[26–31] TATSUNO *et al.* observed that the organ distribution and degree of splenic visualization with ^{99m}Tc-phytate were similar to those obtained with ^{198}Au-colloid in patients, irrespective of the type of liver disease, and that both of these agents exhibited a lower splenic concentration compared to ^{99m}Tc-Sn-colloid.[26] ABE *et al.* reported that ^{99m}Tc-phytate has a higher accumulation in the spleen but a slower blood clearance than ^{198}Au-colloid.[27] The degree of splenic concentration of ^{99m}Tc-phytate was found to be very similar to that of ^{198}Au-colloid by KUBO *et al.*[29] They also investigated the metabolic fate of the administered ^{99m}Tc-phytate and determined the average $T\frac{1}{2}$(biol) to be 13.5 ± 11.0 days with 11% of the ^{99m}Tc activity being excreted in the urine within 24 hr and 0.2% of the activity eliminated in the feces. In a recent article, AKISADA and MIYAMAE compared the use of ^{99m}Tc-phytate and ^{198}Au-colloid in both dynamic and static scintiphotography.[31] Their attempt to alter the organ specificity of phytate by *in vitro* incubation with calcium ion for six hours led to the formation of macroaggregates and subsequent lung uptake following injection of the preformed colloid.

Due to our interest in increasing the splenic uptake of ^{99m}Tc-phytate to a level similar to that obtained with ^{99m}Tc-sulfur colloid, a comparison of the *in vivo* biologic properties of ^{99m}Tc-stannous phytate and ^{99m}Tc-stannous phytate precipitated *in vitro* by the addition of calcium ion was undertaken in our laboratory and the findings are reported here.

EXPERIMENTAL PROCEDURE

Two colloids commonly used for external visualization of the liver and reticulo-endothelial system, ^{99m}Tc-sulfur colloid and ^{99m}Tc-stannous phytate, were evaluated in mice, rats and primates. In the rodents, the biodistribution was determined by the sequential sacrifice of animals followed by excision of organs and tissues of interest and subsequent counting in a well-type gamma-scintillation counter. The studies in primates were performed by obtaining scintigraphic images at specific times after injection of the agent with a Searle Radiographics Pho-Gamma HP Scintillation Camera.

The gamma camera is connected to a dedicated computer system (Gamma-11 (PDP 11/20), Digital Equipment Corp., Maynard, Mass.) with which scintigraphic images are recorded and stored on magnetic disk in a 64×64 matrix format. Images of the syringe containing the radiopharmaceutical are collected before and after injection for calculation of the injected activity. Upon completion of the study, the images of liver and spleen are displayed on a persistence oscilloscope. These images are smoothed with a 9-point weighted technic and the lowest 15% of counts are not displayed so as to enhance organ borders. Regions of interest are drawn around the liver and spleen at the edge of these enhanced borders, and the total activity in each region of interest, number of matrix cells in each region, and the average counts per cell in each region are automatically displayed on the oscilloscope (Fig. 1). The average counts per cell value is used to determine the spleen/liver activity ratio. The average counts per cell value is then divided by the injected activity (initial syringe activity minus final syringe activity as determined by the previously collected and stored images of the syringe before and after injection) to provide the per cent uptake of injected dose in the liver and spleen as seen by posterior scintigraphy.

MATERIALS AND METHODS

Agent preparation

Stannous phytate in lyophilized "kit" form and sulfur colloid were obtained from the New England Nuclear Corp. Both agents were prepared according to the manufacturer's instructions using 4 ml of $^{99m}TcO_4^-$ generator eluate. A modified phytate was prepared by the dropwise addition of varying amounts of a

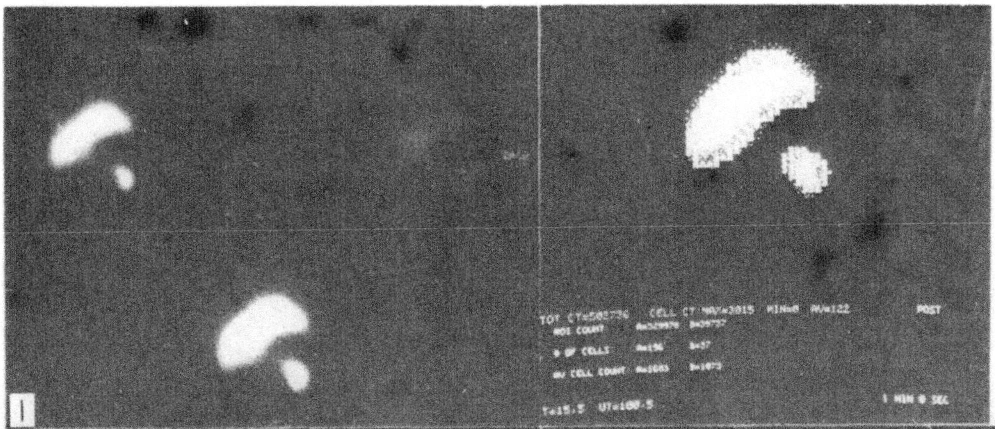

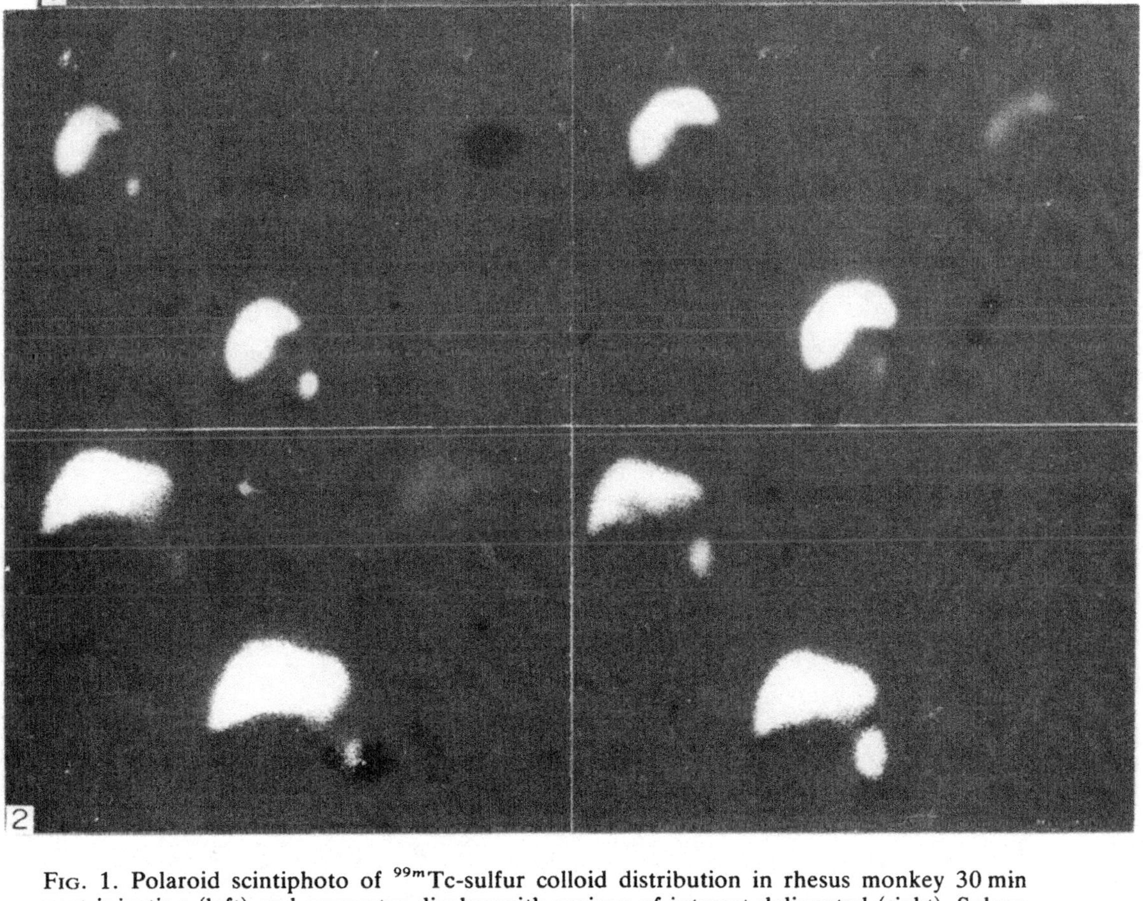

FIG. 1. Polaroid scintiphoto of ^{99m}Tc-sulfur colloid distribution in rhesus monkey 30 min post injection (left) and computer display with regions of interest delineated (right). Spleen to liver ratio was determined to be 0.67 in this animal.

FIG. 2. Scintiphoto of ^{99m}Tc-sulfur colloid (upper left) and ^{99m}Tc-phytate (upper right) in rhesus monkey 30 min after administration. Comparison of commercially available ^{99m}Tc-phytate (lower left) and modified phytate colloid by addition of 2 mg Ca^{2+} and 50 mg gelatin *in vitro* (lower right) in a baboon.

$CaCl_2$ solution (10 mg Ca^{2+}/ml) and 50 mg of either gelatin (1 ml of a 5% solution) or albumin (0.25 ml of 20% salt-poor solution) stabilizer. Calcium phytate colloids labeled with ^{99m}Tc and containing 0.5, 1.0, 2.0, 5.0 and 10.0 mg of calcium per kit were prepared in this manner.

Quality control

The level of unbound $^{99m}TcO_4^-$ was determined by ascending chromatography using either Whatman 3MM or Gelman ITLC strips with saline and/or 85% methanol as eluant. Colloid size was approximated by filtration through a series of Nuclepore filters as previously described.[32] Formation of macroaggregates was monitored by filtration through 3.0 and 5.0 μm nucleopore filters in addition to optical microscopy.

Biologic distribution

The agents were administered intravenously to laboratory mice (Charles River CD-1), rats (Charles River CD, Sprague–Dawley derived), rhesus monkeys and baboons. The rodents were sacrificed by a combination of ether asphyxiation and cardiac exsanguination 30 or 60 min after injection. The organs and tissues of interest were removed, weighed, and counted in a well-type gamma scintillation counter (Nuclear Chicago Model 4230). The primates were tranquilized with either Sernylan™ (Bio-Ceutic Laboratories, St. Joseph, Missouri) or Ketalar™ (Parke–Davis Laboratories, Detroit, Michigan) and anesthetized with Nembutal™ (Abbott Laboratories, North Chicago, Illinois). Approximately 2 mCi of the test agent were administered intravenously and 1-min images were taken and stored on the disk at 5, 15, 30, 45 and 60 min after injection. The spleen to liver ratios were determined at each time interval using the region of interest (ROI) program previously described.

RESULTS AND DISCUSSION

Chromatographic analysis of sulfur colloid, stannous phytate and calcium-tin-phytate indicated that less than 3% of the original ^{99m}Tc activity remained as pertechnetate. Table 1 shows the concentration of ^{99m}Tc labeled sulfur colloid and stannous phytate in the liver, spleen and lungs of mice and rats. There appears to be no difference in the % I.D./organ or in the liver/spleen ratio for either agent in either species. These results are in good agreement with previous findings[24] but do not suggest the lack of splenic visualization encountered with phytate in the nuclear medicine clinic which strongly implies a species difference. This species specificity is clearly illustrated in Table 2 and Fig. 2. Table 2 shows that the concentration of ^{99m}Tc-phytate in the spleens of rhesus monkeys and baboons is only one-third that of ^{99m}Tc-sulfur colloid. This does not agree with the data presented in Table 1 where the uptake of both agents in rodent spleen is essentially the same.

TABLE 1. Organ distribution of ^{99m}Tc-sulfur colloid and Tc-stannous phytate in mice and rats 30 min after intravenous administration

	^{99m}Tc-sulfur colloid		^{99m}Tc-stannous phytate	
Organ	Mice (5)	Rats (5)	Mice (15)	Rats (7)
	%I.D./organ			
Liver	82 ± 4	78 ± 8	78 ± 9	85 ± 3
Spleen	2.8 ± 1.1	3.0 ± 1.7	2.0 ± 0.8	2.1 ± 0.6
Lung	0.8 ± 0.2	0.7 ± 0.2	1.0 ± 2	0.8 ± 0.9
Blood	0.7 ± 0.3	0.9 ± 0.3	1.5 ± 0.5	1.8 ± 0.8
Liver/Spleen	29	25	39	40
	%I.D./g tissue			
Liver	51.2	7.1	48.7	7.7
Spleen	28.0	4.3	20.0	3.0
Lung	4.0	0.5	5.0	0.6
Blood	0.5	0.05	1.0	0.1
Liver/Spleen	1.8	1.6	2.4	2.5

TABLE 2. Spleen : liver ratios in rhesus monkeys and baboons for 4 hepatic radiopharmaceuticals determined by computer region of interest program.

	Spleen : liver (CPM/Unit Cell)			
	^{99m}Tc-phytate	Modified phytate (2 mg Ca^{2+}, 50 mg gelatin)	Modified phytate (2 mg Ca^{2+}, 50 mg albumin)	^{99m}Tc-sulfur colloid
Rhesus	0.21	0.51	0.56	0.65
Baboon	0.23	0.55	0.80	0.67

The excellent splenic visualization obtained with ^{99m}Tc-sulfur colloid (spleen/liver = 0.67) compared to the almost complete absence of visualization with ^{99m}Tc-phytate (spleen/liver = 0.23) is shown scintigraphically in Fig. 2. The addition of 2 mg calcium ion to ^{99m}Tc-phytate prior to injection produced a milky-white, opaque suspension which upon administration exhibited an increased splenic concentration essentially equal to that obtained with sulfur colloid (Table 2). This microaggregate suspension tended to form larger aggregates upon standing but this could easily be prevented by including either gelatin or albumin stabilizers immediately prior to or subsequent to the addition of the calcium ion.

The effect on the biologic distribution of varying the calcium ion concentration can be seen in Fig. 3. It is evident that the addition of calcium ion has no effect upon splenic uptake in mice whereas the uptake in primates is increased 3 to 4 fold. Thus, a species difference is clearly demonstrated and explains why the initial rodent data which suggested ^{99m}Tc-phytate to be

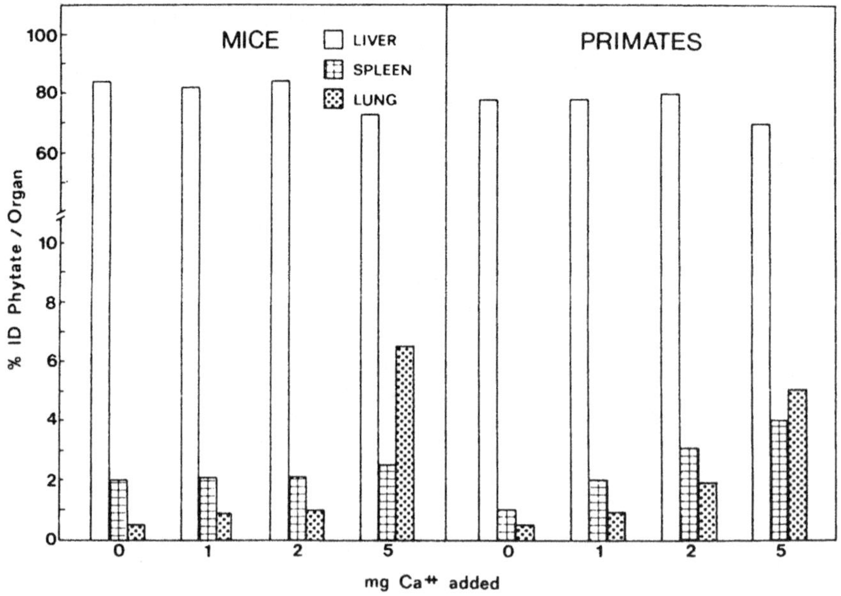

FIG. 3. Organ distribution of ^{99m}Tc-stannous phytate in mice and primates as a function of calcium ion concentration. Note that in mice the splenic uptake is unaffected by the varying calcium levels in contrast to the proportional increase in splenic uptake with changing calcium concentration in primates. Lung activity is appreciable only at the 5 mg level.

nearly identical in distribution to ^{99m}Tc-sulfur colloid[23,24] cannot be extrapolated to the patient situation.

Nuclepore filtration of the preformed phytate colloid correlated well with the observed biologic distribution shown in Fig. 3. Colloid prepared with 1 or 2 mg calcium ion had an average size of 0.3 μm with 90% of the activity between 0.1 and 1.0 μm. This is nearly identical to the activity-size distribution previously found for sulfur colloid.[32] At the 5 mg calcium level, approximately 10% of the activity is associated with colloid greater than 1 μm, 5% of the activity with colloid greater than 3 μm and 1% with colloid greater than 5 μm. This explains the rise in lung activity noted at the 5 mg calcium level in Fig. 3.

Numerous attempts to prepare a freeze-dried one-step kit were unsuccessful. Upon completion of this investigation it was called to our attention that two commercial manufacturers (Ackerman Nuclear and Frosst) had recently introduced modified phytate kits incorporating the concept of preformed colloid through the use of calcium ion. Neither of the kits is in a single-step freeze-dried form indicating the manufacturers may have encountered difficulties similar to those experienced in our laboratory.

CONCLUSIONS

The results of this experimental study can be summarized as follows:

(1) The similarity in the organ distribution values of ^{99m}Tc-phytate and ^{99m}Tc-sulfur colloid is species related (rodents) and decreased splenic uptake of the phytate colloid is found in other primates and man.

(2) The gamma scintillation camera with computer ROI capability is an extremely useful technique for studying the biodistribution and pharmacokinetics of radiodiagnostic agents in primates, where the cost of the animals prohibits the use of the sequential sacrifice technique.

(3) The addition of calcium ion to the ^{99m}Tc-phytate complex produces an *in vitro* colloid with a particle size and biologic distribution very similar to that obtained with ^{99m}Tc-sulfur colloid.

Acknowledgements—We wish to express our thanks to Mrs. Alice Carmel and Mr. Peter Aschaffenburg, for their expert technical assistance and to Mrs. Rebekah Taube for her editorial work on the manuscript. This work was supported in part by NIH Diagnostic Radiology Center Grant #USPHS GM 18674.

REFERENCES

1. Anderson R. J. *J. Biol. Chem.* **17,** 171 (1914).
2. Hoff-Jorgensen E. *Biochem. J.* **40,** 189 (1946).
3. Jenkins N. K. *Nature* **205,** 89 (1965).
4. Hoff-Jorgensen E., Andersen O., Begtrup H. *et al. Biochem. J.* **40,** 453 (1946).
5. Hoff-Jorgensen E., Andersen O. and Nielsen G. *Biochem. J.* **40,** 555 (1946).
6. Widdowson E. M. *Biol. Intern. Nutr. Nutritio et Dieta* **15,** 38 (1970).
7. *The Merck Index*, 8th Edition, p. 829. Merck and Company, Rahway, New Jersey (1968).
8. Hoff-Jorgensen E. *Kgl Danske Vid Seiskab.* **21,** 1 (1944).
9. Mollgaard H. *Biochem. J.* **40,** 589 (1946).
10. Likuski H. and Forbes R. *J. Nutr.* **85,** 230 (1965).
11. Latham M. C. *Nutr. Rev.* **25,** 215 (1967).
12. Bruce H. and Callow R. *Biochem. J.* **28,** 517 (1934).
13. Harrison D. and Mellanby E. *Biochem. J.* **33,** 1660 (1939).
14. Walker A. *Lancet* **261,** 244 (1951).
15. Vermeulen C., Lyon C., Gill W. *et al. J. Urol.* **82,** 249 (1959).
16. Vagelos P. R., Henneman P. H. *N. Engl. J. Med.* **256,** 773 (1957).
17. Nordbo H., Rolla G. *Scand. J. Dent. Res.* **79,** 507 (1971).
18. Dewanjee M. K., Fliegel C. P., Treves S. *et al. J. nucl. Med.* **15,** 176 (1974).
19. Davis M. A. and Jones A. G. *Semin. Nucl. Med.* **6,** 19 (1976).
20. Adler N., Camin L. L. and Shulkin P. *J. nucl. Med.* **17,** 203 (1976).
21. Lin T. H., Khentigan A. and Winchell H. S. *J. nucl. Med.* **15,** 34 (1974).
22. Atkins H. L., Cardinale K. G., Eckelman W. C. *et al. Radiology* **98,** 674 (1971).
23. Subramanian G., McAfee J. G., Mehter A. *et al. J. nucl. Med.* **14,** 459 (1973).
24. Sewatkar A. B., Noronha O. P. D., Ganatra R. D. *et al. Nucl. Med.* (*Stuttg.*) **14,** 46 (1975).
25. Sewatkar A. B., Noronha O. P. D., Ganatra R. D. *Nucl. Med.* (*Stuttg.*) **14,** 293 (1975).
26. Tatsuno I., Michigishi T. and Kato S. *Radioisotopes* **23,** 620 (1974).

27. ABE M., MATSUI K., CHIBA K. *et al. Radioisotopes* **24,** 31 (1975).
28. KANEKO M., WATANABE M., SASAKI T. *et al. Radioisotopes* **24,** 63 (1975).
29. KUBO A., KINOSHITA F., ISOBE Y. *et al. Radioisotopes* **24,** 186 (1975).
30. YAMAGISHI Y., HONDA K., WATANABE H. *et al. Radioisotopes* **24,** 354 (1975).
31. AKISADA M. and MIYAMAE T. *Radioisotopes* **24,** 626 (1975).
32. DAVIS M. A., JONES A. G. and TRINDADE H. *J. nucl. Med.* **15,** 523 (1974).

International Journal of Applied Radiation and Isotopes, 1977, Vol. 28, pp. 131–147. Pergamon Press. Printed in Northern Ireland

Bromine-77 and Iodine-123 Radiopharmaceuticals

G. STÖCKLIN
Institut für Chemie der KFA Jülich GmbH, Institut 1: Nuklearchemie, D-5170 Jülich, FRG

(*Received* 2 *February* 1976)

The neutron deficient isotopes bromine-77 ($T_{1/2} = 56$ hr) and particularly iodine-123 ($T_{1/2} = 13.3$ hr) have the decay properties required for diagnostic *in vivo* studies. Among the radiohalogens these two isotopes have at present the greatest potential. Their production methods are critically reviewed from the viewpoint of medical applications, and recent developments in labelling procedures are outlined in the context of use in nuclear medicine.

1. INTRODUCTION

THE NEUTRON deficient halogen radioisotopes play a special role in nuclear medicine. The major reason for their great potential lies primarily in the fact that they are also "organic" isotopes which can replace a hydrogen atom in almost every organic compound. Depending on the biochemical pathway and the halogen of interest, the biochemical behaviour changes to a greater or lesser extent and may even lead to a blocking of active sites, which in some cases can make a radiohalogen labelled compound or metabolite even more interesting. Thus, the advantages and potential clinical applications of radiohalogen labelled radiopharmaceuticals have been topics of interest for some time (for a recent review cf. ref. 1). Fortunately, most of the neutron deficient halogen radioisotopes have nuclear properties which are suitable for *in vivo* studies and external radiation detection. This is particularly true for the avantgardist's radionuclides such as ^{18}F ($T_{1/2} = 110$ min), ^{34m}Cl ($T_{1/2} = 32$ min), ^{77}Br ($T_{1/2} = 56$ hr) and ^{123}I ($T_{1/2} = 13.3$ hr). While these nuclides are suitable for diagnosis and pharmacokinetics, the heaviest known halogen astatine, in particular ^{211}At ($T_{1/2} = 7.2$ hr), has also received renewed interest, as far as potential radiobiological and -therapeutic applications are concerned.[2–4]

In this paper we shall review bromine-77 and iodine-123, the two radiohalogens which obviously at present have the greatest potential. Although from a chemical standpoint compounds in which a hydrogen atom is substituted by the positron emitter fluorine-18 would be favourable in many cases, due to the great stability of this label and the almost unchanged structure, there are a few very practical reasons why ^{77}Br and, in particular ^{123}I, are presently preferred:

(1) Efficient high speed positron camera are generally not available to date.

(2) Labelling with fluorine is far more difficult than with bromine or iodine, particularly when dealing with practically carrier-free amounts, which are often needed, e.g. in cases of high toxicity.

(3) Both ^{77}Br ($T_{1/2} = 56$ hr) and ^{123}I ($T_{1/2} = 13.3$ hr) have convenient and not too short half-lives, and their major γ-lines are suitable for detection with present day γ-cameras, particularly in the case of iodine-123.

Even though the vast majority of radiopharmaceuticals to date have been prepared with radioiodine, in many cases bromine would be preferable. The C–Br bond is about 10–15 kcal/mole stronger than that of the corresponding C–I bond, and hence the label is more stable. From the neutron deficient bromine isotopes the nuclides with mass numbers 74 to 77 are potentially useful (cf. Table 1).[5–11] The main γ-lines from ^{77}Br (239 keV (30%) and 521 keV (24%)) can be used, although the 521 keV γ-energy is too high for good collimation with a common scintillation camera. The whole body dose delivered by ^{77}Br in the form of a plasma label would be 0.4 mrads/μCi and any ^{77}Br$^-$ which is released would not accumulate in the thyroid.[12] The 56 hr half-life is long enough to allow even long term studies such as the use of

TABLE 1. Some potentially useful neutron deficient bromine isotopes

	^{74m}Br [5]	^{74}Br [6]	^{75}Br [7]	^{76}Br [8,9]	^{77}Br [10,11]
Half-life	28.0 min	41.5 min	101 min	15.9 hr	56 hr
Decay mode	β^+ (85%) E.C. (15%)	β^+ (91%) E.C. (9%)	β^+ (76%) E.C. (24%)	β^+(57%) E.C. (43%)	β^+ (0.7%) E.C. (99.3%)
Conversion electrons*	<1	<1	<1	<1	<1.2
Energies and abundances of the main γ-lines*	511 keV (170) 635 keV (57) 219 keV (17) 634 keV (12)	511 keV (181) 635 keV (89) 728 keV (34) 634 keV (17)	511 keV (152) 287 keV (91)	511 keV (115) 559 keV (78) 657 keV (17)	239 keV (22.4) 521 keV (20.3)
X-Ray (av. energy and abundances)*	11.5 keV (8.5)	11.5 keV (5.1)	11.5 keV (13.4)	11.5 keV (24)	11.5 keV (55)
High energy photons*	53 ≧800 keV	95 ≧750 keV	15.5 ≧550 keV	16 ≧880 keV	4.8 ≧600 keV
Half-thickness in water of the main γ-line	7.2 cm	7.2 cm	7.2 cm	7.2 cm	5.5 cm
Half-thickness in lead of the main γ-line	4.3 mm	4.3 mm	4.3 mm	4.3 mm	1.1 mm

*per 100 decays.

labelled fibrinogen for clot localization.[13,117] Shorterlived bromine isotopes down to ^{74}Br ($T_{1/2} = 42$ min) might be useful in a positron camera because of their dominant 511 keV annihilation radiation. With the advent of efficient high speed positron cameras the interest in these isotopes for the labelling of radiopharmaceuticals should increase.

Among the numerous radioisotopes of iodine the nuclide with mass number 123 is the most suitable for *in vivo* studies, as was pointed out by MYERS and ANGER as early as 1962.[14] The absence of β^--particle emission, the relatively short half-life ($T_{1/2} = 13.3$ hr) together with its convenient major (86%) γ-line of 159 keV make it an almost ideal radioisotope for most diagnostic uses. In Table 2 we compare the major decay properties of six pertinent iodine isotopes.[15–19] The virtue of ^{123}I is the low radiation dose to the patient. Consequently this nuclide is the radioiodine of choice for requested tests in children or pregnant women.[20] Because of the relatively short half-life it can be administered at frequent intervals. Recently ATKINS *et al.*[21] have compared high-purity ^{123}I as sodium iodide and ^{99m}Tc as pertechnetate for thyroid imaging and have concluded that ^{123}I is the agent of choice when utilized with the gamma camera. While others[115,116] confirm the superiority of ^{123}I with regard to imaging and uptake measurements, ARNOLD and PINSKY,[110] on the other hand, come to the conclusion that ^{123}I provides no advantage over ^{99m}Tc for routine thyroid imaging except in patients with very poor thyroid uptake or with suspected retrosternal thyroid tissue. The greatest potential of iodine-123 lies undoubtedly in the field of labelled compounds. Many of the problems encountered in diagnostic procedures requiring radioiodine labelled radiopharmaceuticals, which often could only poorly be solved with iodine-125 and iodine-131 due to the high radiation dose and poor counting statistics, respectively, can now be avoided by using iodine-123. At present the difficulty lies as much in educational as in organizational problems. Nevertheless, a variety of ^{123}I-labelled radiopharmaceuticals have already been in clinical use and the improvement could be clearly demonstrated, e.g. in renography with ^{123}I-hippuran,[22,113,114] urography with ^{123}I-labelled contrast agents[23] and in

TABLE 2. Decay properties of medically important iodine radionuclides

	^{121}I	^{123}I	^{124}I	^{125}I	^{131}I	^{132g}I
Half-life	2.12 hr(16)	13.3 hr(17)	4.15 d(18)	60 d(19)	8.03 d	2.3 hr
Decay mode	β^+ (13%)(16) E.C. (87%)	E.C.(17)	β^+ (27%)(18) E.C. (73%)	E.C.(19)	β^-	β^-
Conversion electrons*	8	15.4(17)	5	93(19)	6.3	~4
Energy and intensity* of the main γ-line	213 keV (86)(16)	159 keV (86)(18)	603 keV (59.5)(18)	35.5 keV (6.4)(19)	364 keV (82)	668 keV (100) 773 keV (93)
X-Ray (av. 28 keV)*	72(15)	92(17)	49(15)	136(19)	82	?
High energy photons*	230–1300 keV incl. 511 keV (33)	274–530keV(3)(17)	511–3000 keV (84.5)(18)	none(19)	637 keV (6.8) 723 keV (1.6)	1–2.2 MeV (50)
Half-thickness of main γ-rays in water	5.1 cm	4.7 cm	7.8 cm	1.7 cm	6.3 cm	8.5 cm
Half-thickness of main γ-rays in lead	0.79 mm	0.37 mm	6.0 mm	0.016 mm	2.4 mm	6.5 mm
Relative number of photons detected per min per rad ($^{123}I = 100$)		100(1)	2(1)		4(1)	46(1)
Relative absorbed dose (rads/μCi in hr/man) ($^{123}I = 1$)		1(1)	69(1)	70(1)	100(1)	1(1)
Remark	$^{121}Te(T_{1/2} = 16.8d)$-daughter					$^{132}Te(T_{1/2} = 78hr)$-daughter

*per 100 decays.

hepatobiliary imaging with ^{123}I-Rose Bengal.[24,105–108] Many other applications have been suggested or are being evaluated (*vide infra*).

2. PRODUCTION OF ^{77}Br AND ^{123}I

Bromine-77

Nuclear data on the production of bromine-77 are still extremely scarce.[25–30] Potential production methods are listed in Table 3. The reactions on bromine, selenium and arsenic can be used also in principle for the production of the lighter isotopes with A from 76 to 74 if higher excitation energies are involved and hence more neutrons are emitted. Due to the lack of relevant nuclear data little can be said about the relative yields and radionuclidic purities for most of the reactions. Extensive studies have, however, been carried out in the case of the ^{75}As(α,2n)^{77}Br process, with 28 MeV α-particles on As_2O_5 by SKRABEL *et al.*[26] and in more detail by HELUS[27]. The excitation function from 14 to 28 MeV reveals a flat maximum around 26 MeV with a cross section of about 800 mbarn.[28] In a 1 hr run at 20μA a ^{77}Br yield of 3.2 mCi was obtained.[27] The bromine was separated from the As_2O_5 powder target material by distillation after dissolution and oxidation to the elemental state in $H_2SO_4/K_2Cr_2O_7$. The preparation of arsenic metal targets was described by NUNN,[31] and the group at the Hammersmith Hospital, London regularly produces ^{77}Br via the ^{75}As(α,2n)^{77}Br-process from As_2O_5, As_2O_3 and As-metal targets with yields of 160, 290 and 171 μCi/μAhr, respectively.[32]

BLUE and BENJAMIN[29] suggested the indirect production via the ^{77}Kr ($T_{1/2}$ = 1.2 hr) parent by bombarding enriched ^{76}Se-targets with 42 MeV α-particles. Recently, the Heidelberg group[30] reported on the production of ^{77}Br via high-energy α-particle induced reactions (cf. Table 3). When bombarding a thick NaBr-target with 100 MeV α-particles a yield of 322 μCi/μAhr was obtained, only 12% of which was formed indirectly via ^{77}Kr. For small compact cyclotrons the low-energy reactions on selenium such as ^{77}Se(p,n)^{77}Br, ^{78}Se(p,2n)^{77}Br, ^{76}Se(d,n)^{77}Br, ^{75}As(^{3}He,n)^{77}Br and the indirect process ^{76}Se(^{3}He,2n)^{77}Kr seem to be promising. The indirect reactions via the ^{77}Kr parent may be preferable not only because the removal of ^{77}Kr provides a purification step which improves the radionuclidic purity, but also because the ^{77}Kr can be used for excitation labelling procedures (see below). More nuclear data, in particular the relevant excitation functions, are needed before conclusions on the suitability of the individual processes can be drawn.[121,122]

Iodine-123

The situation is somewhat different for ^{123}I,

TABLE 3. Nuclear reactions leading to ^{77}Br

Nuclear reaction	Q-value [MeV]	Y*[mCi/μAhr]	Reference
^{79}Br(p, 3n)^{77}Kr $\xrightarrow{\beta^+}$ ^{77}Br	−22.8	0.8	121
^{79}Br(d, 4n)^{77}Kr $\xrightarrow{\beta^+}$ ^{77}Br	−25.0	0.64	122
^{79}Br(α, 6n)^{77}Rb $\xrightarrow{\beta^+}$ ^{77}Kr $\xrightarrow{\beta^+}$ ^{77}Br	−58.7		30
^{79}Br(α, p5n)^{77}Kr $\xrightarrow{\beta^+}$ ^{77}Br	−51.1		30
^{79}Br(α, 2p4n)^{77}Br	−47.3		30
^{79}Br(d,p3n)^{77}Br	−21.2	4.6	122
^{77}Se(p, n)^{77}Br	− 2.1	0.5	25, 121
^{78}Se(p, 2n)^{77}Br	−12.6		—
^{76}Se(d, n)^{77}Br	+ 3.04		—
^{78}Se(d, 3n)^{77}Br	−14.9		—
^{76}Se(α, 3n)^{77}Kr $\xrightarrow{\beta^+}$ ^{77}Br	−26.8		29
^{78}Se(α, 5n)^{77}Kr $\xrightarrow{\beta^+}$ ^{77}Br	−44.7		—
^{76}Se(^{3}He, 2n)^{77}Kr $\xrightarrow{\beta^+}$ ^{77}Br	− 6.2		—
^{78}Se(^{3}He, 4n)^{77}Kr $\xrightarrow{\beta^+}$ ^{77}Br	−24.1		—
^{75}As(α, 2n)^{77}Br	−13.5	0.41	26–28, 31, 32
^{75}As(^{3}He, n)^{77}Br	+ 7.1		—

* Calculated thick target yield

where a remarkable amount of data already exists, including the relevant excitation functions (cf. ref. 123). Most of these data, however, have only recently been obtained. The processes presently used for routine production are shown in Table 4.

Some of the low-energy direct processes which can be applied with a small compact cyclotron are not suitable for the production of iodine-123, since they give rise to non-tolerable amounts of ^{124}I ($T_{1/2} = 4.3$ days). This longer lived impurity increases the radiation dose to the patients (comparable to ^{131}I) and also emits relatively hard γ-lines (major γ-transition 603 keV), which with elapsing time builds up a strong radiation background, thus deteriorating the resolution of the scans. Iodine-123 containing such radionuclidic impurity is ill suited for functional diagnosis extending over a long period of time, but it might still be used for studies which are completed within a short time, such as renography with iodohippuric acid or liver tests with Rose Bengal. The ^{124}I contamination should be less than 3–4% at the time of delivery to the patient. In the case of the reactions on Te isotopes the impurity level can be suppressed by using highly enriched targets, particularly when applying the high yield proton induced ^{123}Te(*p*,*n*) ^{123}I or the ^{124}Te(*p*,2*n*) ^{123}I reactions.(38–42) This, however, makes the process more expensive. For the first reaction the price becomes prohibitive due to the low isotopic abundance of ^{123}Te (0.87%). The ^{124}Te (*p*,2*n*) ^{123}I-reaction, on the other hand, seems to be one of the most promising for small compact cyclotrons, if highly enriched ^{124}Te-targets are used.(41,42,124) For economic reasons, however, it is essential to reprocess the target material using an efficient recovery method. Several laboratories (cf. Table 4) have employed the ^{124}Te (*p*,2*n*) ^{123}I-process and excitation functions have been measured for the relevant (*p*,*n*)- and (*p*,2*n*)-processes for varying ^{124}Te-enrichments up to 92% over the energy range of 10–30 MeV by Resmini *et al.*(41), Kondo *et al.*(42) and van den Bosch *et al.*(124) for the energy range of 10–28 MeV using 99.87% enriched ^{124}Te. The initial ^{124}I-impurity is a function of the ^{124}Te-enrichment, the proton energy and the target thickness (energy loss). This is shown in Fig. 1 for two different degrees of enrichment (91.86 and 99.87% ^{124}Te). It can be seen that the higher the ^{124}Te-enrichment of the target material and the smaller the target thickness (energy degradation) the higher is the radio-nuclidic purity for a given incident particle energy. At a target thickness of 10 MeV yields of about 40 mCi can be obtained. According to Resmini *et al.*(41) the impurity level for the range of 25–20 MeV is only about 0.7% even for an 91.86% enrichment of ^{124}Te, while Wolf *et al.*(42) give an impurity level of 0.5% for a 1 MeV thick target with 99.87% ^{124}Te enrichment. The more pessimistic value of Wolf *et al.* seems to be more realistic, since the lower enrichment is expected to give additional ^{124}I via the ^{125}Te (*p*,2*n*) ^{124}I-reaction.

The physical and chemical properties of tellurium make the targetry somewhat difficult. Te-metal has a low thermal conductivity, melts at 450°C and has a high vapour pressure. Efficient cooling is absolutely necessary to avoid melting and vaporization at higher beam currents. Several target designs have been reported.(38,41,42) Removal of the iodine can be achieved discontinuously by wet chemical separation(38,42) or by dry distillation in a carrier

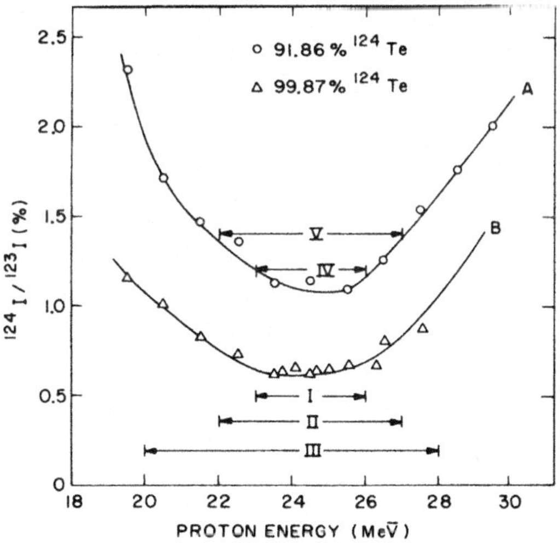

Fig. 1. % iodine-124 impurity formed during iodine-123 production via the ^{124}Te(*p*,2*n*)^{123}I-process as a function of proton energy at different target thicknesses (energy loss ranges) and for two different ^{124}Te enrichments (99.87% lower curve and 91.86% upper curve) after Kondo, Lambrecht and Wolf.(42)

TABLE 4. Processes for routine production of iodine-123

Nuclear reaction	Energy of the incident beam [MeV]	Target material	Thick target yield [mCi/μAh]	Impurities	Reference
$^{121}Sb(\alpha, 2n)^{123}I$	25.1	natural Sb	0.15	0.7% ^{124}I	33
	≦32	natural Sb	?	?	34
	24.5	natural Sb	0.20	0.8% ^{124}I	35
$^{122}Te(\alpha, 3n)^{123}Xe \xrightarrow{\beta^+, EC} {}^{123}I$	42	96% ^{122}Te-metal	0.30	0.4% ^{125}I	36
	42	96% ^{122}Te-metal	0.45	≦0.2% ^{125}I	37, 50
$^{122}Te(d, n)^{123}I$	11	95% ^{122}Te	>1	<1% ^{124}I	125
$^{123}Te(p, n)^{123}I$	15.5	79% ^{123}Te-metal	0.44	0.88% ^{124}I, 0.24% ^{126}I	38
	15	77% ^{123}Te-metal	4	0.7% ^{124}I	39
$^{124}Te(p, 2n)^{123}I$	30	93.4% ^{124}Te	40	?	40
	28	91.9% ^{124}Te	45	0.88% ^{124}I	41
	29	99.87% ^{124}Te	?	0.62% ^{124}I	42
	26	96% $^{124}TEO_2$	?	0.7% ^{124}I	126
	24	91.9% $^{124}TeO_2$	20	0.7% ^{124}I	124
	27	enr. ^{124}Te	8	0.7% ^{124}I	127
	22	96% $^{124}Te/Al$	1.3	0.7% ^{124}I	128
$^{127}I(p, 5n)^{123}Xe \xrightarrow{\beta^+, EC} {}^{123}I$	57.5	I_2	3.0	0.1% ^{125}I	43
	52	CH_2I_2	3.8	<1% ^{125}I	44
	65	KI	14	≦0.1% ^{125}I	45
	60	NaI solution	4.5	0.2% ^{125}I	46
	70	NaI	?	0.1% ^{125}I	47
	58	NaI solution	6	0.13% ^{125}I	129
	58	I_2	5	0.12% ^{125}I	130
$^{127}I(d, 6n)^{123}Xe \xrightarrow{\beta^+, EC} {}^{123}I$	80	NaI	8	0.2% ^{125}I	48, 51
$^{127}I(\alpha, 8n)^{123}Cs$ / $(\alpha, p7n)$ $\rightarrow {}^{123}Xe \xrightarrow{\beta^+, EC} {}^{123}I$	100	NaI	0.53	1.8% ^{125}I	30, 131
$La/Cu(p, x)^{123}Xe \xrightarrow{\beta^+, EC} {}^{123}I$	590	Cu–La alloy	6–10	?	49
$Cs(p, x)^{123}Xe \xrightarrow{\beta^+, EC} {}^{123}I$	450	Cs-metal	?	?	TRIUMF Vancouver planned
$Cs(p, x)^{123}Xe \xrightarrow{\beta^+, EC} {}^{123}I$	590	Cs-metal	?	?	SIN Villigen planned

gas.[41,124] The latter method can also be applied as an on-line technique. Separation yields of 80[41] and 90%[42] are reported for the discontinuous wet chemical separations. For economic reasons, the recovery of the tellurium is just as important. In the wet batch technique, used by Wolf *et al.*, yields of 99% are achieved.

The indirect processes which proceed via the ^{123}Xe ($T_{1/2} = 2.08$ hr) parent, and are therefore often referred to as generator methods, have several advantages:

(1) Higher radionuclidic purity can be obtained due to the Xe-separation.

(2) Continuous on-line methods can easily be applied by sweeping the ^{123}Xe-parent off the target system.

(3) Cheap natural iodine can be used.

(4) Decay-induced labelling (excitation labelling) is possible (see below).

From the indirect reactions listed in Table 4 the high-energy processes on iodine, in particular the ^{127}I $(p,5n)$ ^{123}Xe[43–47,129,130] and the ^{127}I $(d,6n)$ ^{123}Xe[48] are of interest. The ^{122}Te $(\alpha,3n)$ ^{123}Xe reaction, which was frequently used to produce a very pure ^{123}I,[36,37,50] does not seem to be economical because of the poor practical yield. The high energy reactions, on the other hand, require 50–65 MeV protons and 70–85 MeV deuterons, respectively. Machines with such particle energies are rather scarce and quite often have rather low beam intensities. The rather expensive irradiation time on these large machines together with the low beam currents can lead to an uneconomical production. In addition, the often low availability of these large accelerators also imposes an organizational problem. A high-current linear accelerator like the Brookhaven LINAC ISOTOPE SEPARATOR (BLIP), on the other hand, can principally produce Curie-amounts per hour.[45] The major longer-lived radionuclidic impurity encountered in the indirect processes is iodine-125 ($T_{1/2} = 60.14$ d). While this impurity leads to a noticeable increase in dose, its major 35 keV γ-line does not disturb the scans. With the high-energy reactions the initial concentration of this contaminant can be kept at or below 0.2%, provided the beam energy and target thickness are carefully controlled.

For the ^{127}I $(p,5n)$- and ^{127}I $(d,6n)$-reactions a variety of target materials have been tested, such as I_2, CH_2I_2, NaI, KI and $KI \cdot I_2$-solutions. In the case of iodine or its complexes in aqueous solutions, as well as in the case of CH_2I_2, corrosion problems are encountered.[44,46,129,130] Systems are being used in which the iodine solution or liquid is circulated through the target and the ^{123}Xe is stripped off by gas-liquid extraction, using a He-counter current stream, from which the ^{123}Xe is eventually separated in a cooled trap. In other cases[45,47,48] NaI in the form of powder or pressed tablets is used as target material and ^{123}Xe is continuously swept off by He-carrier gas, from which it is then separated in a trap. The latter procedure seems to be the simplest and cleanest way, particularly because experimental yields of 80–90% of the theoretical yields can be achieved.[48,51]

The greatest production capacity lies possibly in the high-energy spallation reactions.[49] At the Los Alamos Meson Physics Facility (LAMPF) it is anticipated that a residual 600–800 MeV proton beam of about 0.5 mA reaches the main beam stop. Since the cost of the prime beam time normally associated with accelerator-produced isotopes is not present, a substantial economy can be realized here, particularly when utilization of high beam currents is possible. The spallation cross section of La at 590 MeV is 57 mbarn for ^{123}I and 36 mbarn for ^{123}Xe.[52] Little is known, however, about the radionuclidic purity of these products.

3. RECENT DEVELOPMENTS IN LABELLING

It is not the intention of this paper to give a complete review on bromination and iodination procedures suitable for labelling but rather to point out a few new trends and developments which are useful, particularly when dealing with relatively short-lived radionuclides in carrier-free form. In many cases, however, modified classical bromination and iodination procedures can be applied and information can be obtained from the well-known labelling methods for ^{82}Br and ^{131}I.

Very frequently the chemical forms of the starting materials are Br^- and I^-, respectively. It is important to point out some difficulties when dealing with such species in carrier-free form. Not only losses in the activity occur, particularly in chromatographic procedures, but also changes of the chemical form are easily observed depending on the environment, especially in the case of iodine. The oxidation state of carrier-free species under the condition of the analytical quality control is meaningless, and stabilization or protection of the halide ions is necessary by adding a reducing or stabilizing agent such as $S_2O_3{}^{2-}$ or a few micrograms of iodide carrier if tolerable.

Most of the applications of Na ^{123}I as an intermediate for further syntheses are practically identical to those of the well-known use of Na ^{131}I. In ^{123}Xe-producing nuclear reactions Na ^{123}I can easily be obtained, after the ^{123}Xe is allowed to decay in a pyrex ampoule containing a small amount of diluted NaOH and traces of $Na_2S_2O_3$.[53,54] The optimum time for ^{123}Xe-decay (6.7 hr) can be conveniently used for shipment. Some groups[36,53] have developed this method to a kit procedure. The time required to prepare Na ^{123}I after the decay of ^{123}Xe precursor is only a few minutes. The major iodine species (95–98%)[54,55] is $^{123}I^-$. The procedure can probably easily be adapted for ^{77}Br if produced via ^{77}Kr.

Decay-induced labelling (excitation labelling)

Hot atom chemists have known this type of labelling *via* radioactive and highly reactive daughter ions formed in a nuclear decay process for many years (for review cf. ref. 56 and 57). β-decay gives rise to different types of excitation of the daughter atom, such as electronic perturbation (shaking) due to the sudden change in Z, change in the chemical identity, kinetic energy from the electron- and neutrino recoil and charge acquisition. The latter can have several sources. The change in Z leads to a charge increase of $+1$ in a β^- and a decrease of -1 in a β^+-decay. In addition, a further electron can be emitted (shake off) due to the above mentioned electronic excitation. In the case of the E.C.-process multiply positive charged daughter ions are formed *via* the Auger electron cascade.

The chemical effects of the β-decay can be usefully employed for labelling in the case of ^{77}Kr- and ^{123}Xe-generators. ^{77}Kr decays with 84% by β^+ emission and 16% by E.C.:[58]

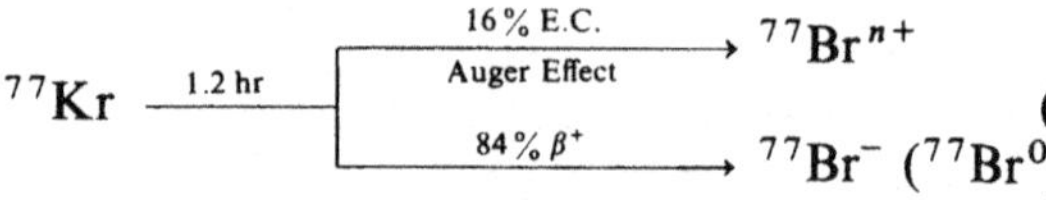

(1)

with maximum recoil energies of 75.8 eV[60] (neutrino recoil from E.C.) and 37.5 eV[60] (β^+ recoil). ^{123}Xe, on the other hand, decays with 22% β^+ and 78% E.C.:[59]

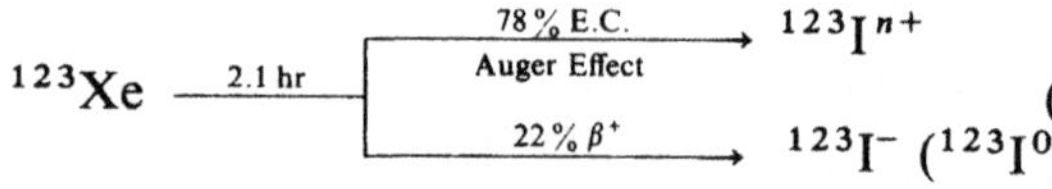

(2)

The iodine ions also receive some kinetic energy resulting from neutrino emission (maximum recoil energy 34 eV)[61] and from β^+ emission (maximum recoil energy 20 eV).[61] When the decay is allowed to occur in the presence of an organic substrate the bromine or iodine species can undergo rapid reactions as singly charged positive ions, as neutral atoms, or as negative ions, i.e. they can undergo electrophilic, homolytic or nucleophilic substitution processes.

It has been shown that in simple gaseous systems such as methane the decay of $^{76,77}Kr$,[60] ^{123}Xe[61] and ^{125}Xe[62] leads to significant bromination and iodination, respectively. In simple aromatic systems it could be demonstrated[63] that decay produced positive bromine and iodine ions exhibit an electrophilic substitution pattern to be expected for unsolvated positive iodine ions. In the liquid phase, however, neutralization seems to occur rapidly and the isomeric *o*-, *m*-, *p*-substitution changes significantly, indicating a substantial contribution from neutral iodine atoms.[64,65,69]

Decay-induced labelling can be accomplished by simple ^{77}Kr- or ^{123}Xe-exposure of the substrate for a time period necessary to obtain the maximum daughter activity (6.9 hr for ^{77}Br and 6.7 hr for ^{123}I). The iodination *via* the decay of ^{123}Xe frozen on serum albumin was first reported by WELCH,[66] who obtained an 80% radiochemical yield of organically bound iodine. WOLF *et al.*[1] reported a 20% yield for the labelling of indocyanine green *via* the decay of ^{123}Xe, which was adsorbed at the dye surface at 77 k. In general, however, decay-induced labelling of solid biomolecules gives only poor yields, ranging from about 1 to 3%,[67] mainly due to the fact that the solubility of Xe in organic solids is very low and the labelling process takes place essentially at the surface of the solid substrate. Direct decay-induced labelling is also inherently unspecific (see below, Fig. 2a) and both hydrogen and halogen atoms can be replaced.[63]

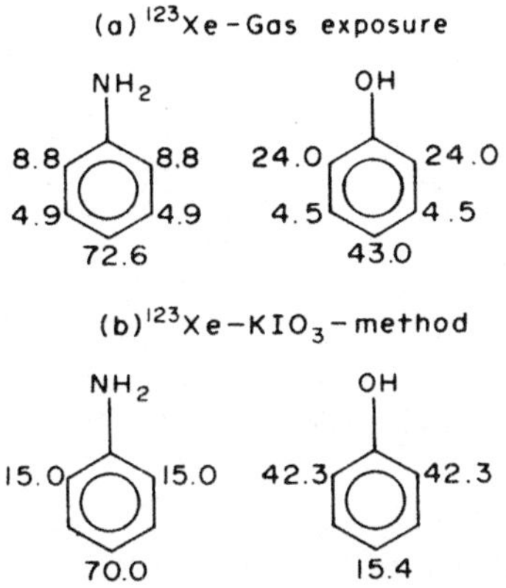

FIG. 2. Isomer distribution in the iodination of aniline and phenol. (a) ^{123}Xe-gas exposure of liquid aniline and phenol.[69] 100 μl of substrate exposed to 2 ml of Xe (^{123}Xe + Xe-carrier) for 6 hr at room temperature in a 300 μl pyrex ampoule. (b) ^{123}Xe-KIO_3-method.[67,69] 1 mg KIO_3-powder exposed to 2 ml Xe (^{123}Xe + Xe-carrier) in a 100 μl pyrex ampoule exposed for 6 hr at room temperature. KIO_3 then dissolved in 1 ml of 0.1 N HCl solution of substrate (1 mg/ml) at room temperature.

A different approach entails the decay-induced preparation of iodinating reagents which can subsequently serve for the final labelling procedure. Two methods of this type have been developed so far. The first was reported by LAMBRECHT *et al.*[68] and consists essentially of the carrier-free preparation of the classical iodination reagent iodine monochloride *via* the decay of ^{123}Xe in chlorine:

$$^{123}\text{Xe} \xrightarrow{\beta^+,\ \text{E.C.}} {}^{123}\text{I}^* \xrightarrow{\text{Cl}_2} {}^{123}\text{ICl} \qquad (3)$$

The second method, reported by EL-GARHY and STÖCKLIN,[67] is based on the preparation of a reactive iodination reagent via the decay of ^{123}Xe on solid KIO_3:

$$^{123}\text{Xe} \xrightarrow{\beta^+,\ \text{E.C.}} {}^{123}\text{I}^* \xrightarrow{\text{KIO}_3} {}^{123}\text{IO}^-\ (?) \qquad (4)$$

In the first method carrier-free ^{123}ICl is obtained with a radiochemical yield of about 90% when the ^{123}Xe is allowed to decay in chlorine gas, followed by the separation of unreacted Cl_2. Of course, carrier containing ICl–^{123}I can also be prepared when the ^{123}Xe is allowed to decay in anhydrous ^{123}ICl, and $^{123}ICl_2^-$ is obtained by dissolution of these species in diluted HCl. The use of the classical but adapted to carrier-free iodinating reagents for labelling radiopharmaceuticals such as Teridex, human serum albumin, tyrosine and diiodosalicylic acid has been discussed.[68]

The second method is also practically a carrier-free iodination procedure.[67,69] The ^{123}Xe exposed KIO_3 is dissolved in an acidic solution of the substrate (e.g. 0.1 N HCl solution). Iodination generally occurs immediately during dissolution. Only aromatic systems are iodinated with good yields, and the selectivity is that expected for an electrophilic substitution *via* positive iodine (Fig. 2b), the highest yields being obtained in the pH region of 1–3.[69] The substrate concentration should be as high as possible but at least 10^{-3} mole/l, in order to prevent competing reactions of the carrier-free iodine species with impurities. Fig. 2 shows the isomer distribution for the iodination of simple aromatic molecules such as aniline and phenol using decay-induced labelling by ^{123}Xe-exposure (Fig. 2a) and the ^{123}Xe–KIO_3 method (Fig. 2b). It can be seen that the ^{123}Xe–KIO_3 method is somewhat more selective and shows a distribution which is also observed in the classical iodide–iodate method[69]. A great variety of compounds have been labelled with the ^{123,125}Xe–KIO_3 method[69,70] (see below, Table 5). Radiochemical yields in the range of 30–90% are obtained. The method is extremely fast and mild. The decay period for the growth of ^{123}I can be used for shipment, and a kit procedure is principally possible. It is particularly useful for cases in which classical procedures show only poor results with carrier-free iodine.

The methods can easily be adapted to the ^{77}Kr–^{77}Br generator system. It has recently been demonstrated by WONG and ACHE[71,132] that the positive bromine ions produced in the isomeric transition of ^{80m}Br also give rise to effective bromination reagents when the decay is allowed to take place in the presence of Cl_2 gas and solid

TABLE 5. ^{123}I-radiopharmaceuticals.

Compound	Labelling method	Application	References
3-acetamido-5-methylcarbamyl-2,4,6-triiodobenzoic acid ("Conray")	isotope exchange ICl	urography	23
iodoantipyrine		brain imaging	96
iodobleomycin	ICl	tumor localization	97, 98
3,5-diacetamido-2,4,6-triiodobenzoic acid ("Hypaque")	isotope exchange ICl	urography	23
5-iododeoxyuridine	^{123}Xe–KIO_3	control of tumor therapy	70
6-iododopamine	$I_2/AgOCOCF_3$		99
2-iodoestradiol	isotope exchange in melt		75, 100
2,4-diiodoestradiol	isotope exchange in melt		75, 100
iodofibrinogen	(a) ICl (b) electrolysis (c) ^{123}Xe–KIO_3	thrombus localization	101 102,103 70
iodofibrin		thrombus localization	111
16-iodo-9-hexadecanoic acid	I-for-Br exchange	myocardial imaging	78, 133
17-iodoheptadecanoic acid	I-for-Br exchange	heart muscle metabolism	69, 134
o-, *m*-, *p*-iodohippuric acid	isotope exchange in melt I-for-Br exchange in melt ^{123}Xe–KIO_3	renography	73, 74 22, 113, 114, 135 70
iodinated hepatitis antibody	^{123}Xe–KIO_3	*in vivo* studies of antibody–antigen reactions	70, 104, 137
iodinated serum albumin	^{123}Xe–exposure, ICl	cardio-pulmonary dynamics	66, 120
iodinated indocyanine green	^{123}Xe-exposure	eye melanoma imaging	1
iodoinsulin	^{123}Xe–KIO_3		67, 70
iodinated oleic acid	^{123}Xe-exposure	myocardial imaging	67, 70
2-iodopalmitic acid	I-for-Br exchange in melt ^{123}Xe-exposure in melt		67, 69, 70
4-iodophenylalanine	isotope exchange in melt	pancreas imaging	76
iodinated Rose Bengal	I^{θ}/H_2O_2	liver function test evaluation of jaundice	24, 105–108
ioglycemic acid		liver function studies	136
3,5-diiodosalicyclic acid	ICl		68, 109
sodium iodide	^{123}Xe-decay in NaOH	thyroid imaging	21, 53, 55, 110, 115, 116
iodostreptokinase	ICl	thrombus imaging	112
2-iodostearic acid	I-for-Br exchange		69, 70, 134
iodotyrosine	ICl, ^{123}Xe–KIO_3		67, 68, 70, 109
iodothyronine	^{123}Xe–KIO_3		68, 70
diiodotyrosine	ICl		68, 109
5- and 6-iodotryptophan	ICl		76

Cf. also ref. 25 and 138

$KBrO_3$, respectively. A variety of biomolecules have been labelled with this method and radiochemical yields of 20–85% were obtained (see below, Table 6).

Halogen exchange labelling

Another important class of methods for introducing radioactive bromine or iodine into a molecule is the halogen exchange. Some new approaches have been reported in this area during the past years. In principle, the exchange can be isotopic and non-isotopic. In the interhalogen exchange carrier-free products can be obtained. The reaction can be either electrophilic *via* positive halogens or nucleophilic *via* halide ions or even a homolytic process *via* halogen atoms; it can proceed in solution or in molten organic systems.

Isotopic exchange in melt is a new approach[72] which can be useful when the exchange in solution cannot conveniently be achieved. ELIAS *et al.*[73] have reported a new technique for iodine labelling of aromatic iodine compounds, which is based on the isotopic iodine exchange in melts of carrier-free sodium iodide and an organic substrate. This method was used to prepare radioiodine labelled *m*-iodohippuric acid which—due to its greater stability—is advantageous for nuclear medical applications when compared with *o*-iodohippuric acid. The method is very efficient, and in this particular case the exchange is completed within about 30 min at a temperature of 161 °C. This procedure has also been applied to prepare ^{123}I-labelled *o*-iodohippuric acids.[74] It has further been reported[75] that ^{123}I-labelled 2-iodo- and 2,4-diiodoestradiol is formed in a melt of estradiol and carrier-free $Na^{123}I$.

The exchange in melt is limited by the necessity of melting the substrate to be labelled without decomposition. In some instances where the compound of interest melts with decomposition, low melting derivatives of the compound can be used. This approach has been utilized by LAMBRECHT *et al.*[76] who applied the decay of ^{123}Xe in the exchange labelling in the melt to prepare ^{123}I-4-iodophenyl-alanine and ^{123}I-5- and -6-iodotryptophan. In another approach ELIAS and LOTTERHOS[77] have bypassed this problem by using molten "unreactive" solvents with low melting points such as acetamide, which has a good solubility for many organic

TABLE 6. ^{77}Br-radiopharmaceuticals and labelled compounds

Compound	Labelling method	Yield (%)	Ref.
brominated albumin	enzymatic, chloroperoxidase	76	93
brominated fibrinogen	enzymatic, chloroperoxidase (acylating agent)	—	117
brominated thyroglobulin	enzymatic, chloroperoxidase	79	93
bromotyrosine	enzymatic, chloroperoxidase	83	93
bromocholesterol	excitation labelling	—	139
bromodeoxyuridine	excitation reagent labelling* ($KBrO_3$, Cl_2)	27	71, 152
bromoguanosine	excitation reagent labelling	52	71
bromotyrosine	excitation reagent labelling	85	71
17-bromoheptadecanoic acid	Br-for-I exchange	45	134
4-bromo-2,5-dimethoxyphenyliso-propylamine	direct bromination	—	94, 118
6β-bromomethyl-19-nor cholest-5(10)en-3β-ol			140
sulfobromophthalein	recoil bromination†	7.5	95

*^{80}Br via isomeric transition, but adaptable to ^{77}Kr–^{77}Br excitation labelling.
†^{82}Br via (n, γ)-recoil, possibly adaptable to ^{77}Kr–^{77}Br excitation labelling.

compounds and also for metal halides. Using molten acetamide solutions the authors also observe an element exchange in simple aromatic systems, e.g. the I-for-Br exchange in *m*-bromobenzoic acid leads to carrier-free *m*-iodobenzoic acid with a radiochemical yield of 92% after 2 hr at 180°C.

Benzoic acid derivatives containing two or three iodine atoms are used routinely in urography as contrast media. Iodine-123 labelled products are desirable for *in vivo* measurements. An efficient labelling method by exchange in solution has been recently reported by THAKUR[23] for the preparation ^{123}I-labelled Conray and Hypaque. In the case of Conray (3-acetamido-5-methyl-carbamyl-2,4,6-tri-iodobenzoic acid with its two electron attracting and one electron releasing groups, nucleophilic substitution with iodide ions was found to be the most efficient method. At elevated temperatures the exchange was completed within less than two minutes. In the case of Hypaque (3,5-diacetamido-2,4,6-triiodobenzoic acid), on the other hand, with its two electron releasing and one electron attracting groups, an electrophilic substitution via iodine monochloride was expectedly the fastest procedure, and the exchange was completed under reflux within 2 min.

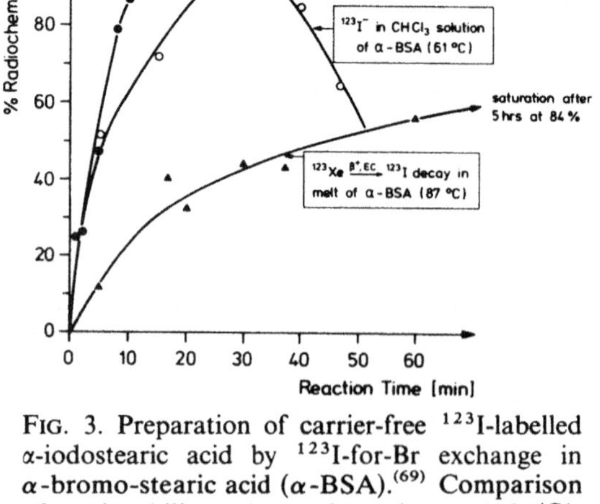

FIG. 3. Preparation of carrier-free ^{123}I-labelled α-iodostearic acid by ^{123}I-for-Br exchange in α-bromo-stearic acid (α-BSA).[69] Comparison of nucleophilic exchange in melt at 87°C (●), nucleophilic exchange in $CHCl_3$ solution at 61°C (○) and decay-induced exchange via ^{123}Xe-gas exposure of melt at 87°C(▲).

While the halogen exchange in aromatic systems is often difficult or impossible to accomplish, even in molten systems, aliphatic halogen generally readily exchanges by a nucleophilic reaction. ROBINSON and LEE[78] and MACHULLA *et al.*[69,134] have prepared ^{123}I-labelled fatty acids, by a nucleophilic ^{123}I-for-Br exchange of the corresponding bromocarboxylic acids in acetone solution. MACHULLA *et al.*[69] have shown that carrier-free ^{123}I-labelled α-iodo fatty acids can be most effectively prepared in a melt. Figure 3 shows the kinetics of the ^{123}I-for-Br exchange in an α-bromo stearic acid (BSA) for the melt (carrier-free $Na^{123}I$ in molten BSA), the solution (carrier-free $Na^{123}I$ and BSA in $CHCl_3$) and for the decay induced labelling (^{123}Xe exposure of molten BSA). It can be seen that at 87°C the exchange in the melt is the fastest process and is completed within about 20 min. Isotopic and non-isotopic exchange in the molten state can easily be adapted to the labelling with ^{77}Br.

For special cases exchange labelling on chromatographic columns is also of interest. The method of labelling volatile compounds by heterogeneous halogen exchange reactions on gas chromatographic columns was first reported by STÖCKLIN *et al.*[79] who prepared carrier-free ^{82}Br-labelled alkyl bromides and ^{18}F-labelled alkyl fluorides. A complete summary of this type of exchange can be found in a review by ELIAS[80]. The exchange on gas chromatographic columns can be described as follows:

$$AX_{mobile} + BY^*_{station.} \rightleftarrows AY^*_{mobile} + BX_{station.} \quad (5)$$

BY* (or BX*), are the exchange partners carrying the label Y* (or X*), in general solids which serve as stationary phase and are adsorbed on supports. The virtue of this method lies in the fact that it can produce carrier-free products in a mild and fast way, combining the exchange and purification step. The procedure is limited to volatile compounds; however, higher boiling compounds such as α-halo fatty acids have been recently prepared via their esters using ion exchange resins as exchanging support.[81] A batch technique can also be applied for liquid-solid systems. Such an approach has been used by ROBINSON[82] for the synthesis of ^{18}F-α-fluoro fatty acid esters. The ^{18}F loaded resin was sealed along with an aqueous solution of the substrates in glass ampoules and then heated. Some of these techniques might also be interesting for ^{77}Br- and ^{123}I-labelling procedures.

Enzymatic labelling

Among the various labelling methods which have been used for many years the enzymatic halogenation is the most gentle and deserves special attention as a fast and mild procedure for introducing radioactive halogen into sensitive biomolecules such as proteins. Radioiodination of proteins is generally carried out by one of three methods: The ICl-method of MacFarlane[83], the chloramine-T oxidation of Hunter and Greenwood[84] or the electrolytic procedure of Rosa[85]. These methods often show disadvantages with respect to denaturation and/or specific activity.[86–88] Enzymatic oxidation was developed for the iodination of proteins[89] and immunoglobulins[86] and has since also been used to label gonadotropin[90] aromatic steroids and other hormones and proteins.[91] This method requires the reaction of lactoperioxidase, carrier-free radioiodine and the compound to be labelled together with nanomolar quantities of H_2O_2. It yields high specific activity products and the biological activity for radioimmunoassay is preserved.[86] Krohn and Welch[88] have reported the carrier-free iodination of fibrinogen, a molecule which is extremely sensitive to chemical effects that can destroy its biological activity.[92] Under optimum conditions the authors found that within 30 min 60–85% binding of the carrier-free radioiodine occurred.

Bromine-77 has several advantages for labelling of proteins. Due to its stronger bond the extent of hydrolysis should be smaller as compared to that of the iodinated products, and the 56 hr half-life also allows studies of longer duration. Welch *et al.*[93] have therefore labelled *L*-tyrosine and proteins with ^{77}Br using the enzyme chloroperoxidase. They obtained yields of 75% bromotyrosine, 80% labelled albumin and 80% labelled thyroglobulin. All the brominated compounds have been shown to be more stable to *in vitro* hydrolysis than the corresponding iodinated compounds, and the authors conclude that ^{77}Br-labelled proteins are superior to their iodinated analogs. Knight, Harwig and Welch[117] have also prepared ^{77}Br-labelled fibrinogen using the intermediate enzymatic labelling of the acylating agent N-succinimidyl-3-(4-hydroxphenyl)propionate) (SHPP) with carrier-free ^{77}Br. Labelled SHPP was then allowed to conjugate with fibrinogen. The bromine labelled fibrinogen was found to be >90% clottable.

4. CONCLUSION

It is obvious that the greatest potential of ^{77}Br and ^{123}I lies in the field of radiopharmaceuticals for *in vivo* studies. In the case of iodine-131 which is in current use, the limited quantity that can be safely administered to the patient often restricts its application. The low photon flux with low counting statistics requires a relatively prolonged imaging time or poor resolution image results. Due to the suitable decay properties of iodine-123, larger activities can be applied. The overall reduction in imaging time and radiation exposure together with the better images greatly improves the diagnostic capabilities. Most of the iodine-131 or iodine-125 radiopharmaceuticals which are presently being used for *in vivo* procedures can advantageously be replaced by iodine-123, unless long term studies require a longer half-life as in some cases of thrombus localization. In fact, a variety of ^{123}I-labelled radiopharmaceuticals are already in clinical use or are being evaluated (cf. also refs. 119, 125), as can be seen from Table 5. Besides the expected improvement, whenever well-known ^{131}I- or ^{125}I-radiopharmaceuticals are replaced by ^{123}I-radiopharmaceuticals, a variety of new approaches for *in vivo* metabolic studies are conceivable and are also listed in column 3 of Table 5. In some instances ^{123}I or its labelled compounds are also superior to ^{99m}Tc, as in the case of thyroid imaging and uptake measurements[21] or for studies of cardiopulmonary dynamics[120] with ^{123}I-labelled albumin in the form of macroaggregates, due to the stronger chemical bond of iodine when compared with technetium. At present the limitation lies not only in the availability. A network of collaborating accelerators seems to be necessary to satisfy the demand. The great potential of ^{123}I-labelled compounds and radiopharmaceuticals also necessitates basic research and development of efficient and mild labelling procedures, both classical and non-classical, with special emphasis on the preparation of practically carrier-free products. Iodine-123 radiopharmaceuticals should eventually be made commercially available to a broader spectrum of physicians just as iodine-131 radiopharmaceuticals. This is ultimately an organizational and economic problem which should be solved, otherwise the benefits of this radioisotope will remain confined to the few hospitals which have the facilities and the chemical know-how.

The situation is somewhat different for bromine-77. Production, labelling and application of this nuclide have hardly started. This is also reflected in Table 6. While the stronger C–Br bond and in some instances the longer half-life can be of considerable advantage for a broader use, the 521 keV γ-line imposes limitations. The shorter lived positron emitting bromine isotopes might eventually be more useful when efficient high speed positron cameras are available. Another problem is the availability of these medically interesting bromine isotopes. At present, too little is known about the relevant nuclear and radiochemical data, but it can be expected that these problems are no more serious than in the case of iodine-123. A considerable amount of basic research and development work is needed in nuclear chemistry, carrier-free labelling, biochemistry and pharmacokinetics in order to explore the great potential of radiobromine as a label for radiopharmaceuticals.

For the development of radiobromine and -iodine labelled radiopharmaceuticals accompanying work in radioanalytical quality control is mandatory. This is particularly true for practically carrier-free products. Efficient chromatographic procedures have to be applied in order to ensure radiochemical purity. High pressure liquid chromatography is frequently the only method of choice, and it has been extensively used as an efficient and rapid method for the quality control of radiopharmaceuticals labelled with short lived isotopes (cf. e.g. ref. 54).

While iodine-123 is the radioisotope of today, bromine-77–74 might be the radioisotopes of tomorrow. It took more than a decade of campaigning, which was started by Myers (for a recent review cf. ref. 119), before iodine-123 was generally recognized, and the ultimate stage of routine production and availability has yet not been reached. It is to be hoped, however, that the tremendous interest in short-lived radioisotopes for nuclear medicine will considerably accelerate this procedure for the other medically interesting radiohalogens such as $^{77-74}$Br, ^{34m}Cl and ^{18}F. Nuclear- and radiochemists may find it challenging to help nuclear medicine open up these new routes.

Acknowledgements—The author is indebted to Dr. H.-J. Machulla and Dr. R. Weinreich for many stimulating discussions. He also wishes to thank Dr. A. P. Wolf and Dr. M. J. Welch for making some of the experimental data available to him prior to publication.

REFERENCES

1. Lambrecht R. M. and Wolf A. P. *Proc. Symp. New Devel. Radiopharm. Labelled Comp., Copenhagen,* Vol. 1, p. 275, IAEA, Vienna (1973).
2. Neirinckx R. D., Myburgh J. A. and Smit J. A. *ibid.* p. 171. IAEA, Vienna (1973); cf. also Smit J. A., Myburgh J. A. and Neirinckx R. D. *Clin. exp. Immunol.* **14,** 107 (1973).
3. Meyer G.-J., Rössler K. and Stöcklin G. *J. Labelled Comp. Radiopharm.* in press; cf. also Rössler K., Tornau W. and Stöcklin G. *J. Radioanalyt. Chem.* **21,** 199 (1974).
4. Aaij C., Tschroots W. R., Lindner L. and Feltkamp T. E. W. *Int. J. appl. Radiat. Isotopes* **26,** 25 (1975).
5. Coban A. *J. Phys.* **A7,** 1705 (1974).
6. Coban A., Lisle J. C., Murray G. and Willmott J. C. *Particles and Nuclei* **4,** 108 (1972).
7. Coban A., Willmott J. C., Lisle J. C. and Murray G. *Nucl. Phys.* **A182,** 385 (1972).
8. Dzhelepov B. S., Dimitriev A. G., Zhelev Zh., Zhukovskii N. N., Moskvin L. N. and Fominykh V. I. *Bull. Acad. Sci. USSR, Phys. Ser.* **34,** 1838 (1971).
9. Dzhelepov B. S., Dimitriev A. G. and Zhukovskii N. N. *Bull. Acad. Sci. USSR, Phys. Ser.* **35,** 2043 (1972).
10. Sarantites D. G. and Erdal B. R. *Phys. Rev.* **177,** 1631 (1969).
11. Braga R. A. and Sarantites D. G. *Phys. Rev.* **C9,** 1493 (1974).
12. Nicholson J. P. and Zilva J. F. *Clin. Sci.* **19,**

391 (1960).

13. Kakkar V. V., Nicolaides A. N., Renney J. T. G., Friend J. R. and Clark M. B. *Lancet* **1**, 540 (1970).
14. Myers W. G. and Anger H. O. *J. nucl. Med.* **3**, 183 (1962); cf. also Myers *et al. Proc. Symp. New Devel. Radiopharm. Labelled Comp., Copenhagen*, Vol. 1, p. 249, IAEA, Vienna (1973).
15. Myers W. G. *Proc. Symp. Radioactive Pharmaceuticals, Oak Ridge*, USAEC CONF-651111, Chapter 12, p. 217 (1965).
16. Gföller D. and Langhoff H. *Z. Physik* **211**, 317 (1968).
17. Ragaini R. C., Walters W. B., Gordon G. E. and Baedecker P. A. *Nucl. Phys.* **A115**, 611 (1968).
18. Bechvarzh F., Gromov K. Ya., Zhelev Zh. T., Zaitseva N. G., Loshchilov M. G., Nazarov U. K. and Sabirov S. S. *Bull. Acad. Sci. USSR, Phys. Ser.* **33**, 1228 (1970).
19. Karttunen E., Freund H. U. and Fink R. W. *Nucl. Phys.* **A131**, 343 (1969).
20. Goolden A., Glass H. I. and Silvester D. *Br. J. Radiol.* **41**, 20 (1968).
21. Atkins H. L., Klopper J., Lambrecht R. M. and Wolf A. P. *Am. J. Roentgenol.* **117**, 195 (1973).
22. Holroyd A. M., Chisholm G. D. and Glass H. I. *Phys. Med. Biol.* **15**, 483 (1970).
23. Thakur M. L. *Int. J. appl. Radiat. Isotopes* **25**, 576 (1974).
24. Serafini A. N., Smoak W. M., Hupf H. B., Beaver J. E., Holder J. and Gilson A. G. *J. nucl. Med.* **16**, 629 (1975).
25. Johnson C. H., Galonsky A. and Inskeep C. N. Report ORNL-2910, 25 (1960).
26. Skrabel F., Arnot R. N., Helus F., Glass H. L. and Joplin G. F. *Int. J. appl. Radiat. Isotopes* **21**, 183 (1970).
27. Helus F. *Radiochem. Radioanal. Lett.* **3**, 45 (1970).
28. Waters S. L., Nunn A. D. and Thakur M. L. *J. inorg. nucl. Chem.* **35**, 3413 (1973).
29. Blue J. E. and Benjamin P. P. *J. nucl. Med.* **12**, 416 (1971).
30. Helus F., Maier-Borst W., Lambrecht R. M. and Wolf A. P. *Proc. 7th Int. Conf. on Cyclotrons and their Applications, Experientia Suppl.* **24**, 474 (1975).
31. Nunn A. D. *Nucl. Instrum. Methods* **99**, 251 (1971).
32. Nunn A. D. and Waters S. L. *Int. J. appl. Radiat. Isotopes* **26**, 731 (1975).
33. Silvester D. J., Sugden J. and Watson I. A. *Radiochem. Radioanalyt. Lett.* **2**, 17 (1969).
34. Neirinckx R. D. *Radiochem. Radioanalyt. Lett.* **5**, 205 (1970).
35. Murakami Y., Akiha F. and Ezawa O. *Proc. Symp. New Devel. Radiopharm. Labelled Comp., Copenhagen*, Vol. 1, p. 257. IAEA, Vienna (1973).
36. Sodd V. J., Blue J. W., Scholz K. L. and Oselka M. C. *Int. J. appl. Radiat. Isotopes* **24**, 171 (1973).
37. Guillaume M., Lambrecht R. M. and Wolf A. P. *Int. J. appl. Radiat. Isotopes* **26**, 703 (1975).
38. Hupf H. B., Eldridge J. S. and Beaver J. E. *Int. J. appl. Radiat. Isotopes* **19**, 345 (1968).
39. Gelbard A. S., Hara T., Tilbury R. S. and Laughlin J. S. *Proc. Symp. New Devel. Radiopharm. Labelled Comp., Copenhagen*, Vol. 1, p. 239. IAEA, Vienna (1973).
40. Cauwe F., Deutsch J. P., Favart D., Prieels R. and Cogneau M. *Int. J. appl. Radiat. Isotopes* **25**, 187 (1974).
41. Acerbi E., Birattari C., Castiglioni M., Resmini F. and Villa M. *Int. J. appl. Radiat. Isotopes* **26**, 741 (1975); cf. also Acerbi *et al. Proc. 7th Int. Conf. on Cyclotrons and their Applications, Experimenta Suppl.* **24**, 461 (1975).
42. Kondo K., Lambrecht R. M., Wolf A. P. *Int. J. appl. Radiat. Isotopes*, in press.
43. Fusco M. A., Peek N. F., Jungermann J. A., Zielinski F. W., Denardo S. J. and Denardo G. L. *J. nucl. Med.* **13**, 729 (1972).
44. Lindner L., Brinkman G. A., Suer T. H. G. A., Schimmel A., Veenboer J. Th., Karten F. H. S., Visser J. and Leurs C. J. *Proc. Symp. New Devel. Radiopharm. Labelled Comp., Copenhagen*, Vol. 1, p. 275. IAEA, Vienna (1973).
45. Richards P., Lebowitz E. and Stang L. G. *Proc. Symp. New Dev. Radiopharm. Labelled Comp., Copenhagen*, Vol. 1, p. 341. IAEA, Vienna (1973).
46. Bievelez P., Cogneau M., Charlier R. and Jongen Y. *J. Radioanalyt. Chem.*, **30**, 67 (1976)
47. Loepfe E. and Huszar I. RIS Würenlingen, Switzerland. Private communication.
48. Weinreich R., Schult O. and Stöcklin G. *Int. J. appl. Radiat. Isotopes* **25**, 535 (1974).
49. O'Brien H. A., Jr. *Proc. Symposium on Practical Appl. of Accelerators*, LA-5535 (CONF-730849). p. 93 (1973).
50. Lambrecht R. M. and Wolf A. P. *Rad. Res.* **52**, 32 (1972).
51. Weinreich R., Bräutigam W., Machulla H.-J., Probst H. J., Qaim S. M. and Stöcklin G. *Proc. 7th Int. Conf. on Cyclotrons and their Applications, Experientia Suppl.* **24**, 245 (1975); cf. also Weinreich R., Qaim S. M., Michael H. and Stöcklin G. *J. Radioanal. Chem.* **30**, 53 (1976).
52. O'Brien H. A., Jr. and Schllaci M. E. *Isotopes and Radiation Technology* **9**, 105 (1971).

53. Lambrecht R. M., Norton E. and Wolf A. P. *J. nucl. Med.* **14,** 269 (1973).
54. Machulla H.-J., Laufer P. and Stöcklin G. *J. Radioanalyt. Chem.* **32,** 381 (1976).
55. Neirinckx R. D., Lambrecht R. M., Norton E. F. and Wolf A. P. *Int. J. appl. Radiat. Isotopes* **25,** 387 (1974).
56. Wexler S. *Action Chimiques et Biologiques des Radiations* (Haissinsky ed), huitième série, chapter III, Masson et Cie (1965).
57. Stöcklin G. *Chemie heißer Atome,* Verlag Chemie, Weinheim FRG (1969); cf. also revised French translation "*Chimie des Atomes Chauds*". Masson et Cie (1972).
58. Houdayer A., Mark S. K., Kitching J. E. and Paradellis T. *Phys. Fennica* **9,** 85 (1974).
59. Gföller D., Schöneberg R. and Flammersfeld A. *Z. Physik* **208,** 299 (1968).
60. Frost J. J. and Welch M. J. *8th Int. Hot Atom Chem. Symp.,* Spa, Belgium, 1975, abstract.
61. Loberg M. D. and Welch M. J. *J. Am. Chem. Soc.* **95,** 1075 (1973).
62. Schroth F. and Adloff J. P. *J. Chim. Phys.* **61,** 1373 (1964).
63. Knust E. J., Halpern A. and Stöcklin G. *J. Am. Chem. Soc.* **96,** 3733 (1974).
64. Vernois J., Abdel Ghani A. H. and Muxart R. *Radiochem. Radioanal. Lett.* **12,** 7 (1972).
65. Muxart R., Vernois J. and Strapelias H. *Radiochimica Acta* **21,** 88 (1974).
66. Welch M. J. *J. Am. Chem. Soc.* **92,** 408 (1970).
67. El-Garhy M. and Stöcklin G. *Radiochem. Radioanal. Lett.* **18,** 281 (1974).
68. Lambrecht R. M., Mantescu C., Redvanly C. S. and Wolf A. P. *J. nucl. Med.* **13,** 266 (1972).
69. Machulla H.-J., Shanshal A. and Stöcklin G. *Institut für Chemie,* KFA Jülich GmbH, Jülich, unpublished results.
70. Machulla H.-J. and Stöcklin G. *Quality Factors in Nuclear Medicine,* p. 144.1, Proc. 8th Int. Annual Meeting of the Society of Nuclear Medicine, Copenhagen (Fadl.s Forlag Copenhagen 1975).
71. Wong St. and Ache H. *Int. J. appl. Radiat. Isotopes* **27,** 19 (1976).
72. Elias H. *Proc. 9th Japan Conf. on Radioisotopes, Tokio,* p. 538 (1969).
73. Elias H., Arnold Ch. and Kloss G. *Int. J. appl. Radiat. Isotopes* **24,** 463 (1973).
74. Thakur M. L., Chauser B. M. and Hudson R. F. *Int. J. appl. Radiat. Isotopes* **26,** 319 (1975).
75. Ghandadian R., Waters S. L., Thakur M. L. and Chisholm G. D. *Int. J. appl. Radiat. Isotopes* **26,** 343 (1975).
76. Lambrecht R. M., Atkins H., Elias H., Fowler J. S., Lin S. S. and Wolf A. P. *J. nucl. Med.* **15,** 863 (1974).
77. Elias H. and Lotterhos F. *Chemische Berichte,* in press.
78. Robinson G. D., Jr. and Lee A. W. *J. nucl. Med.* **16,** 17 (1975).
79. Stöcklin G. *Proc. Symp. Preparation and Biomedical Application of Labelled Molecules, Venice,* EUR 2200.e Brussels, p. 481 (1964); cf. also Schmidt-Bleek F., Stöcklin G. and Herr W. *Angew Chem.* **72,** 778 (1960).
80. Elias H. *Advances in Chromatography,* Vol. 7, p. 243. Marcel Dekker Inc., New York (1968).
81. Karim H. M. A. Report Jül-1145-NC (1974).
82. Robinson G. D., Jr. *Proc. Symp. New Dev. Radiopharm. Labelled Comp., Copenhagen,* Vol. 1, p. 423. IAEA, Vienna (1973).
83. MacFarlane A. S. *Nature, Lond.* **182,** 53 (1958).
84. Hunter W. M. and Greenwood F. C. *Nature, Lond.* **194,** 495 (1962).
85. Rosa U., Scasselai G., Pennisi F., Riciccioni N., Giagnoni P. and Giordani R. *Biochim. Biophys. Acta* **86,** 519 (1964).
86. Marchalonis J. J. *Biochem. J.* **113,** 299 (1969).
87. MacFarlane A. S. *J. clin. Invest.* **42,** 346 (1963).
88. Krohn K. A. and Welch M. J. *Int. J. appl. Radiat. Isotopes* **25,** 315 (1974).
89. Nunez J., Pommier J., El Hilali and Roche J. *J. Labelled Compounds* **1,** 128 (1965).
90. Matkovics B., Rakonczay Z. and Rajki S. *Steroidologia* **2,** 77 (1971).
91. Morrison M. and Bayse G. *Biochemistry* **9,** 2995 (1970).
92. Regoeczi E. *Br. J. Haematol.* **20,** 649 (1971).
93. Knight L., Krohn K. A., Welch M. J., Spomer B. and Hager L. P. *Proc. Int. Symposium Radiopharmaceuticals.*
94. Sargent T., Abbo Kalbhen D., Shulgin A. T., Stauffer H. and Kusubov N. *J. nucl. Med.* **16,** 243 (1975).
95. Schumacher J. and Maier-Borst W. *Int. J. appl. Radiat. Isotopes* **23,** 519 (1972).
96. Uszler J. M., Bennett L. R., McDonald N. S. and Messa I. *J. nucl. Med.* **15,** 540 (1974).
97. De Nardo G. L., De Nardo S. J., Meyers J. and Krohn K. A. *J. nucl. Med.* **16,** 524 (1975).
98. Meyers J., Krohn K. A., Janshoff A. L. and De Nardo G. L. *J. nucl. Med.* **16,** 550 (1975).
99. Fowler J. S., McGregor R. R. and Wolf A. P. *J. Medicinical Chem,* in press.
100. Kirschner A. S. and Ice R. D. *J. nucl. Med.* **16,** 542 (1975).
101. De Nardo S. J., De Nardo G. L., O'Brien T., Peek N. F., Zielinski F. W. and Jungerman J. A. *J. nucl. Med.* **15,** 487 (1974).
102. De Nardo S. J., De Nardo G. L., Caretta R. F., Jansholt A. L., Krohn K. A. and Peek N. F. *J. nucl. Med.* **16,** 524 (1975).
103. Coleman R. E., Harwig S. S. L., Harwig J. T.,

SIEGEL B. A. and WELCH M. J. *J. nucl. Med.* **16,** 370 (1975).
104. STÜRNER K. H. and MACHULLA H.-J. *Abstract 14th Cong. Int. Soc. Blood Transfusion,* July 1975, Helsinki.
105. CHRISTY B., KING G. and SMOAK W. M. *J. nucl. Med.* **15,** 484 (1974).
106. KAVULA M. P., ROGGE J. D., ANGER R. T., JR., WELLMAN H. N. and FITZGERALD J. S. *J. nucl. Med.* **16,** 540 (1975).
107. SERAFINI A. N., HUPF H. B., LINDBERG D., SMOAK W. M. and GILSON A. J. *J. nucl. Med.* **16,** 567 (1975).
108. SERAFINI A. N., SMOAK W. M., HUPF H. B., BEAVER J. E., HOLDER J. and GILSON A. J. *J. nucl. Med.* **16,** 629 (1975).
109. LAMBRECHT R. M., MANTESCU C., ATKINS H. L. and WOLF A. P. *Trans. Am. nucl. Soc.* **15,** 131 (1972).
110. ARNOLD J. E. and PINSKY S. *J. nucl. Med.* **17,** 261 (1976).
111. COLEMAN R. E., HARWIG S. S. L., HARWIG J. T., SHERMAN L. A. and WELCH M. J. *J. nucl. Med.* **16,** 521 (1975).
112. COATES G., DE NARDO S. J., DE NARDO G. L. and TROY F. A. *J. nucl. Med.* **16,** 136 (1975).
113. WELLMAN H. N., BERKE R. A., ROBBINS P. J. and ANGER R. T. *J. nucl. Med.* **12,** 405 (1971).
114. KENNY R. W., ACKERY D. M., FLEMING J. S., GODDARD B. A. and GRANT R. W. *J. Radiol.* **48,** 481 (1975).
115. CEDERQUIST E., NAVERSTEN Y., PERSSON R. B. R., STRAND S. E. and SVENSSON L. *Quality Factors in Nuclear Medicine,* p. 166.1, Proc. 8th Int. Ann. Meeting Soc. Nucl. Med., Copenhagen (Fadl.s Forlag Copenhagen 1975).
116. BUSEMANN SOKOLE E. and VAN DER SCHOOF J. B. *Quality Factors in Nuclear Medicine,* p. 167.1, Proc. 8th Int. Ann. Meeting Soc. Nucl. Med. Copenhagen (Fadl.s Forlag Copenhagen 1975).
117. KNIGHT L. C., HARWIG S. S. L. and WELCH M. J. Abstract, 47th Session of the American Heart Association. *Nat. Conf. on Thrombosis and Hemostasis,* Dallas Nov. (1974).
118. SARGENT T., KALBHEN D. A., SHULGIN A. T., BRAUN G., STAUFFER H. and KUSUBOV N. *Neuropharmacol.* **14,** 165 (1975).
119. MYERS W. G. In *Recent Advances in Nuclear Medicine* (Edited by JOHN H. LAWRENCE), Vol. IV, p. 131. Grune & Stratton, New York (1974).
120. DROUET J., GOUTHERAUD R., AMOUCH P.-J., DE VERNEJOUL P., BARRITAULT L. and KELLERSOHN C. *J. Biol. Med. Nucl.* **4,** 21 (1969).
121. LAMBRECHT R. M., NORTON E., WOLF A. P. and TREVES S. 1st Int. Symp. Radiopharm. Chem., Brookhaven, Sept. 1976, Abstr. No. 76 in *J. Lab. Comp. Radiopharm.*, in press.
122. WEINREICH R., QAIM S. M. and STÖCKLIN G. 1st Int. Symp. Radiopharm. Chem., Brookhaven, Sept. 1976, Abstr. No. 66 in *J. Lab. Comp. Radiopharm.*, in press.
123. WEINREICH R., *Proc. Panel Disc. Iodine-123 in Western Europe,* Jülich, Febr. 1976, Report Jül-Conf. 20, p. 49 (1976).
124. VAN DEN BOSCH R., DE GOEIJ J. J. M., VAN DER HEIDE J. A., TERTOOLEN J. F. W., THEELEN H. M. J. and ZEGERS C. *Int. J. appl. Radiat. Isotopes,* in press.
125. FLEISCHER A. A. *Proc. Conf. Appl. Iodine-123 in Nucl. Med.*, Rockville 1975, HEW(FDA)76–8033, p. 44 (1976).
126. MICHEL F., MÜNZEL H., SCHULZ F. and SCHWEICKERT H. *Proc. Panel Disc. Iodine-123 in Western Europe,* Jülich, Febr. 1976, Rept. Jül-Conf. 20, p. 91 (1976).
127. BEAVER J. E. *Proc. Conf. App. Iodine-123 in Nucl. Med.*, Rockville 1975, Report HEW (FDA) 76-8033, p. 48 (1976).
128. ZIELINSKI F., ROBINSON G. D. JR., MACDONALD N. S., BOCKSRUCKER A. V., EASTON M. and LEE A. W. 1st Int. Symp. Radiopharm. Chem., Brookhaven, Sept. 1976, Abstr. No. 64 in *J. Lab. Comp. Radiopharm.*, in press.
129. CUNINGHAME J. D., MORRIS B., NICHOLS A. L. and TAYLOR N. K. *Int. J. appl. Radiat. Isotopes* **27,** 597 (1976).
130. GODART J., BARAT J. L., GIZON A. and BRIÈRE J. *Proc. Panel Disc. Iodine-123 in Western Europe,* Jülich, Febr. 1976, Report. Jül-Conf 20, p. 129 (1976).
131. LAMBRECHT R. M., WOLF A. P., HELUS F., UNTUCHT M. and MAIER-BORST W. *Int. J. appl. Radiat. Isotopes* **27,** 675 (1976).
132. WONG St. H., MUSTAKLEM J. F. and ACHE H. J. *Int. J. appl. Radiat. Isotopes* **27,** 379 (1976).
133. ROBINSON G. D. JR. and ZIELINSKI F. N. 1st Int. Symp. Radiopharm. Chem., Brookhaven, Sept. 1976, Abstract No. 76 in *J. Lab. Comp. Radiopharm.*, in press.
134. MACHULLA H.-J., KUPFERNAGEL CH. and STÖCKLIN G. *Proc. 9th Int. Ann. Meeting Soc. Nucl. Med.*, Berlin, Sept. 1976, cf. also Abstracts of the 1st Int. Symp. Radiopharm. Chem., Brookhaven, Sept. 1976 in *J. Lab. Comp. Radiopharm.*, in press.
135. BUTTERMANN G., WOLF I. H., HÖR G. and PABST H. W. Quality Factors in Nuclear Medicine, P. 9.1, *Proc. 8th Int. Annual Meeting of the Society of Nuclear Medicine,* Copenhagen (Fadl.s Forlag Copenhagen 1976).
136. BUTTERMANN G., WOLF I. H., HÖR G. and

Pabst H. W., *Proc. Panel Disc. Iodine-123 in Western Europe*, Jülich, Febr. 1976, Rept. Jül-Conf 20, p. 19 (1976).

137. Feinendegen L. E. *Proc. Panel Disc. Iodine-123 in Western Europe*, Jülich, Febr. 1976, Rept. Jül-Conf 20, p. 35 (1976).

138. DeNardo G. L., Krohn K. A., Jansholt A. L., DeNardo S. J., Lagunas Solar M., Jungerman J. A. *IAEA Symp. on Medical Radionuclide Imaging*, Los Angeles, U.S.A. (Oct. 1976).

139. Nozaki T., Fukushi K. and Irie T. 1st Int. Symp. Radiopharm. Chem., Brookhaven, Sept. 1976, Abstract No. 59 in *J. Lab. Comp. Radiopharm.*, in press.

140. Kojima M., Maeda M., Ogawa M., Nitta K. and Ito T. 1st Int. Symp. Radiopharm. Chem., Brookhaven, Sept. 1976, Abstract No. 60 in *J. Lab. Comp. Radiopharm.*, in press.

International Journal of Applied Radiation and Isotopes, 1977, Vol. 28, pp. 149–156. Pergamon Press. Printed in Northern Ireland

Synthesis of ^{123}I-16-iodo-9-Hexadecenoic Acid and Derivatives for Use as Myocardial Perfusion Imaging Agents*

G. D. ROBINSON, JR.†

Department of Radiological Sciences, Center for the Health Sciences, and Laboratory of Nuclear Medicine and Radiation Biology, University of California, Los Angeles, U.S.A.

(*Received* 6 *March* 1976)

Radioiodinated long chain ω-iodocarboxylic acids have been proposed for use as fatty acid analog tracer molecules. Fatty acids serve as fuel for myocardial metabolism and, since ^{131}I-16-iodo-9-hexadecenoic acid has been shown to simulate the heart uptake and turnover of oleic and stearic acids, the ^{123}I labeled compound has been suggested as a radiopharmaceutical for use in myocardial perfusion imaging.

^{123}I-16-iodo-9-hexadecenoic acid was synthesized by reacting the brominated precursor with ^{123}I-iodide in refluxing anhydrous acetone or MEK. ^{123}I-iodide was extracted into the reaction medium from aqueous 4 N NaOH. Single extraction transfer into acetone was more efficient than into MEK, but labeling yields were higher using MEK. In routine syntheses, using MEK as the solvent, 20 mCi of ^{123}I-16-iodo-9-hexadecenoic acid at a specific activity of 20 mCi/mg are produced.

The label of radioiodinated 16-iodo-9-hexadecenoic acid shows rapid metabolic turnover within the heart. This limits its usefulness for myocardial imaging when multiple projection images are required. The heart uptake and turnover of labeled 16-iodohexadecenoic acid were found to be indistinguishable from those of the unsaturated molecule.

INTRODUCTION

LONG CHAIN carboxylic acids are a significant nutrient source for myocardial metabolism.[1,2] A large fraction of the free fatty acids which enter the coronary circulation are taken up by the myocardium.[3–5] As a result, long chain carboxylic acids, labeled with a variety of gamma-emitting radionuclides, have been proposed as radiopharmaceuticals for imaging the regional distribution of myocardial perfusion.[5–11]

Initially, EVANS and GUNTON used ^{131}I labeled oleic acid as an agent for imaging myocardial infarcts in experimental animals and man.[6,7] Subsequently, other investigators have reported the use of ^{11}C, ^{18}F, ^{131}I and ^{99m}Tc labeled carboxylic acids for heart perfusion imaging.[4,5,8–11] These studies often had limited success due either to the poor biological behavior of the labeled molecule or to the unfavorable physical characteristics of the radionuclide.

While reinvestigating the use of radioiodinated fatty acids for myocardial perfusion imaging, we demonstrated that ICl addition to unsaturated molecules, such as oleic acid, dramatically reduces the level of their extraction by the myocardium.[4,5] This reduction is presumably caused by the alteration of structure which results when bulky iodine and chlorine atoms

*Work supported by ERDA Contract No. E(04-1) GEN-12 and USPHS (NHLI) Contract No. HV-12491.

†Address: G. D. Robinson, Jr., Ph.D. Nuclear Medicine Research Laboratory, 900 Veteran Avenue, Los Angeles, CA 90024, U.S.A.

are added to the molecule. Our observations were subsequently confirmed by BEIERWALTES *et al.*[10] In contrast, labeling of ω-bromocarboxylic acids by interhalogen replacement with radioiodide produces molecules which are structural analogs of naturally occurring carboxylic acids.[8] In these molecules, the iodine atom occupies the position and approximate volume of the terminal methyl group in the corresponding C_{n+1} "physiologic" molecule. Such compounds were shown to retain the myocardial extraction characteristics of the parent molecules.[4,5] An analog of 9-heptadecenoic acid, 16-iodo-9-hexadecenoic acid, has been evaluated extensively in experimental animals.[12–14]

The preparation, *in vitro*, and *in vivo* behavior of ^{131}I-16-iodo-9-hexadecenoic acid have been described in detail.[4,5,8,14,15] After intravenous administration the labeled compound is promptly taken up by the heart, with a regional distribution pattern which is essentially identical to that of ^{43}K.[14,15] The myocardial clearance half-time is approximately 20 min, which is similar to those of ^{11}C labeled stearic and oleic acids.[15] Since ^{123}I is acknowledged as an attractive radionuclide for use with scintillation camera imaging systems,[16–18] a method for routine production of large quantities (> 20 mCi) of high specific activity ^{123}I-16-iodo-9-hexadecenoic acid was developed. This agent is now being evaluated for use in external quantification of relative regional myocardial blood flow in man.

The major apparent limitation to use of this labeled compound for such quantitative studies is its rapid metabolic turnover within the myocardium, resulting in liberation of radioiodide.[15] Small variations in the molecular structure of fatty acids can produce changes in myocardial specificity and/or metabolism.[19,20] Variation in degree of unsaturation, increase in carbon atom chain length ($\geq C_{20}$), or inclusion of non-radioactive substituents, such as fluorine atoms, may be used to alter the rate of metabolic degradation of the analog compounds without loss of myocardial specificity. With this possibility in mind, more general synthetic routes to radioiodine labeled ω-iodocarboxylic acids, their derivatives, and their unlabeled precursors are being devised.

MATERIALS AND METHODS

16-*Bromo-9-hexadecenoic acid*

16-Bromo-9-hexadecenoic acid (Aldrich Chemical Co.; #B6,820–8; m.p. 35–40°C) was obtained as either a clear viscous liquid or a white solid. Gas chromatographic analysis of the methyl esters indicated that the solid material was slightly more impure than the liquid. However, with regard to labeling and/or biological behavior of the labeled product, no significant differences were noted.

16-*Bromohexadecanoic acid*

16-Bromohexadecanoic acid was synthesized from 16-bromo-9-hexadecenoic acid by hydrogenation in the presence of a nickel boride catalyst. The catalyst was prepared by reaction of sodium borohydride with aqueous nickel acetate.[21]

Four gm of 16-bromo-9-hexadecenoic acid were dissolved in 50 ml of ethanol containing 0.35 gm of freshly prepared NiB catalyst. The hydrogenation proceded smoothly while H_2 was passed through the stirred solution for 1.5 hr. The catalyst was filtered from the reaction mixture and ethanol in the filtrate was removed by evaporation. The organic residue was dissolved in CCl_4 and bromine was added to convert unreacted 16-bromo-9-hexadecenoic acid to the tribromide. The brominated CCl_4 solution was extracted with 10% aqueous $Na_2S_2O_3$, and the organic layer was isolated and dried with $MgSO_4$. Carbon tetrachloride was removed by evaporation, the residue was redissolved in a minimum volume of acetone and the resulting solution was cooled overnight at 4°C. The crude 16-bromohexadecanoic acid which precipitated from the cold acetone was isolated and purified 3 times by reprecipitation from ethanol by addition of water. The m.p. of the white product was 64–65°C, and NMR spectroscopy verified the absence of ethylenic hydrogens.

^{123}I-16-*iodo-9-hexadecenoic acid*

^{123}I-16-iodo-9-hexadecenoic acid was prepared by interhalogen replacement of the bromine atom of 16-bromo-9-hexadecenoic acid by ^{123}I-iodide in refluxing anhydrous 2-butanone (MEK).

^{123}I was produced by the ^{124}Te$(p, 2n)^{123}$I

reaction using 22 MeV protons from the UCLA CS-22 biomedical cyclotron. The irradiated ^{124}Te (96% enriched) targets were dissolved in acidic hydrogen peroxide, ^{123}I was distilled into dilute NaOH, and the distillate was concentrated to one ml. After one ml of 8 N NaOH was added, ^{123}I-iodide was extracted from the resulting 4 N NaOH solution with seven successive 3 ml portions of freshly distilled MEK. The combined MEK extracts, containing 85–95% of the total isolated ^{123}I, were dried by shaking them with one gm of $CaSO_4$ and the volume of the dried MEK solution was reduced to 2 ml by gentle heating.

The concentrated ^{123}I/MEK solution was placed in a 10 ml round bottom flask and a reflux condenser was attached. One milligram of 16-bromo-9-hexadecenoic acid was added and the labeling reaction proceeded in the refluxing solution. After 90 min a labeling yield of 75–85%, as determined by TLC,[8] was usually achieved. The MEK was then removed with low heat under a N_2 gas stream and the residue was taken up in one ml of 25% human serum albumin (HSA). ^{123}I-iodide was removed by passing the HSA solution through a 1 cm diameter × 3 cm anion exchange column ("Dowex 1-8X," chloride form) eluted with 3 ml of 0.9% NaCl. Terminal sterilization was by membrane filtration (0.22 μm).

^{131}I-ω-*Iodocarboxylic acids*

For preliminary studies of the myocardial extraction and/or turnover of radioiodine labeled ω-iodocarboxylic acids, the compounds were labeled with ^{131}I in a manner similar to that reported previously.[8] Two milliliters of freshly distilled MEK containing the desired quantity of ^{131}I-iodide and one mg of the appropriate ω-bromocarboxylic acid were placed in a 10 ml round bottom flask. A reflux condenser was attached to the flask and labeling proceded under MEK reflux. Radiochemical yields, measured as described previously,[8] of 80–90% were achieved within 2 hr. After being allowed to cool, the flask was removed from the condenser, capped with a ground glass stopper, and stored at 4°C. Aliquots of the reaction solution were removed and processed for use as needed. Evaporation of MEK, dissolution in aqueous albumin (human or canine), removal of ^{131}I-iodide, and terminal sterilization were as described for ^{123}I-16-iodo-9-hexadecenoic acid.

RESULTS AND DISCUSSION

16-iodo-9-hexadecenoic acid was chosen for preliminary biological studies of the ω-iodocarboxylic acids because of the high myocardial extraction of fatty acids and because the unlabeled precursor, 16-bromo-9-hexadecenoic acid, was commercially available. 11-Iodo undecanoic acid and 6-iodohexanoic acid were also synthesized[8] but the heart uptakes of these medium chain length molecules was not expected to be as great as for the 16 carbon analog compound. Subsequently it was discovered that hydrolytic liberation of radioiodide from ω-iodocarboxylic acids in 6% HSA solution is inversely related to carbon atom chain length.[8] Only 16-iodo-9-hexadecenoic acid was found to be stable enough to be used in aqueous solution.

Our initial work on preparation of ^{131}I-16-iodo-9-hexadecenoic acid used 10 mg of the brominated precursor and 10 ml of refluxing acetone as the labeling medium.[8] For clinical applications using the ^{123}I-labeled molecule, the potential toxicity of the early preparations was of concern. Beta-oxidation is a primary metabolic route for free long chain fatty acids in animals and man.[22] Total metabolism of a 10 mg preparation by beta-oxidation would ultimately produce the equivalent of 4.2 mg of bromoacetate. This is because the labeled product consists almost entirely of 16-bromo-9-hexadecenoic acid on a mass basis. The LD_{50} of bromoacetate, which is a potent enzyme inhibitor, is $\simeq$ 50 mg/kg in a variety of animals.[23] When ^{123}I-16-iodo-9-hexadecenoic acid is used in humans, the highest specific activity is required to minimize the amount of the potentially toxic ω-bromocarboxylic acid which is administered. Simply substituting 1 mg of the precursor in 10 ml of solvent gave unacceptably low labeling yields with 90 min of refluxing. Higher yields, similar to those obtained in our earlier ^{131}I work, were achieved using 1 mg of 16-bromo-9-hexadecenoic acid in a 2 ml volume of MEK. These results are illustrated in Fig. 1. The result was an effective 10 fold increase in the specific activity of the labeled product.

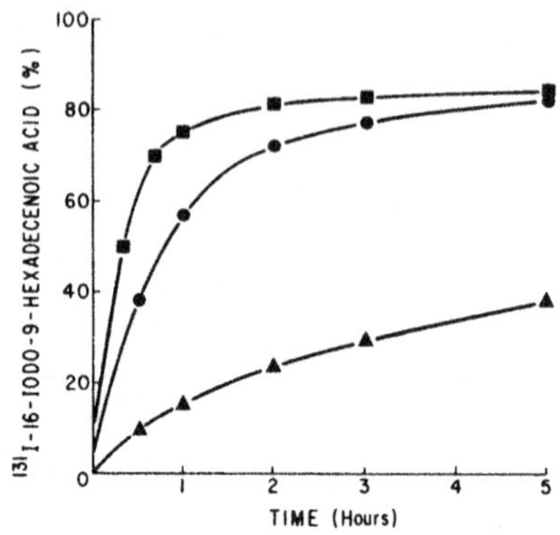

FIG. 1. Radiochemical yield of ^{131}I-16-iodo-9-hexadecenoic acid as a function of time of refluxing ^{131}I-iodide with (●) 10 mg of 16-bromo-9-hexadecenoic acid in 10 ml of freshly distilled, anhydrous MEK, (▲) 1 mg in 10 ml, and (■) 1 mg in 2 ml.

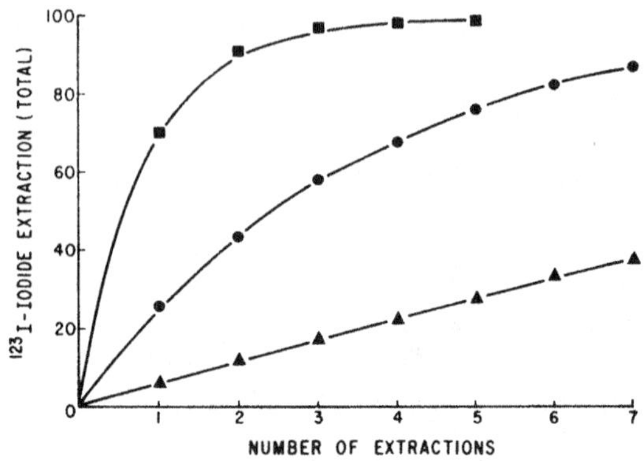

FIG. 2. Cumulative extraction of ^{123}I-iodide from 2 ml of 4 N NaOH into successive 3 ml aliquots of (■) freshly distilled acetone, (●) freshly distilled MEK, and (▲) undistilled MEK. The latter results varied, but those shown are representative.

Since labeling must take place in an anhydrous solvent, ^{123}I-iodide was extracted from aqueous base into the desired organic reaction medium. Details of our method for production of ^{123}I-iodide will be published in a separate report. Using 2 ml of 4 N NaOH, a single extraction with 3 ml of acetone gave 60–80% transfer into the organic phase. After 2 or 3 extractions, quantitative overall extraction yields were usually obtained. Using freshly distilled MEK, transfer was only 25–30% per extraction. As many as 7 extractions were required reliably to achieve 85–95% overall transfer of ^{123}I-iodide into MEK. If freshly distilled MEK was not used, results were variable but could be as low as 2–3% per extraction. These results are shown in Fig. 2.

As was previously reported, the labeling yield of ^{131}I-16-iodo-9-hexadecenoic acid in refluxing acetone was 60% after 90 min and 75% after 4 hr.[8] Using the higher specific activity method described here, similar values were obtained for the ^{123}I-labeled compound. In refluxing MEK, the yield after 90 min was 75–85%. These results are shown in Fig. 3. The higher 90 min yields in the MEK system are presumably the result of the higher reaction temperature because of the higher boiling point of MEK (80°C for MEK compared with 56° for acetone). The overall radiochemical yield after refluxing for 90 min in acetone was approximately 60%. ([Extraction Yield][Labeling Yield] = [1.00][0.60] = 0.60.) With MEK for the same refluxing period, an overall yield of over 70% ([0.90][0.80] = 0.72) was achieved. The latter route was chosen for routine production runs because of the higher mCi yields and greater specific activities which result. Subsequent processing into injectable form is identical for both methods. Twenty mCi of ^{123}I-16-iodo-9-hexadecenoic acid in isotonic 6% HSA solution are produced from a target containing 50 mCi of ^{123}I at EOB. The specific activity of the sterile, pyrogen-free product, based on total fatty acid content, is 20 mCi/mg. The time required from

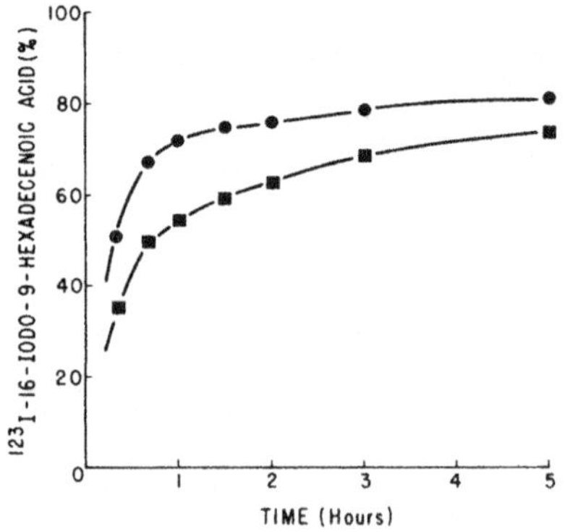

FIG. 3. Radiochemical yield of ^{123}I-16-iodo-9-hexadecenoic acid synthesized from 1 mg of 16-bromo-9-hexadecenoic acid and ^{123}I-iodide as a function of time of refluxing in 2 ml of anhydrous, freshly distilled (■) acetone and (●) MEK.

target processing through preparation of the final dosage form is 5 hr.

We have recently been investigating the use of sealed ampoules to rapidly achieve higher yields by labeling at temperatures above the solvent boiling point. Preliminary results indicate that, using quantities identical to those described for the refluxing systems, a 90% yield of ^{131}I-16-iodo-9-hexadecenoic acid can be obtained after 10 min at 120°C with MEK in a flame sealed 3 ml glass ampoule. For longer heating times, yields decrease as a result of thermal decomposition. Results from our preliminary studies are shown in Fig. 4. If this method becomes routine, the higher extraction yields of acetone might be combined with the high yield, rapid labeling in the closed system. Overall preparation time for ^{123}I-16-iodo-9-hexadecenoic acid could then be reduced to 3–4 hr.

The literature contains conflicting reports regarding the relative extractions of saturated and unsaturated long chain fatty acids by the heart. Some investigators have found little difference in the differential concentrations of saturated and unsaturated fatty acids before and after passage through the coronary circulation.(24,25) WILLEBRANDS, however, found that the percentage difference in oleic acid concentrations in arterial compared with venous coronary blood samples was nearly twice that of stearic acid;(3) thus implying a factor of two difference

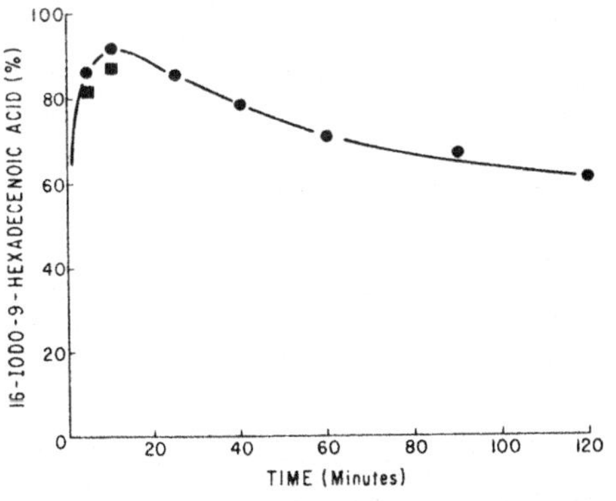

FIG. 4. Radiochemical yield of labeled 16-iodo-9-hexadecenoic acid synthesized from 1 mg of 16-bromo-9-hexadecenoic acid and radioiodide as a function of time at 120°C in 2 ml of anhydrous, freshly distilled MEK in a sealed glass ampoule: (●) ^{131}I and (■) ^{123}I.

in myocardial extraction for the two compounds. In an effort to study the role of unsaturation on the myocardial extraction of ω-iodocarboxylic acids, ^{131}I-16-iodohexadecanoic acid was synthesized. Nickel boride was chosen as the catalyst for preparation of the required 16-bromohexadecanoic acid because it had been used for hydrogenation of a variety of compounds under relatively mild conditions.(26) After hydrogenation of 4 g of the unsaturated brominated compound, 0.35 g of saturated product was isolated. Although the overall synthetic yield was low (≃9%), it was adequate for preliminary studies of labeling and heart uptake. Radiochemical yields, were similar to those for the unsaturated molecule. Using a collimated NaI(Tl) detector and direct intracoronary administration of the labeled compounds,(5) the canine myocardial extractions of 16-iodo-9-hexadecenoic acid and 16-iodo-hexadecanoic acid were found to be $77 \pm 11\%$ and $70 \pm 7\%$, respectively. For these molecules, the presence of a double bond in the carbon atom chain appears to be of little consequence with regard to myocardial extraction. Details of our biological comparisons of these compounds will be published in the near future.

The rapid turnover of 16-iodo-9-hexadecenoic acid within the heart is a potential limitation to use of the ^{123}I-labeled compound for multiple projection imaging of regional myocardial blood flow. As is shown in Fig. 5, the high quality images which can be obtained initially degrade severely within 20 min post injection, and loss of detailed perfusion information results. The residence time of the label within the heart may be prolonged by structural modification of the labeled fatty acid. As we have shown with ICl-labeled oleic, linoleic and linolenic acids, some alterations lead to reduction in myocardial uptake.(5) However, other modifications, such as substitution of a fluorine atom for a hydrogen atom attached along the carbon atom skeleton, can lead to initial retention of a more "physiologic" biological behavior, with subsequent alteration of metabolism.(27) Total length of the carbon atom chain can affect the rate of metabolic turnover of fatty acids within the heart. Erucic acid, which is a naturally occurring, unsaturated C_{22} fatty acid, is concentrated within the myocardium but is metabolized very slowly.(28–30) Since liberation of radioiodide from long chain ω-iodocarboxylic acids is the result of active metabolism, rather than simple hydrolysis,[8] use of labeled compounds longer than C_{16} may result in slower turnover. With these possibilities in mind, we are currently investigating some structure-activity relationships which may affect myocardial uptake and turnover of ω-iodocarboxylic acids.

The rapid clearance of activity from the heart may, in fact, be an advantage of applying ^{123}I-16-iodo-9-hexadecenoic acid and similar molecules to biological studies. We have demonstrated the feasibility of using such compounds for rapid sequential imaging to evaluate the effects of surgical and/or drug intervention upon regional myocardial blood flow.[31] In such studies, where multiple sequential single projection images are used, rapid clearance is essential. In addition, since the myocardial clearance is related to metabolism, it may be possible to obtain information concerning the metabolic status or fate of myocardial tissue by analysis of regional clearance rates. This approach is not practical with ionic tracers such as ^{43}K, ^{129}Cs or ^{201}Tl.

Acknowledgement—The author wishes to acknowledge the technical assistance furnished by A. W. LEE during initial studies of the labeling of ω-iodocarboxylic acids and by M. P. EASTON during synthesis of 16-bromohexadecanoic acid, synthesis of labeled 16-iodohexadecanoic acid and routine preparations of ^{123}I-16-iodo-9-hexadecenoic acid. F. W. ZIELINSKI developed our current method for isolation of ^{123}I-iodide in anhydrous solvents. Biological evaluations and imaging were performed in collaboration with N. D. POE.

REFERENCES

1. BLAIN J. M., SCHAFER H., SIEGEL A. L. and BING R. J. *Am. J. Med.* **20,** 820 (1956).
2. BALLARD F. B., DANFORTH W. H., NAEGLE S. and BING R. J. *J. Clin. Invest.* **39,** 717 (1960).
3. WILLEBRANDS A. F. *Clin. Chim. Acta.* **10,** 435 (1964).
4. POE N. D., ROBINSON G. D. and MACDONALD N. S. *J. nucl. Med.* **14,** 440 (1973).
5. POE N. D., ROBINSON G. D. and MACDONALD N. S. *Proc. Soc. Exp. Biol. Med.* **148,** 215 (1975).
6. EVANS J. R., GUNTON R. W., BAKER R. G., BEANLANDS D. S. and SPEARS J. C. *Circ. Res.* **16,** 1 (1965).
7. GUNTON R. W., EVANS J. R., BAKER R. G., SPEARS J. C. and BEANLANDS D. S. *Am. J. Cardiol.* **16,** 482 (1965).
8. ROBINSON G. D. and LEE A. W. *J. nucl. Med.* **16,** 17 (1975).
9. BONTE J. F., GRAHAM K. D., MOORE J. G., PARKEY R. W. and CARRY G. C. *J. nucl. Med.* **14,** 381 (1973).
10. BEIERWALTES W. H., ICE R. D., SHAW M. J. and RYO U. Y. *J. nucl. Med.* **16,** 842 (1975).
11. WEISS E. S., HOFFMAN E. J., AHMED S. A., PHELPS M. E., WELCH M. J., TER-POGOSSIAN M. M. and SOBEL B. E. *Circ.* **II-52** (1975).
12. POE N. D., ROBINSON G. D., MACDONALD N. S., LEE A. W. and SELIN C. S. *J. nucl. Med.* **15,** 524 (1974).
13. ROBINSON G. D., POE N. D., LEE A. W. and SELIN C. S. *J. nucl. Med.* **15,** 528 (1974).
14. POE N. D., ROBINSON G. D., GRAHAM L. S. and MACDONALD N. S. *J. nucl. Med.* (submitted).
15. POE N. D., ROBINSON G. D. and ZIELINSKI F. W. *J. nucl. Med.* (1976) Abstract.
16. SODD V. J. In *Radiopharmaceuticals* (Edited by SUBRAMANIAN G., RHODES B. A., COOPER J. F. *et al.*), p. 125. Society of Nuclear Medicine, New York (1975).
17. MYERS W. G. In *Radioactive Pharmaceuticals* (Edited by ANDREWS G. A., KINSELEY R. M., WAGNER H. N.) p. 217. AEC Symposium Series 6, CONF-651111, Springfield, VA. National Bureau of Standards (1966).
18. RHODES B. A., WAGNER H. N. and GERRARD M. *Isot. Rad. Tech.* **4,** 275 (1967).
19. HARPER H. A. In *Review of Physiological Chemistry*, p. 189. Lange Medical Publications, Los Altos, CA (1957).
20. HOLUM J. R. In *Elements of General and Biological Chemistry*, p. 263. Wiley, New York (1962).
21. BROWN C. A. *J. org. Chem.* **35,** 1900 (1970).
22. BALDWIN E. In *The Nature of Biochemistry*, p. 73. University Press, Cambridge (1962).
23. WEBB L. J. In *Enzyme and Metabolic Inhibitors*, Vol. 3, p. 1. Academic Press, New York (1966).
24. CARLSTEN A., HALLGREN B., JAGENBURG R., SVANBORG A. and WERKO L. *Scan. J. Clin. Lab. Invest.* **13,** 418 (1961).
25. STEIN O. and STEIN Y. *Biochim. Biophys. Acta.* **70,** 517 (1963).
26. BROWN C. A. and BROWN H. C. *J. Org. Chem.* **31,** 3989 (1966).
27. SMITH F. A. *Chem. Tech.* **3,** 422 (1973).
28. BEARE-ROGERS J. L. and NERA E. A. *Lipids* **9,** 365 (1974).
29. CHRISTOPHERSON B. O. and BREMER *J. Biochem. Biophys. Acta.* **280,** 506 (1972).
30. HOUTSMULLER U. M. T., STRUIJK C. B. and VAN DER BEEK A. *Biochem. Biophys. Acta* **218,** 564 (1970).
31. POE N. D. and ROBINSON G. D. (to be published).

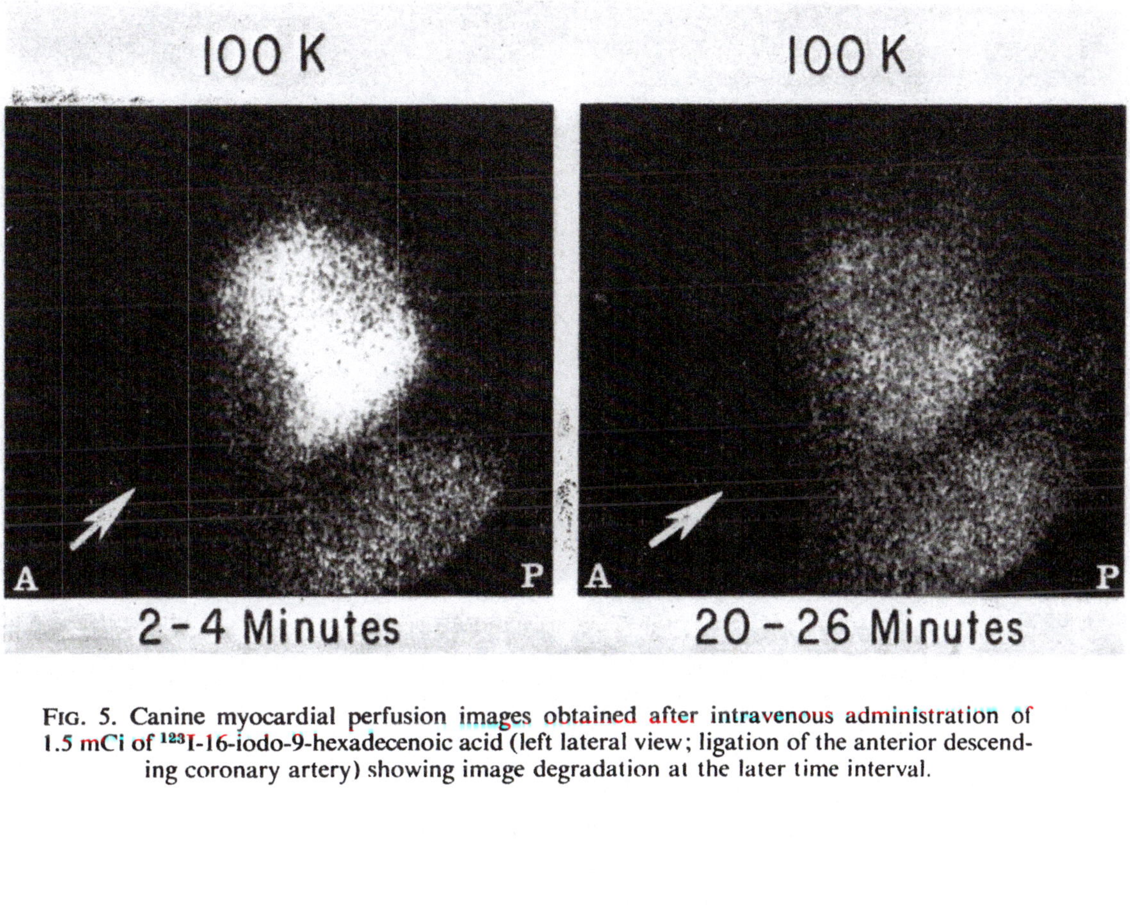

FIG. 5. Canine myocardial perfusion images obtained after intravenous administration of 1.5 mCi of ^{123}I-16-iodo-9-hexadecenoic acid (left lateral view; ligation of the anterior descending coronary artery) showing image degradation at the later time interval.

International Journal of Applied Radiation and Isotopes, 1976, Vol. 28. pp. 157–162. Pergamon Press. Printed in Northern Ireland

^{123}I-Labeled Soluble Fibrin: Preparation and Comparison with Other Thrombus Imaging Agents*

JOHN F. HARWIG,† SYLVIA S. L. HARWIG, JOHN O. EICHLING, R. EDWARD COLEMAN and MICHAEL J. WELCH

Edward Mallinckrodt Institute of Radiology, Washington University School of Medicine, 510 South Kingshighway Blvd., St. Louis, MO 63110, U.S.A.

(*Received* 5 *April* 1976)

Radioiodinated soluble fibrin, a derivative of radioiodinated fibrinogen, has been investigated for thrombus localization. Soluble fibrin can be labeled with ^{123}I as the radionuclide for imaging thrombi with a scintillation camera. Protein iodination grade ^{123}I is produced by a distillation and extraction procedure from the product of the ^{121}Sb (α, $2n$) ^{123}I reaction in the Washington University 54-in. cyclotron. Labeling with these ^{123}I preparations is effected by the iodine monochloride method. Induced 4–6-hr-old femoral vein thrombi in dogs can be visualized beginning 2 hr after injection of 1 mCi of ^{123}I-labeled soluble fibrin. This new radiopharmaceutical may be useful for imaging actively-propagating thrombi in patients, especially in areas of large blood pool. Its potential appears to be similar to that of ^{123}I-labeled highly iodinated fibrinogen, another new thrombus imaging agent.

INTRODUCTION

RADIOIODINATED soluble fibrin, a protein derived from fibrinogen by the action of thrombin, has been investigated as a thrombus localizing agent.[1] These studies, which were performed with ^{131}I as the radionuclide, indicated that soluble fibrin is incorporated into fresh thrombi in dogs to the same extent as conventional radioiodinated fibrinogen. Soluble fibrin, however, clears considerably faster from the blood, producing significantly higher thrombus: blood activity ratios than obtained with fibrinogen. The thrombus: blood ratios are similar to those obtained with highly iodinated fibrinogen, another new agent evaluated for thrombus detection.[2] Radiolabeled soluble fibrin may be more useful than labeled fibrinogen for thrombus detection in areas of large blood pool and may allow scintiscanning to be performed sooner after administration of the radiopharmaceutical. These studies have now been extended to soluble fibrin labeled with ^{123}I, an isotope whose physical properties are more compatible with a scintillation camera for short-term imaging procedures in nuclear medicine. This paper reports the preparation of ^{123}I-labeled soluble fibrin and its application to imaging recently-induced thrombi in dogs.

MATERIALS AND METHODS

Sodium iodide-^{123}I suitable for protein iodination was produced by modification of the procedure previously reported.[3] High-purity (99.999%) antimony shot of natural isotopic abundance* was fused in a mold to form an oval slug which was soft-soldered to a copper target-backing plate. The antimony target was bombarded with a 20-μA beam of 26-MeV alpha particles in the Washington University 54-in. cyclotron. Following bombardment the target was milled (Fig. 1) to remove a 0.015-in. layer of antimony.

The irradiated antimony scrapings were placed along with 25 ml concentrated sulfuric acid in the still pot of a distillation apparatus (Fig. 2). The mixture was heated to dissolve the antimony, then allowed to cool briefly before addition of sodium iodide carrier (25 μg I^-). Twenty ml

*Supported by NIH SCOR in Thrombosis 1 P17 HL-14147.

†Address all reprint requests to this author at address above.

*Alfa-Ventron Corp., Danvers, MA.

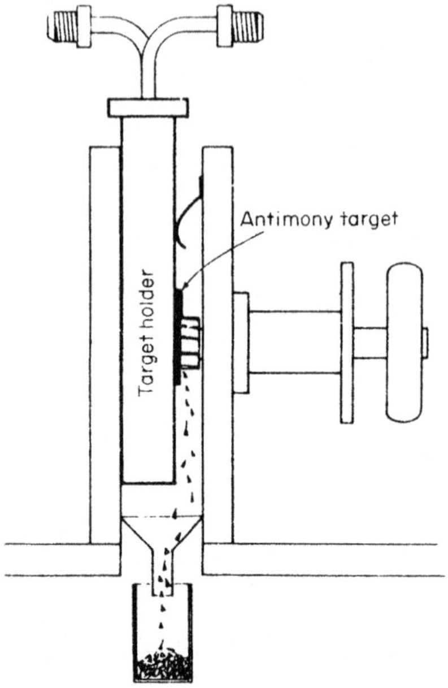

FIG. 1. Device for milling the surface of the antimony target following cyclotron bombardment.

0.5 M sodium nitrite was then added slowly, after which the solution was heated to boiling. The distillate (5 ml) was combined with the solution from the scrubber (Fig. 2) and extracted with 2 portions (5 ml each) of carbon tetrachloride. The combined light pink CCl_4 solution was back-extracted with 2 ml 0.1 N sodium hydroxide. The NaOH extract was evaporated to dryness on a rotary evaporator at 50°C *in vacuo*. The residue was reconstituted and neutralized with 200 μl 1N HCl-2M NaCl.

Canine fibrinogen, isolated from pooled canine plasma as the Blombäck 1-2 fraction, was iodinated with these ^{123}I preparations by the iodine monochloride method.[4,5] The ^{123}I-labeled fibrinogen was then treated as previously described[1] to form ^{123}I-labeled soluble fibrin. The radionuclidic purity of the ^{123}I-labeled soluble fibrin at the time of injection was determined by gamma spectroscopy as previously reported.[3] Radiation dosimetry estimates for man of ^{123}I-labeled soluble fibrin and ^{123}I-labeled highly iodinated fibrinogen were calculated by the absorbed fraction method[6] and utilized tissue distribution data from rabbits. The extrapolation from rabbit to man was achieved by normalizing the tissue concentration data by the ratio of whole body weights.[7]

A thrombus was induced by intimal injury in a femoral vein of each of 6 dogs.[8] One mCi of ^{123}I-labeled soluble fibrin was injected intravenously 4 hr after thrombus induction in 4 of the dogs and 5 and 6 hr after induction in 1 dog each. Each dose of labeled soluble fibrin was prepared from an individual cyclotron run and individual isolation and labeling procedures. Imaging of both hind legs, as well as the abdomen and thorax, was performed with a scintillation camera and mid-energy collimator beginning 2 hr after injection. Eighteen to 24 hr later a venogram was obtained to demonstrate the location of the thrombus.

RESULTS

Yields of ^{123}I with the new target system reported here range from 500–600 μCi/μAhr. The modified ^{123}I distillation and purification procedure requires about 3 hr, with at least 80% recovery corrected for decay. Iodination of fibrinogen with these ^{123}I preparations by the ICl method proceeds with an efficiency of 50–60%, and 60–70% of the activity which initially binds to the fibrinogen remains protein-bound during the preparation of soluble fibrin. These latter two steps require a total of about 1 hr. The physicochemical properties and *in vivo* clearance behavior of radioiodinated soluble

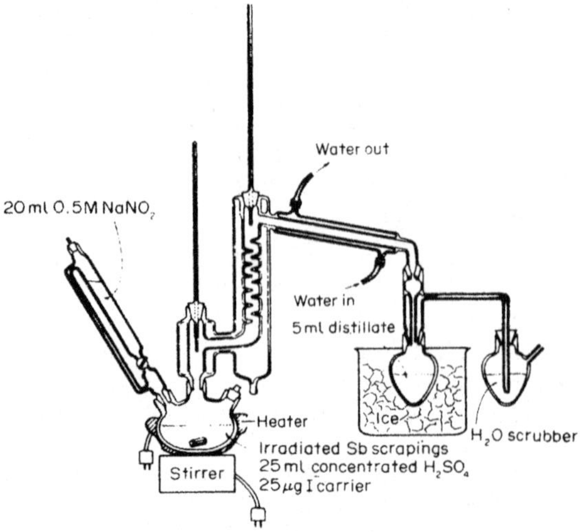

FIG. 2. Distillation apparatus for isolating the ^{123}I from the irradiated antimony scrapings.

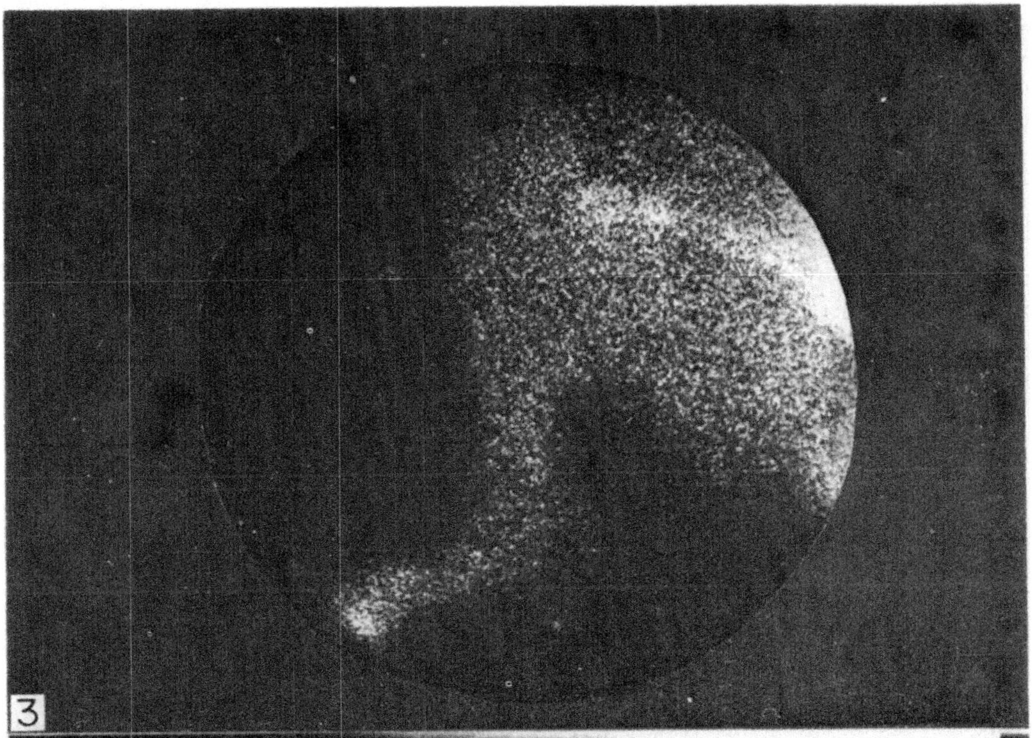

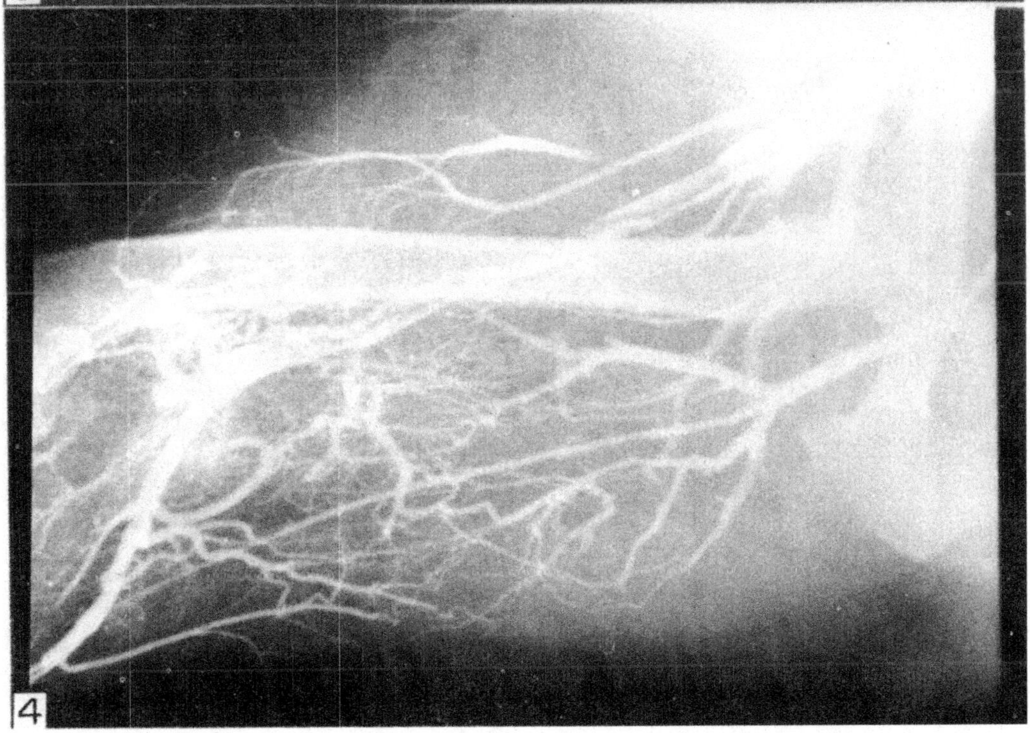

FIG. 3. Scintigram obtained 2 hr after injection of 1 mCi ^{123}I-labeled soluble fibrin in a dog with a 4-hr-old thrombus induced in the right femoral vein. The large thrombus is clearly visualized.

FIG. 4. Venogram of right thigh of same dog as in Fig. 2. The thrombus is seen as a large filling defect in the femoral vein.

fibrin have been extensively studied.[1] The radiation dose to man from an intravenous injection of ^{123}I-labeled soluble fibrin, based on the observed radionuclidic purity of 98% ^{123}I and 2% ^{124}I, was determined to be 22 mrad/mCi whole body and 64 mrad/mCi critical organ (liver). If radioisotopically pure ^{123}I were used these dose estimates would be reduced to 19 mrad/mCi whole body and 57 mrad/mCi critical organ. Dosimetry of ^{123}I-labeled highly iodinated fibrinogen, which has not been previously reported, was computed to be 25 mrad/mCi whole body and 122 mrad/mCi critical organ (blood) for 98% ^{123}I–2% ^{124}I, and 23 mrad/mCi whole body and 99 mrad/mCi critical organ for pure ^{123}I.

In each of the 6 dogs the thrombus was clearly visualized with the scintillation camera and documented by venography. Figure 3 is a typical scintigram of a 4-hr-old thrombus, while the corresponding venogram in Fig. 4 confirms the size and position of the thrombus. Comparable scintigrams were obtained with the 5-hr-old and 6-hr-old thrombi. The other area of localized activity in Fig. 3 is the bladder. In the remainder of the body only the liver showed significant accumulation.

DISCUSSION

The modified ^{123}I production and isolation procedure described here is faster and more efficient than the method previously reported.[3] Although less carrier iodide is added in the distillation step to maintain a lower ratio of iodide: protein in the subsequent labeling reaction, there is no adverse effect on ^{123}I recovery. Labeling of fibrinogen with these ^{123}I preparations is rapid and simple and yields a product suitable for use in preparing soluble fibrin. Although the overall preparation time of 4 hr is not prohibitive, it could be reduced by 75% if protein iodination grade ^{123}I were routinely available commercially.

The whole body radiation dose resulting from ^{123}I-labeled soluble fibrin is very low, due to its very rapid biologic clearance and the good physical properties of ^{123}I. Although significant liver accumulation occurs, as might be expected from a protein which probably consists of various-sized oligomers of monomeric fibrin,[9] the dose to this organ is still quite low. Radiation dosimetry of ^{123}I-labeled soluble fibrin is lower than ^{123}I-labeled highly iodinated fibrinogen and presumably lower than ^{123}I-labeled fibrinogen, another agent which has been used for thrombus imaging.[10,11] For human use, administration of up to 5 mCi of ^{123}I-labeled highly iodinated fibrinogen and up to 10 mCi of ^{123}I-labeled soluble fibrin seems reasonable to provide a high count rate without undue radiation exposure. The 2% ^{124}I in these ^{123}I preparations increases the dose only slightly, due to the rapid clearance of the labeled protein. Therefore, labeling with high purity ^{123}I would reduce high energy scatter but would not significantly affect radiation dose.

In four of the dogs the ^{123}I-labeled soluble fibrin was administered 4 hr after thrombus induction. In agreement with previous studies showing maximum thrombus: blood ratios with soluble fibrin in 4-hr-old thrombi,[1] the present thrombi were clearly delineated in the scintigrams (Fig. 3). These results suggest that ^{123}I-labeled soluble fibrin may be an ideal agent for imaging thrombi which are actively propagating. The 5-hr- and 6-hr-old thrombi were equally well visualized in the present study. The images of the 6-hr thrombus were comparable to those of 6-hr thrombi obtained previously with ^{123}I-labeled highly iodinated fibrinogen.[3] Radioiodinated soluble fibrin has been shown to produce higher thrombus : blood ratios than conventional radioiodinated fibrinogen in thrombi as old as 24 hr.[1] The usefulness of soluble fibrin may be limited, however, in the case of older thrombi in which net propagation has slowed, since the rate of blood clearance of soluble fibrin may be too fast to allow sufficient accumulation of these thrombi. The relationship between thrombus age in this animal model and thrombus age in patients in a clinical situation remains to be assessed. The efficacy of ^{123}I-labeled soluble fibrin for imaging arterial thrombi also remains to be determined.

Acknowledgements—The authors wish to express their appreciation to the staff of the Washington University cyclotron, especially Mr. Roland Head who designed and constructed the antimony target milling machine. The authors also thank Mr. Robert Feldhaus for assistance with the animal studies, Mr. James Frost for determining the gamma spectra, and Mr. Julius Hecht for preparing the illustrations.

REFERENCES

1. Coleman R. E., Harwig S. S. L., Harwig J. F., Sherman L. A. and Welch M. J. *Circ. Res.* **37**, 35 (1975).
2. Harwig J. F., Coleman R. E., Harwig S. S. L., Sherman L. A., Siegel B. A. and Welch M. J. *J. nucl. Med.* **16**, 756 (1975).
3. Harwig J. F., Welch M. J. and Coleman R. E. *J. nucl. Med.* **17**, 397 (1976).
4. McFarlane A. S. *Nature, Lond.* **182**, 53 (1958).
5. Harwig S. S. L., Harwig J. F., Coleman R. E. and Welch M. J. *Thromb. Res.* **6**, 375 (1975)
6. Loevinger R. and Berman M. *J. nucl. Med.* Suppl. No. 1, 7 (1968).
7. Kirschner A. S., Ice R. D. and Beierwaltes W. H. *J. nucl. Med.* **16**, 248 (1975).
8. Coleman R. E., Harwig S. S. L., Harwig J. F., Siegel B. A. and Welch M. J. *J. nucl. Med.* **16**, 370 (1975).
9. Sherman L. A., Harwig S. and Lee J. *J. Lab. Clin. Med.* **86**, 100 (1975).
10. DeNardo S. J., DeNardo G. L., O'Brien T., Peak N. F., Zielinski F. W. and Jungerman J. A. *J. nucl. Med.* **15**, 487 (1974).
11. DeNardo S. J., DeNardo G. L., Carretta R. F., Jansholt A. L., Krohn K. A. and Peek N. F. *J. nucl. Med.* **16**, 524 (1975).

International Journal of Applied Radiation and Isotopes, 1977, Vol. 28, pp. 163–168. Pergamon Press. Printed in Northern Ireland

A Case for 77-Bromine Labelled Radiopharmaceuticals

JAY A. SPICER,* DAVID F. PRESTON, RALPH G. ROBINSON, DIANA L. BRADSHAW, STEVEN H. STERN, R. DALE DEAN,† NORMAN L. MARTIN and BUCK A. RHODES

Division of Nuclear Medicine, Department of Diagnostic Radiology, University of Kansas Medical Center, Kansas City, KS 66103, U.S.A.

(*Received* 2 *February* 1976)

The use of radiobromine as a label for radiopharmaceuticals was demonstrated by the synthesis of ^{82}Br-2,4-dibromoestrone, using the procedure of SLAUNWHITE and NEELY. Structure determination of the brominated steroid was accomplished by the use of melting point data, paper chromatography and mass spectra analysis. Tissue distribution studies in animals demonstrated rapid blood clearance and excretion in the bile. Calculated radiation exposure dose for ^{77}Br-2,4-dibromoestrone is less than ^{131}I-Rose Bengal. A significant difference in tissue distribution was found when the data for ^{82}Br-2,4-dibromoestrone was compared to literature values for ^{82}Br-7,8-dibromoestrone. Comparison with ^{131}I-Rose Bengal and ^{99m}Tc-pyridoxylideneglutamate suggest ^{77}Br-2,4-dibromoestrone would be a tracer of potential value for liver and gallbladder function studies.

INTRODUCTION

AT THE present time the enterohepatic circulation cannot be investigated by a noninvasive method. TWOMBLY *et al.*[1,2] suggested that ^{82}Br-7,8-dibromoestrone when injected intravenously into humans is recirculated by the enterohepatic system. Since ^{131}I is not a suitable label for estrone,[3] ^{82}Br was used to prepare ^{82}Br-2,4-dibromoestrone (^{82}Br-2,4-DBE, Fig. 1) for animal distribution studies and to demonstrate the procedure for preparing ^{77}Br-2,4-DBE. Bromine-82's decay scheme makes it unsuitable for scintillation scanning, however, radiation dosimetry calculations reveal that if the 2,4-DBE were labelled with ^{77}Br a new radiopharmaceutical for studying the enterohepatic circulation could result. Theoretical tissue distribution and radiation dose estimates for ^{77}Br-2,4-DBE compare favorably with ^{131}I-Rose Bengal. Uptake in the gallbladder of rabbits is higher than either ^{131}I-Rose Bengal or ^{99m}Tc-pyridoxylideneglutamate (^{99m}Tc-PG).

MATERIALS AND METHODS

The radionuclide ^{82}Br was obtained from ICN Life Sciences in the form of KBr in a minimal amount of water (1–2 ml). Estrone (99% pure) was purchased from Nutritional Biochemicals as a dry powder. The following procedure was adapted from that of SLAUNWHITE and NEELY[3,4].

1. *Chemicals*

(1) ^{82}Br–KBr, (2) estrone, (3) glacial acetic acid (reagent grade), (4) manganese dioxide (prepared by method in Ref. 3), (5) conc. sulfuric acid (reagent grade), and (6) absolute ethanol (sterile, pyrogen-free).

2. *Method*

The procedure was performed in a well-ventilated hood and behind lead shielding.

(1) The $^{82}Br_2$ generator system consisting of a generator flask with air flow through a cold trap into the bromination vessel was assembled as shown by SLAUNWHITE and NEELY[3].

*For reprints contact: JAY A. SPICER, M.S., Department of Diagnostic Radiology, University of Kansas Medical Center, 39th & Rainbow Boulevard, Kansas City, KS 66103, U.S.A.

†Department of Radiation Oncology, University of Kansas Medical Center, Kansas City, KS 66103, U.S.A.

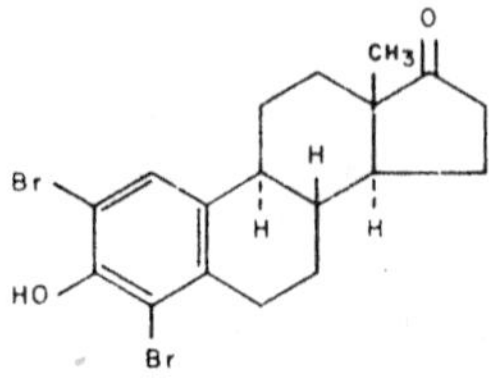

FIG. 1. Chemical structure of 2,4-DBE.

(2) 5 mg of estrone was dissolved in 2 ml of glacial acetic acid (heating is required) and placed in the reaction vessel.

(3) To the generator flask the following were added in the order listed:

(1) 10 mg MnO_2
(2) 5 ml of conc. H_2SO_4
(3) 6–7 mCi ^{82}Br–KBr.

(4) The air sweep through the system into the reaction vessel was started and the contents of the generator flask were gently boiled for 10 min.

(5) Saturated sodium bicarbonate solution was cautiously added to the glacial acetic acid mixture until a cloudy solution occurred.

(6) Next, 5–10 ml of sterile, distilled water was added to the neutralized solution and allowed to stand for 5–10 min while the ^{82}Br-2,4-DBE precipitated.

(7) The mixture was centrifuged for 5 min and the supernate was drawn off leaving the ^{82}Br-2,4-DBE as a solid.

(8) The solid was washed 2–3 times with sterile distilled water to remove any unreacted bromine or bicarbonate solution.

(9) The yield was 2–4 mCi of ^{82}Br-2,4-DBE.

3. *Preparation of product for injection*

(1) The solid was dissolved in 2 ml of absolute ethanol and passed through a 0.22 μm Solvenert Millipore® filter.

Fifty mCi of ^{82}Br-2,4-DBE dissolved in absolute ethanol were added dropwise (approximately 3–5 drops) with shaking to 1.5 ml of rabbit serum. Each of 26 New Zealand rabbits (12 females and 14 males) weighing 1.3–3.6 kg were injected in a marginal ear vein with 1.5 ml of the radioactive serum. In each case two identical syringes of the ^{82}Br-2,4-DBE were drawn up and counted on a Baird Atomic well counter (#810C) and spectrometer (#530). Care was taken not to exceed the count rate capacity of the instrument by counting the syringe at an appropriate distance from the crystal. One syringe was injected into the animal and the other injected into a 100 ml volumetric flask containing water so that aliquots could be taken for counting standards. The residual activity and net counts injected were determined. The ratio of counts of the injection syringe to the counts of the standard syringe yielded the injection factor.

The rabbits were sacrified at intervals between 10 min and 24 hr with i.v. injections of sodium pentobarbital. Two blood samples of 3 ml each were taken immediately before the animal's death and after death specific tissue samples were routinely removed, dissected free of fat and connective tissue, rinsed and/or blotted dry of all blood or fluid. The tissues were then placed in preweighed culture tubes, weighed, and radioactive content assayed. Percent of injected dose was calculated by comparing a 1% std(1 ml aliquot from 100 ml volumetric flask with the ^{82}Br-2,4-DBE added) with the tissue samples counted on an automatic gamma scintillation well counter with a Nuclear Data ND 812 computer attachment. Corrections for radioactive decay and volume differences were applied and results were expressed in percent of injected dose/g tissue.

RESULTS

A radiochromatograph using silica gel paper and distilled water as a carrier solvent was scanned using a Nuclear Chicago Actigraph® III radiochromatographic scanner. The R_f of the compound was 0 and the R_f of free bromide ions was 0.95. The final product showed the labelled estrone to be greater than 98% labelled for as long as 9 days. Mass spectra analysis of estrone and brominated estrone showed a dibromoestrone product with a molecular weight of 430 (calculated value 430). The literature value[4] for the melting point of 2,4-dibromoestrone was 226°C. Our product had a melting point of 221–223°C.

Animal tissue distribution data (Table 1) showed rapid collection of ^{82}Br-2,4-DBE in the gallbladder and bile. Disappearance of the ^{82}Br-2,4-DBE from the blood system is rapid as shown by a gallbladder (including bile) to blood ratio of 22:1 at 20 min. The maximum concentration of activity in the gallbladder was

TABLE 1. Tissue distribution of ^{82}Br-2,4-dibromoestrone in rabbits
% injected dose/g tissue: mean values ± S.D.

	20 min ($N = 6$)	50 min ($N = 6$)	150 min ($N = 2$)	180 min ($N = 3$)	360 min ($N = 3$)	1440 min ($N = 6$)
Gallbladder and bile	0.770 ± 0.292	2.400 ± 1.19	3.476 ± 1.079	3.078 ± 1.745	2.521 ± 1.969	2.153 ± 1.840
Liver	0.104 ± 0.038	0.083 ± 0.046	0.051 ± 0.016	0.046 ± 0.010	0.046 ± 0.028	0.014 ± 0.005
Blood	0.032 ± 0.008	0.037 ± 0.010	0.037 ± 0.018	0.034 ± 0.001	0.028 ± 0.001	0.015 ± 0.007
Pancreas	0.115 ± 0.041	0.057 ± 0.015	0.032 ± 0.005	0.028 ± 0.007	0.017 ± 0.007	0.007 ± 0.005
Adrenals	*0.185 ± 0.057	0.124 ± 0.052	0.100 ± 0.054	0.024 ± 0.004	0.013 ± 0.003	0.006 ± 0.002
Kidneys	0.346 ± 0.105	0.191 ± 0.092	0.088 ± 0.037	0.088 ± 0.019	0.083 ± 0.040	0.028 ± 0.011

*For females the value was 0.217 ± 0.061 and for males the value was 0.152 ± 0.033. The difference was significant at the 95% confidence level.

reached at approx 150 min (Fig. 2). The graph in Fig. 2 showed a relatively slow clearance of the ^{82}Br-2,4-DBE from the gallbladder-bile with more than 2% of the injected dose remaining in the organ at 24 hr.

Graphic displays of gallbladder to liver ratios (Fig. 3) and gallbladder to blood ratios (Fig. 4) comparing ^{82}Br-2,4-DBE with the results obtained by KUBOTA *et al.*[5] for ^{131}I-Rose Bengal and ^{99m}Tc-PG indicate that higher ratios are obtained for ^{82}Br-2,4-DBE in 1 hr than either the ^{131}I-Rose Bengal or the ^{99m}Tc-PG. The work of KUBOTA *et al.* shows that both the ^{131}I-Rose Bengal and the ^{99m}Tc-PG reached their maximum activity in the gallbladder and their maximum gallbladder to liver ratio at 60 min. From the curve in Fig. 3 it can be seen that ^{82}Br-2,4-DBE did not reach its point of maximum activity in the gallbladder at 60 min, however at this time a higher gallbladder to liver ratio as compared to ^{131}I-Rose Bengal or ^{99m}Tc-PG was obtained.

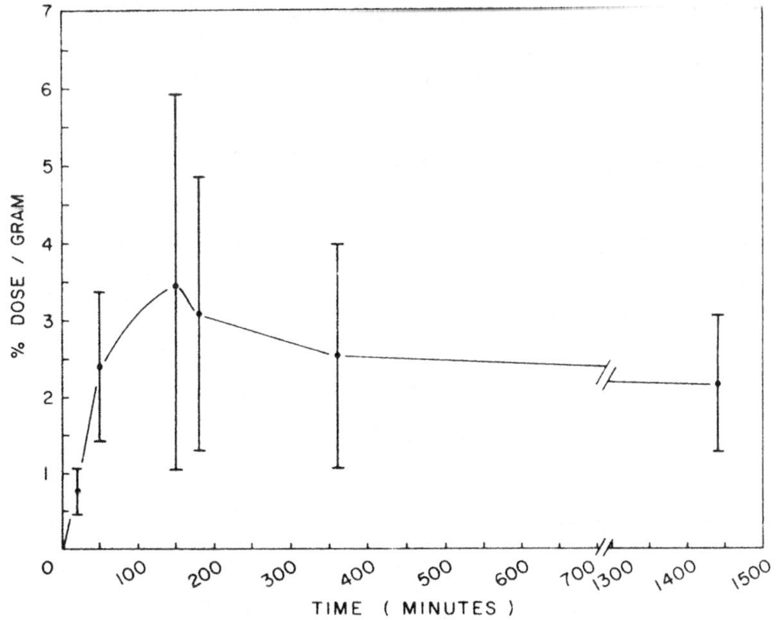

FIG. 2. ^{82}Br-2,4-DBE uptake in the gallbladder (including bile). Brackets represent the standard error of the mean.

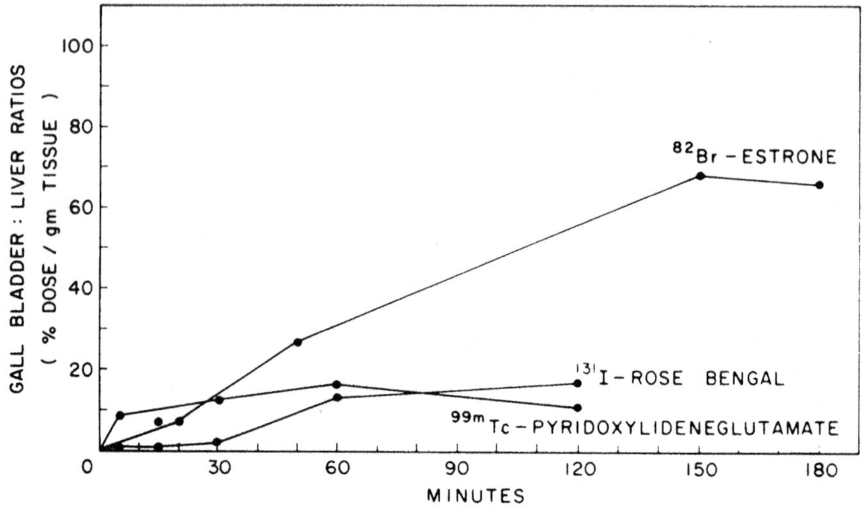

FIG. 3. Comparison of gallbladder: liver ratios.

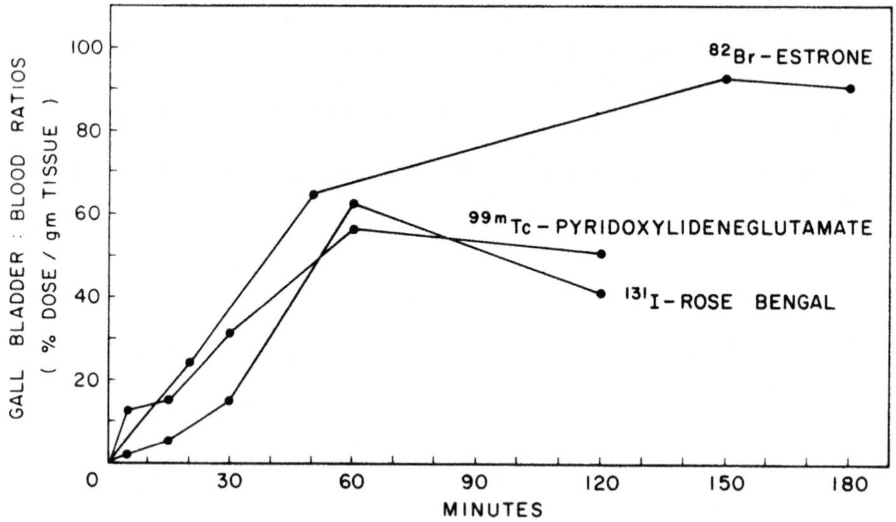

FIG. 4. Comparison of gallbladder: blood ratios.

TABLE 2. Estimated radiation absorbed dose for ^{77}Br-2,4-dibromoestrone in man: mrem/mCi

Organ	$t_{1/2}$eff measured	$t_{1/2}$eff = $t_{1/2}$ physical
Gallbladder wall	1250	3187
Liver	250	740
Pancreas	151	508
Adrenals	163	447
Kidneys	206	838

Calculated radiation dosimetry data (Table 2) for ^{77}Br-2,4-DBE was compared with those for ^{131}I-Rose Bengal.[6] The long effective half-life of ^{77}Br-2,4-DBE causes the absorbed radiation dose to the gallbladder wall to be approximately the same as ^{131}I-Rose Bengal, but the overall total body absorbed radiation dose for ^{77}Br-2,4-DBE was found to be considerably less than the total body dose for ^{131}I-Rose Bengal (0.11 vs 0.26 rads/mCi).

DISCUSSION

Earlier work by TWOMBLY *et al.*[1] showed that ^{82}Br-7,8-DBE was excreted into the bile, although little or none of the ^{82}Br-7,8-DBE was found in the bile at 30 min. A second paper by TWOMBLY *et al.*[2] showed a marked decrease of ^{82}Br-7,8-DBE in the gallbladder of human subjects with liver damage. As these patients returned to normal the amount of ^{82}Br-7,8-DBE in the gallbladder increased, indicating that a labelled dibromoestrone agent has potential for use in hepatobiliary function studies. The rapid accumulation of ^{82}Br-2,4-DBE in the gallbladder as compared to ^{82}Br-7,8-DBE suggests that if labelled with ^{77}Br, the 2,4-DBE analog of estrone is more appropriate for use as a hepatobiliary function agent.

When estrogens are excreted into the intestine, they are absorbed and recirculated. Two, 4-DBE, if labelled with ^{77}Br, may be a new radiopharmaceutical for studying enterohepatic circulation. It has not been shown that any of the current ^{131}I-labelled or ^{99m}Tc-labelled gallbladder agents are absorbed and recirculated. Because of its recirculation ^{77}Br-2,4-DBE could be used to measure parameters of hepatobiliary function not measurable by currently used gallbladder agents.

The results presented in this paper suggest other advantages of ^{77}Br-2,4-DBE over ^{131}I-Rose Bengal and ^{99m}Tc-PG besides the potential to study the enterohepatic circulation. The first advantage of ^{77}Br-2,4-DBE is the higher gallbladder to liver ratios obtained at early times. A second advantage is its longer effective half-life, allowing external measurements to be made over 2–3 days with reduced whole body radiation when compared with ^{131}I-Rose Bengal. Absorption from the gut and recirculation prolong the residence time in the body and result in the higher gallbladder to liver and gallbladder to blood ratios at 6–24 hr. A third advantage is the excellent stability of ^{82}Br-2,4-DBE when compared with ^{131}I-Rose Bengal. Bromine-82-2, 4-DBE exhibited an efficient tag *in vitro* and when rabbit urine (collected at 24 hr) was chromatographed, no free bromide was indicated. Tissue distribution data showed little activity in the thyroid with ^{82}Br-2,4-DBE, thus eliminating the need of a thyroid blocking agent and unnecessary radiation exposure to the thyroid as in the case of ^{131}I-Rose Bengal.

Another area of potential development is *in vivo* estimation of estrone receptor sites in the intact female breast and whole body. The *in vitro* assay of ^{3}H and ^{14}C labelled estrogen to predict responsiveness of a breast tumor to endocrine therapy is a recently discovered predictive parameter of breast carcinoma.[7] Estrogen receptor sites throughout the body have been measured.[8] Bromine-77-2,4-DBE may provide an analogous whole body measure of estrogen binding sites *in vivo* by external scanning.

Although ^{77}Br does not have ideal physical characteristics for imaging (principle energy 240 keV), its use as a radiolabel to give satisfactory images has already been shown by other investigators. The preparation and use of the ^{77}Br as a protein label has been studied by KNIGHT *et al.*[9] at The Edward Mallinckrodt Institute of Radiology, Washington University School of Medicine in St. Louis, MO. A group, SARGENT *et al.*[10,11], at Donner Laboratories, University of California in Berkeley, California, has recently published work involving the use of 4-^{77}Br-2,5-Dimethoxyphenylisopropylamine as a potential brain scanning agent.

Preliminary experiments at Los Alamos Scientific Laboratory have shown that ^{77}Br is formed during the irradiation of molybdenum targets with high energy protons in sufficiently high yields that the production of large amounts of this isotope for clinical use appears feasible in the near future.[12]

REFERENCES

1. TWOMBLY G. H., MCCLINTOCK L. and ENGELMAN M. *Am. J. Obstet. & Gynecol.* **56,** 260 (1948).

2. TWOMBLY G. H. and SCHOENEWALDT E. F. *Cancer* **3,** 601 (1950).
3. SLAUNWHITE W. R., JR. and NEELY L. *Anal. Biochem.* **5,** 133 (1963).
4. SLAUNWHITE W. R., JR. and NEELY L. *J. Org. Chem.* **27,** 1749 (1962).
5. KUBOTA H., ECKELMAN W. C., POULOSE K. P. and REBA R. C. *J. nucl. Med.* **17,** 36 (1976).
6. MIRD/Dose Estimate Report No. 7. *J. nucl. Med.* **16,** 1214 (1975).
7. BAULIEU E. E. *JAMA* **234,** 404 (1975).
8. GOLOMB H. M. and THOMSEN S. *Arch. Intern. Med.* **135,** 942 (1975).
9. KNIGHT L., KROHN K. A., WELCH M. J., SPOMER B. and HAGER L. P. In *Radiopharmaceuticals* (Edited by SUBRAMANIAN G., RHODES B. A., COOPER J. F. and SODD V. J.), pp. 149–154. Soc. of Nucl. Med., New York (1975).
10. SARGENT T., III, KALBHEN D. A., SHULGIN A. T., STAUFFER H. and KUSUBOU N. *J. nucl. Med.* **16,** 243 (1975).
11. SARGENT T., III, KALBHEN D. A., SHULGIN A. T., BRAUN G., STAUFFER H. and KUSUBOU N. *Neuropharmacology* **14,** 165 (1975).
12. O'BRIAN H. A. Personal communication, 1976.

International Journal of Applied Radiation and Isotopes, 1977, Vol. 28, pp. 169–181. Pergamon Press. Printed in Northern Ireland

A Review of Radiopharmaceutical Development with Short-lived Generator-produced Radionuclides Other than ^{99m}Tc

D. J. HNATOWICH
Massachusetts General Hospital, Boston, MA 02114, U.S.A.

(*Received* 6 *February* 1976)

In addition to the technetium generator there are approximately eight generators of short-lived radionuclides in which the daughter, in some chemical form, has seen human use. This article reviews progress in the development of imaging agents prepared with these eight short-lived radionuclides. The bulk of the report deals with ^{113m}In and ^{68}Ga radiopharmaceuticals.

INTRODUCTION

THE OBJECT of this article is to review radiopharmaceutical development with short-lived generator-produced radionuclides other than ^{99m}Tc. It is evident from the literature that the majority of effort in the development of imaging agents with these generator products has concentrated on ^{113m}In and, to a lesser extent, on ^{68}Ga. Interest in ^{113m}In is partly the result of its conveniently long parent half life; the indium generator is especially attractive in those countries in which the technetium generator is not available because of its short shelf life. Increasing interest in the ^{68}Ga generator results from improvements in positron annihilation detection devices for imaging. Many of the remaining generators are relatively unattractive for reasons of short parent half-life, low photon yield, poor photon energy, etc. Thus, although 118 potential radionuclide generators have been listed by BRUCER,[1] only eight generator-produced short-lived radionuclides in addition to technetium have been used clinically or in preliminary human studies.[2] The important decay properties of these eight are given in Table 1.

As a consequence of the above, the bulk of this report will deal with indium and gallium radiopharmaceuticals. The intent is not to review the many important applications of radiopharmaceuticals prepared with these nuclides but rather to review work which has contributed significantly towards radiopharmaceutical development and radiochemical analysis.

Both indium and gallium are unusual in that several radioisotopes of each element have found applications in nuclear medicine imaging. ^{111}In ($T_{1/2}$ 2.8 days), ^{114m}In ($T_{1/2}$ 50 days) and ^{67}Ga ($T_{1/2}$ 3.3 days) are not generator products, nevertheless since labeling and analysis techniques often apply equally well to different isotopes, this discussion of ^{113m}In and ^{68}Ga radiopharmaceuticals will include important developments in which the longer-lived isotopes have been used as label.

INDIUM RADIOPHARMACEUTICALS

COOPER and WAGNER[3] have described the preparation and quality control for the ^{113m}In labeled radiopharmaceuticals used in their laboratory in 1970. The number of ^{113m}In labeled agents in use there and elsewhere is impressively large considering that the indium generator was introduced just four years earlier.[4,5] Use of the radionuclide as a blood label resulted from the reports of STERN *et al.*[6] and HOSAIN *et al.*[7] that indium binds to plasma proteins when administered intravenously at low pH. The same agent was used for kidney and bone imaging after plasma transferrin was saturated with stable indium,[8] stable iron,[9] or stable gallium.[8] Lung images were obtained with ^{113m}In labeled macroaggregates of indium

TABLE 1. Important decay properties for several generators of short-lived radionuclides*

Generator	Parent $T_{1/2}$	Daughter $T_{1/2}$	Daughter E_γ (%)
^{68}Ge–^{68}Ga	275 days	68 min	511 keV (176)
^{81}Rb–^{81m}Kr	4.7 hr	13 sec	190 (65)
^{82}Sr–^{82}Rb	25 days	1.3 min	511 (192)
^{87}Y–^{87m}Sr	3.3 days	2.8 hr	388 (80)
^{113}Sn–^{113m}In	115 days	1.7 hr	393 (64)
^{132}Te–^{132}I	3.2 days	2.3 hr	many
^{137}Cs–^{137m}Ba	30 years	2.6 min	662 (89)
^{191}Os–^{191m}Ir	15 days	4.9 sec	129 (25)

*Taken from ref. (2).

hydroxide,[(10)] ferric hydroxide,[(10–13)] aluminium hydroxide,[(14)] and albumin.[(15,16)] Labeled cation exchange resin particles have been used for lung imaging.[(17)] Iron-containing human serum albumin microspheres had been introduced several years earlier and a method was developed for their labeling with ^{113m}In.[(18)] Smaller sized particles were developed for imaging the organs of the reticuloendothelial system: ^{113m}In labeled colloids of indium hydroxide,[(13,19)] indium sulphide,[(20,21)] indium phosphate,[(3,22,23)] each with carrier indium, albumin,[(15)] hydroxyapatite,[(24)] rhenium sulfide,[(25)] and stannous oxide[(26)] have been used to obtain liver, spleen, or bone marrow images. A colloidal form of ^{113m}In has been used for imaging the subarachnoid space.[(27)] Chelates of ^{113m}In with Fe-DTPA,[(6)] DTPA,[(28,29)] Fe-EDTA,[(6)] Fe-DTPA-ascorbic acid,[(30)] and others[(29,31)] have been used for brain tumour and kidney imaging. Finally, ^{113m}In aerosols have been used for lung ventilation measurements.[(32,33)]

In the several years since these developments, the number of indium-induced agents in clinical use has increased considerably; a recent article by ALVAREZ[(34)] summarizes the present status of ^{113m}In radiopharmaceuticals. The remainder of this section will deal with these recent developments. For convenience, indium and gallium radiopharmaceuticals will be discussed under the headings: Preparation and Analytical Testing; Chelates and Complexes; and Particles, Colloids and Aerosols.

Preparation and analytical testing

Indium generators in clinical use are designed to be eluted with dilute HCl solution. PENKOSKE *et al.*[(35)] have determined that indium in the eluant is mainly in the In $(H_2O)_5Cl^{2+}$ chemical form. This was accomplished by separating the chloride species by a batch cation exchange method. Indium is stable in these chemical forms at pH below about 4; however at higher pH insoluble indium hydroxide and, in the presence of phosphate ion, insoluble indium phosphate will form. When administered intravenously at low pH, indium binds to serum transferrin and may be used for blood pool imaging. If the eluant is first neutralized prior to injection, insoluble and colloidal forms of indium result such that the activity localizes in organs of the reticuloendothelial system.

These properties of indium must be considered in the preparation of indium complexes. The complexing agent should be added to the generator eluant before neutralization. Depending on the agent and on its concentration, upon neutralization indium may bind preferentially to the complexing agent rather than to hydroxyl or phosphate ions. HILL *et al.*[(28)] have developed a one step method for the preparation of In-DTPA at neutral pH by the use of acetate ion to form an indium acetate complex. This intermediate complex prevents the formation of indium hydroxide as the pH is raised above 4 but is not sufficiently stable to prevent the ultimate formation of In-DTPA. Suitable brain scans were obtained with the agent prepared in this fashion. In an excellent discussion of indium radiopharmaceutical chemistry, WELCH and WELCH[(36)] have explained on the basis of equilibrium constants and rates of reaction the observed *in vivo* behavior of this and other indium-labeled agents.

Initially, indium-labeled chelates of DTPA and EDTA were prepared in the presence of ferric ions;[6] ascorbic acid has been added to prevent the formation of insoluble ferric hydroxide which may result from this practice.[30] However the iron will compete for chelate binding sites and can serve no useful purpose.[37] Successful iron-free preparations of indium labeled DTPA have been reported;[28,29] iron or other metals are no longer used in the preparation of indium chelates. Further improvements in the indium labeling procedure have been described by ADATEPE *et al.*[38]; a closed eluting system and sterile materials were employed to help insure the safety of the product, and by COLOMBETTI *et al.*[11] in which thymol blue was used as an acid-base indicator to control preparation pH.

Only infrequent use has been made of analytical techniques to establish radiopharmaceutical purity of indium-labeled agents. An ammonia:ethanol:water solvent was used with ascending paper chromatography by MANI and NARASIMHAN[39] to analyze preparations of ^{113m}In labeled DTPA. These authors were able to analyze an ^{113m}In-gelatin complex for ionic indium by dialysis against distilled water. A similar solvent system and ascending paper chromatography as well as gel column chromatography has been used by BESNARD *et al.*[40] to analyze ^{113m}In-DTPA preparations. ^{113m}In labeled DTPA, gelatin complex, and an indium hydroxide colloid for liver and spleen imaging were prepared by JAISWAL *et al.*[41] and analyzed by ascending paper chromatography on Whatman No. 1 paper using several different solvent systems. Several additional authors have adopted radiochemical test procedures for indium labeled radiopharmaceuticals as mentioned below.

Chelates and complexes

Following the introduction into nuclear medicine of ^{111}In[42] radiopharmaceutical preparation techniques developed for ^{113m}In were applied to this longer-lived isotope. As such, both ^{111}In transferrin complex and ^{111}In phosphate colloid were prepared and used clinically for cisternographic studies.[42] The indium binding capacity of patient serum was investigated by electrophoresis; the stability of the indium-transferrin complex was tested by immunoelectrophoresis. Improved procedures for the preparation of ^{111}In labeled transferrin, EDTA, and DTPA have been described by GOODWIN *et al.*[43] in which the compounds were tested by column chromatography.

GOODWIN *et al.*[44] have prepared ^{111}In-8-hydroxy-quinoline and ^{111}In-8-thio-quinoline as well as ^{111}In labeled citrate and chloride for a comparison study of tumor and organ distribution. MERRICK *et al.*[45] investigated the influence of ligand on the biodistribution of ^{111}In compounds in rats. They prepared labeled bleomycin, tetraphenylporphine, hydroxy-ethyl ethylenediamine triacetic acid (HEDTA) as well as labeled citrate, lactate and acetate. Their preparation procedure for the majority of the chelates was to add the chelating agent to dilute HCl presumably followed by injection without prior neutralization. With the exception of the chloride and tetraphenylporphine chemical forms, there was no significant difference in the distributions of the compounds other than in the rate of excretion.

GOODWIN *et al.*[46] have prepared ^{111}In labeled citrate, nitrilotriacetic acid (NTA), ortho-phenylene diaminetetraacetic acid, 1,2-cyclohexanediaminetetraacetic acid and EDTA and using column chromatography, radioimmunoelectrophoresis, and perturbed angular correlation, they found that in the chloride, citrate, or NTA chemical forms, indium binds *in vivo* to transferrin whereas a lack of protein binding is evident when indium is injected in the other chemical forms.

Interest in bleomycin for tumor scanning followed the observation that labeled with ^{57}Co, the polypeptide localizes to a useful extent in tumors. The long half life of ^{57}Co makes this radiolabel unattractive and for this reason a ^{111}In label for bleomycin was developed. Two chromatographic methods were used by GROVE *et al.* to establish compound purity:[47] Whatman No. 1 paper with 10% NH_4Cl and silica gel thin layer chromatography with a 1:1 mixture of 10% ammonium acetate and methanol. In both systems, indium hydroxide remains at the origin while the bleomycin fractions travel with the solvent front. Essentially the same preparation and thin layer chromatography procedures were used by ROBBINS *et al.*[48] These latter authors investigated the thermal stability of indium bleomycin and its chemical stability towards various metal ions as well as the dependence of tagging yield on the quantity of bleomycin used.

Recently a novel approach to the labeling of large molecules such as proteins with indium has been developed. PRICHARD *et al.* have described a method of labeling antibody with ^{111}In by first preparing an antibody conjugate using gluteraldehyde as part of the linking molecule.[49] The conjugate may be stored at low temperature and may be labeled when required. The labeling step itself is rapid and thus the procedure has application towards the use of ^{113m}In. An alternative method is that of SUNDBERG *et al.*[50] in which an EDTA analogue, azo-ϕ-EDTA, is used to form a link between EDTA and proteins and polypeptides via a diazo bond. Once the conjugate is prepared, the labeling with indium by chelation is rapid. Human serum albumin, bovine fibrinogen, and bleomycin have been labeled with ^{111}In following this procedure.

Another EDTA analogue (EDTMP) and an analogue of DTPA (DTPMP) in which the acetate groups have been replaced with methylene phosphate groups, have been labeled with ^{113m}In by SUBRAMANIAN *et al.*[51] and found to be useful bone imaging agents. These compounds combine the chelating ability of EDTA and DTPA with the bone seeking properties of disphosphonic acid. Several other indium labeled multidentate chelating agents were also prepared and tested. The amount of unbound indium in the preparations was determined by paper electrophoresis. In preliminary work, these authors studied indium-labeled polyphosphate, pyrophosphate and other compounds that had been used with ^{99m}Tc for bone imaging. Although these indium-labeled compounds appeared to have been prepared successfully based on *in vitro* testing, very little skeletal localization occurred, presumably because of poor *in vivo* stability. JONES *et al.*[52] and DEWANJEE and KAHN[53] have studied the same ^{113m}In-labeled compounds for bone and myocardial infarct imaging respectively. In both cases thin layer chromatography with 0.1 M NH_4OH as solvent was used to establish preparation purity.

Additional chelates and complexes which have been prepared include ^{113m}In-labeled red blood cells and plasma protein.[54] The labeling was accomplished by mixing the ^{113m}In generator eluant with whole blood or its constituents followed by incubation. The labeled protein fractions were identified by block electrophoresis and radioimmunoelectrophoresis. Fibrinogen has been labeled by incubation with ^{113m}In and has been found to accumulate in actively forming thrombosis.[55] The labeling of potassium chromate-treated erythrocytes with ^{113m}In has been accomplished by BURAGGI *et al.*[56] Following the labeling, the red blood cells were heat denatured and used for spleen imaging. The labeling efficiency and stability of the label are high, however, the preparation time of 90–100 min is long with respect to the ^{113m}In half life. Erythrocytes have also been labeled with ^{113m}In by SINN *et al.*[57]

Particles, colloids and aerosols

It was mentioned above that the injection of indium chloride at low pH results in the *in vivo* binding of indium to blood serum proteins, mainly transferrin, such that the label is essentially nondiffusible and may be used for blood pool imaging. If, however, the indium chloride is first neutralized prior to injection, colloidal particles of indium form which accumulate in the organs of the reticuloendothelial system. ADATEPE *et al.*[58] have studied the biodistributions of ^{113m}In colloids prepared in the usual manner with NaOH and prepared by using a phosphate buffer for neutralization. They conclude that the difference in organ distribution which they observed may be related to differences in the size of colloids prepared with the two bases.

Ferric hydroxide macroaggregates labeled with indium continue to receive attention despite reports indicating that the lung clearance rate of the ferric hydroxide might be considerably slower than previously suspected. GOODWIN[59] investigated lung retention in mice of ^{114m}In-labeled ferric hydroxide macroaggregates. Similar work was reported by BARKER and GUSMANO[60]. The conclusion from these studies is that, depending on "hardness" and average particle size of the preparation, an appreciable fraction of the injected particles may still remain in the lungs after 20 days.

Another particle type consists of ^{113m}In-labeled stannous hydroxide. The preparation of macroaggregates with this material for lung imaging and colloids for liver and bone marrow imaging have both been described by THOMAS and WIENER.[(61)] A particle type consisting of macroaggregated human serum albumin containing stannous tin has been labeled with ^{113m}In for use in the study of myocardial perfusion.[(62)] A ^{113m}In-labeled indium oxinate particle type (containing stable indium) for lung scanning has also been introduced[(63)] to provide an alternative to iron-containing particles. The long term lung retention was established by using ^{114m}In as label. Ascending paper chromatography using a chloroform solvent was used to determine that only a few percent of the activity was in a soluble form. One further macroaggregate consists of ^{113m}In sulfide particles free of carrier metal.[(64)] The influence of zirconium (from the ^{113m}In generator packing) on the labeling of albumin macroaggregates has been investigated by CARO *et al.*[(65)] The observation that the *in vivo* stability of the label was variable led to this study which determined that small amounts of Zr(IV) are necessary to fix ^{113m}In to the macroaggregates.

Human serum albumin microspheres have the advantage over the macroaggregates that the particles are spherical and are prepared and sized prior to labeling. Albumin microspheres containing small amounts of iron hydroxide to facilitate labeling with ^{113m}In for capillary perfusion studies have been described by BUCHANAN *et al.*[(18)] Iron-free microspheres have been prepared and labeled with ^{113m}In by BUCHANAN *et al.*[(66)] The various parameters affecting labeling efficiency such as pH, reaction time, weight of particles, etc. were investigated. An alternative method for the labeling of iron-free microspheres was developed by RABAN *et al.*[(67)] Smaller iron-containing microspheres (less than 1 μm dia.) have been labeled with ^{113m}In and used in studies of the reticuloendothelial system.[(68)]

Recent work on the use of ^{113m}In labeled aerosols for lung ventilation studies includes that of ISAWA *et al.*[(69)] in which the aerosolized preparation was assumed to be a complex of indium hydroxide and human serum albumin. LIN *et al.*[(70)] aerosolized the eluant of a ^{113m}In generator both without prior neutralization and following neutralization to pH of about 6 with sodium phosphate. They did not observe significant differences in the lung images with either form. The generator eluant has also been used without prior neutralization for inhalation lung scanning by COOK and LANDER[(71)].

GALLIUM RADIOPHARMACEUTICALS

Only a few positron-emitting radionuclides are available as products of radionuclide generators; among the generator-produced radionuclides considered in this review only ^{68}Ga and ^{82}Rb fall in this category. Other positron emitters may become available as generator products in the near future. ^{68}Ga is in many ways the most attractive of these because of its favorable chemical and physical properties. It appears likely that this nuclide will play a roll in positron scintigraphy which is comparable to that played by ^{99m}Tc in gamma-ray scintigraphy. As an example of the use of ^{68}Ga, Fig. 1 shows images obtained with the Massachusetts General Hospital Positron Camera in a dog following intracoronary injection of ^{68}Ga labeled microspheres and sacrifice. The image quality of the posterior–anterior and left lateral images shown on the left are entirely comparable with those obtained with gamma-ray scintigraphic methods. However, the use of a positron-emitting radionuclide has permitted the reconstruction of emission tomographic sections through the activity distribution. Several sections are shown on the right and correspond to the positions shown by the horizontal lines in the figures in the middle.

Gallium occupies a position in the periodic table just above that of indium (group IIIA) and consequently the solution chemistry of the two elements are quite similar. Both, for example, exhibit only one stable oxidation state (+3) in solution, form chloride complexes in HCl solutions, and form insoluble hydroxides and phosphates.

The first gallium generator was developed by GLEASON[(72)] and required that ^{68}Ga be separated from its parent by solvent extraction. The activity was back extracted into a dilute HCl solution from which an injectable preparation could be made. An EDTA chelate of gallium was prepared in this fashion. An improved generator was developed by GREENE and TUCKER[73] in which the parent was loaded on an activated alumina column and the ^{68}Ga extracted with 0.005 M NaEDTA solution. Gallium generators are now commercially available and like the generator of GREENE and TUCKER provide the activity as ^{68}Ga-EDTA chelate in neutral solution. Some modest success has been achieved in the development of gallium generators in which the gallium is eluted in ionic form;[74] however, because of the large concentration of aluminum extracted into the eluant, these generators must be improved before they become practical.

Preparation and analytical testing

The eluant from gallium generators contain ^{68}Ga-EDTA in neutral solution. This chemical form is useful for brain tumor imaging; however the preparation of other gallium-labeled compounds is complicated by the need to dissociate Ga-EDTA in the first step of any labeling procedure.

The earliest method of separating ^{68}Ga from EDTA is that of YANO and ANGER;[75] stable gallium was added in excess to an acid solution containing the complex such that upon precipitating gallium hydroxide only a small fraction of the total gallium remained in chelated form in solution. The precipitate was then washed free of EDTA. HAYES *et al.*[76] have extracted ^{68}Ga as the chloride complex from HCl solutions free of EDTA using isopropyl ether; the gallum chloride complex was then back extracted into distilled water. Additional methods include ashing the complex in a crucible[77] and the addition of carrier gallium[78] and carrier iron[79] to exchange with ^{68}Ga chelated to EDTA.

None of these methods has the simplicity of the anion exchange method of CARLTON and HAYES[80]. The Ga-EDTA complex is dissociated in conc. HCl media and the mixture passed through an anion exchange column. The anionic chloride complex of gallium is retained on the resin while the protonated and therefore uncharged EDTA is washed off. Lowering the chloride ion concentration in the eluant by eluting with dilute HCl converts the complex to the neutral trichloride which is then released from the resin. This approach has been extended by GOULET *et al.*[81] and by HNATOWICH[82] so that the final eluant is contained in a small volume and is collected in sterile vials for subsequent radiochemical processing.

As is the case with other radionuclides, the possibility exists that several chemical forms of gallium may be present as contaminants in the preparation of gallium compounds. HNATOWICH[82] has described his use of ascending paper chromatography with three solvent systems to determine preparation purity for two ^{68}Ga complexes. The analysis was found to be sensitive to changes in labeling procedure which effect preparation quality. WAXMAN *et al.*[83] have used silica gel thin layer chromatography with 85% methanol in water to analyze commercial ^{67}Ga citrate preparations. Differences in the preparations were observed and may be related to the concentration of citrate.

Chelates and complexes

Compared to indium, the development of gallium-labeled radiopharmaceuticals has not been extensive.[84,85] The clinical use of gallium is still largely restricted to the citrate with ^{67}Ga as label. This agent continues to enjoy popularity primarily as a soft tissue tumor imaging agent. The preparation procedure employed by EDWARDS and HAYES in their pioneer work on the agent has been described.[86] When injected as the citrate, carrier-free gallium has been found to bind to serum constituents. As is the case with indium, if these serum binding sites are first saturated, for example with stable gallium, the activity is free to localize in bone.[87]

Several additional chemical forms of ^{68}Ga were investigated as potential brain tumor imaging agents by SHEALY *et al.*[88]. ANGHILERI[89] has prepared a ^{68}Ga labeled polymetaphosphate-Mg-polymetaphosphate compound for kidney scanning. More recently, KONIKOWSKI *et al.*[90] have described their procedure for the preparation of ^{67}Ga-labeled chloride, citrate, lactate, and Fe-DTPA for use in a study of the kinetics of these compounds in brain sarcomas and in kidneys of mice. HNATOWICH[82] has described a method for the preparation of ^{68}Ga labeled citrate and adenosine triphosphate in which phenol red, an injectable acid–base indicator, is

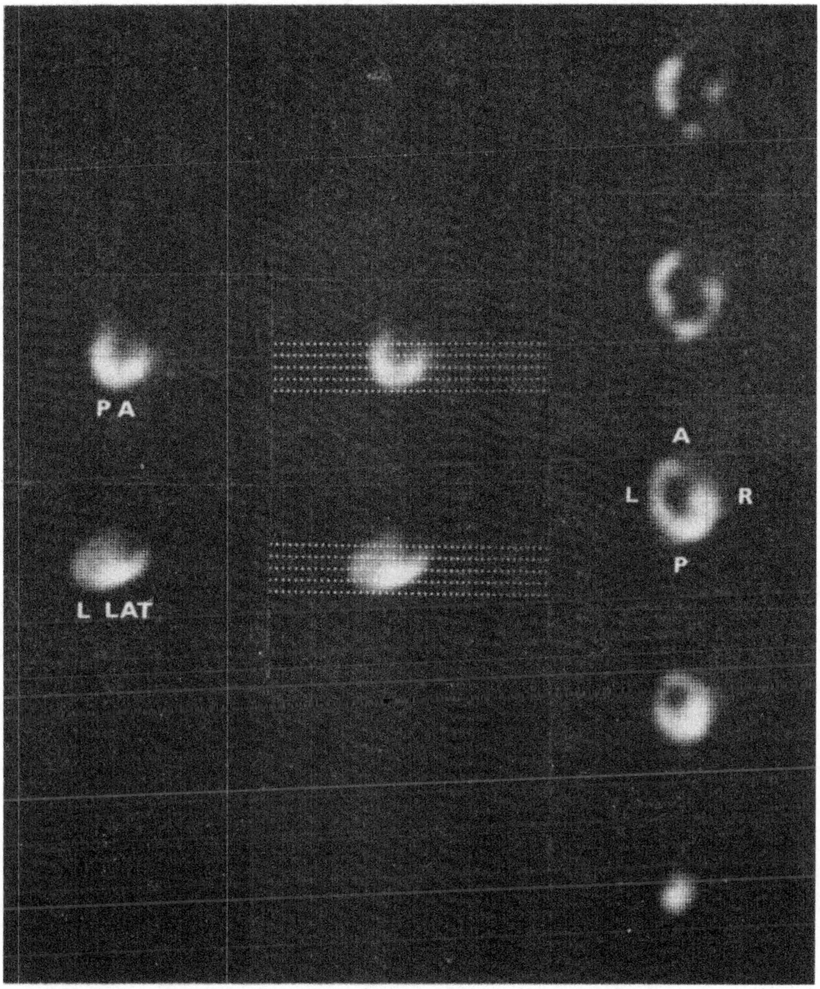

FIG. 1. Images obtained with the MGH Positron Camera in a dog following intracoronary injection of ^{68}Ga-labeled microspheres and sacrifice. *Left:* Posterior–Anterior and left lateral conventional images. *Right:* Emission tomographic sections through the heart, position of the sections shown by horizontal lines in the middle figures.

used for pH adjustment. GOULET *et al.*[81] have labeled tripolyphosphate with ^{68}Ga and determined the tissue distribution in mice and rabbits. The agent was found to localize in bone to a useful extent particularly in the presence of carrier gallium. Phytic acid has been labeled with ^{67}Ga by DEWANJEE *et al.*[91] and studied in rabbits as a potential lymph node scanning agent. Preparations were analyzed by paper chromatography in 85% methanol. Labeling of rabbit leukocytes by incubation with ^{67}Ga citrate has been reported by BURLESON *et al.*;[92] the agent has been suggested for use in abscess detection. The effect of iron on the labeling of leukocytes with ^{67}Ga citrate has been investigated by HILL *et al.*[93] Bleomycin has been labeled with ^{67}Ga by REBA *et al.*[94] and the product tested with silica gel thin layer chromatography.

Particles, colloids and aerosols

HAYES *et al.*[95] first reported the use of ^{68}Ga labeled colloids by preparing hydrous ferric oxide colloid for imaging liver, spleen, and bone marrow in animals. ANGHILERI and PRPIC introduced a ^{68}Ga-chromic phosphate colloid for liver scanning.[96] Gallium was separated from EDTA by co-precipitation with magnesium hydroxide from basic solution. The precipitate was washed, dissolved in HCl, and ^{68}Ga co-precipitated a second time to form the chromic phosphate colloid.

As mentioned above, carrier iron has been used to displace ^{68}Ga from EDTA. The iron is present in sufficient concentration to co-precipitate ^{68}Ga if the solution is neutralized. Thus, COLOMBETTI *et al.*[97] have prepared ferric hydroxide macroaggregates labeled with ^{68}Ga for lung studies. A similar method in which the iron was encouraged to form a ^{68}Ga-labeled hydrous ferric oxide colloid for liver scintigraphy has also been reported.[98] These methods are rapid but suffer from the disadvantage that about 20% of the gallium is present in the final preparation as the original EDTA chelate.[97] CHESLER *et al.*[99] have performed lung tomographic studies in dogs using ^{68}Ga labeled albumin microspheres containing lanthanum to facilitate labeling. HNATOWICH[100] has described a procedure for the labeling of tin-soaked albumin microspheres with ^{68}Ga and has shown the preparation to be stable in the lungs of dogs over a period of several hours. Finally, aerosols labeled with ^{68}Ga have been used in ventilation studies.[101]

RADIOPHARMACEUTICALS FROM OTHER GENERATOR PRODUCTS

Much less effort has been devoted to radiopharmaceutical development with the remaining short-lived generator products listed in Table 1. This section will deal with these generator products.

Krypton-81m

As an isotope of an insoluble gas, ^{81m}Kr has application principally for lung ventilation measurements by inhalation of the activity in gaseous state and for lung perfusion measurements by intravenous injection of the activity in solution. The krypton generator was introduced in 1969[102,103] and was designed so that the activity could be obtained in solution by eluting the generator with bubble-free water or in the gaseous phase by forcing air through the apparatus. Later YANO *et al.*[104] added a mixing chamber so that 2% NaCl solution would mix with the water eluant to provide an isotonic injectate (present generators may not be eluted with saline without extracting appreciable ^{81}Rb). MAYRON *et al.*[105] improved the generator such that it may be eluted with an isotonic dextrose water solution to eliminate the need to mix hypertonic saline with the generator eluant.

A novel application of this nuclide has been described by JONES *et al.*[106] in which heat-denatured red cells were labeled with ^{81}Rb and were intravenously administered to localize in the spleen. The continuous generation of ^{81m}Kr and its continuous elution by perfusing blood offered a means of obtaining spleen blood flow.

Rubidium-82

Like ^{68}Ga, ^{82}Rb is a positron emitting radionuclide (Table 1) and as such is attractive for use with positron annihilation detection devices. The first rubidium generator was developed by YANO and ANGER[107] and employed a cation exchange resin to retain ^{82}Sr while permitting ^{82}Rb to be eluted in dilute ammonium acetate or sodium chloride solution. GRANT *et al.*[108] have developed a generator in which a chelating resin is used as the solid support. Generator produced rubidium has been used only as a monovalent cation in which form it achieves useful localization in myocardial and kidney tissue.[107]

Strontium-87m

The strontium generator was introduced in 1965[109,110] for use in imaging malignant bone lesions. Early versions required that the eluant be evaporated to dryness and taken up in sterile saline for administration. The generator was subsequently improved by HILLMAN *et al.*[111] such that it became possible to elute the generator with 0.005% citric acid solution at pH 5 to provide an eluant suitable for injection. Details of the strontium generators have been described by TROTT and POPHAM.[112]

^{87m}Sr has been used chiefly for bone imaging in which it is administered presumably in ionic form as Sr^{2+}. Only a few radiopharmaceuticals in which the strontium is in other chemical forms have been reported. CHAUDHURI *et al.*[113] have prepared a colloidal form of ^{87m}Sr by adjusting the pH of the generator eluant to 7.4 with a phosphate buffer. The strontium phosphate colloid accumulates to a high degree in the liver followed by gradual release as ionic strontium to accumulate in bone. A much larger particle type has been prepared[114] by introducing calcium gluconate to the phosphate buffer. The range of particle sizes is extreme and as a result combined lung and liver images may be obtained. Aggregates of strontium carbonate have been prepared and found to localize initially in lungs and, at longer times, in the liver.[115]

Iodine-132

The half life of ^{132}I is sufficiently long that this radionuclide may be tagged to the majority of the compounds which have been labeled with the more popular isotopes of iodine. However, ^{132}I possesses a particularly disadvantageous decay scheme for scintigraphy and as a consequence, it has not achieved popularity. GANATRA[116] has reviewed the generator development for this nuclide and its use as a radiopharmaceutical label.

Barium-137m

The first use of the barium generator in medicine was that of BLAU *et al.*[117] The eluant consisted of a dilute $HCl–NH_4Cl$ solution and required neutralization with NaOH prior to intravenous use. CASTRONOVO *et al.*[118] developed a generator in which ^{137m}Ba is eluted as the chloride at a pH of 1.0. The authors developed a rapid method of converting barium to the EDTA chelate at neutral pH. Subsequently, CASTRONOVO[119] added phenol red indicator to the eluant and a probe type colorimeter to the apparatus in order to monitor continuously the pH of the infusate. NAGAI and WATARI[120] have developed a generator which is eluted directly with physiological saline solution.

Iridium-191m

The sole report appearing in the literature on the development and medical use of the iridium generator is that of YANO and ANGER.[121] Their eluant consisted of a 16–18% NaCl solution. They have performed continuous infusion of the eluant in studies of human lung disease.

DISCUSSION

Along with technetium, the eight generators discussed herein provide the short-lived radionuclides which are either in use clinically or seriously considered for such.[2] Consideration of the properties of these radionuclides will illustrate several important aspects of radiopharmaceutical development. As an isotope of krypton, ^{81m}Kr is chemically inert; consequently it is unlikely that numerous radiopharmaceuticals will be developed with this nuclide particularly considering its short half life. The same conclusion applies to ^{191m}Ir, despite more favorable chemical properties, because of its exceedingly short half life. Similar considerations also apply to other radionuclides of Table 1. Rubidium, as an alkali metal, does not form stable chelates and this alone will limit the development of radiopharmaceuticals with ^{82}Rb. To some extent, this is also true of ^{87m}Sr and ^{137m}Ba. ^{113m}In and ^{68}Ga, however, have half lives sufficiently long to allow extensive chemical manipulation in agent development and, furthermore, both are isotopes of elements with versatile chemical properties. Consequently, both isotopes have been incorporated into numerous compounds of medical value.

An observation which may be apparent from this review is that radiochemical analysis of imaging agents labeled with these generator products is too infrequently performed. The half lives of ^{113m}In and ^{68}Ga for example are sufficiently long that rapid analytical techniques such as paper chromatography may be used to establish preparation purity. As a supplement to bioassay, the *in vitro* methods offer the advantage of greater reproducibility and are often simpler to perform.

Acknowledgements—I wish to thank MARSHA LEVINE for her assistance in the completion of this manuscript; in particular for finding most of my mistakes and introducing few of her own.

REFERENCES

1. BRUCER M. *Isotopes Radiat. Tech.* **3**, 1 (1965).
2. YANO Y. In *Radiopharmaceuticals* (Edited by SUBRAMANIAN G. *et al.*), p. 236. Soc. Nucl. Med., New York (1975).
3. COOPER J. F. and WAGNER H. N., JR. In *Radiopharmaceuticals from Generator-Produced Radionuclides*, p. 83. IAEA-PL-392 (1971).
4. KRAMER H. H. and STERN H. S. *J. nucl. Med.* **7**, 365 (1966) Abstract.
5. SUBRAMANIAN G. and MCAFEE J. G. *Int. J. appl. Radiat. Isotopes* **18**, 215 (1967).
6. STERN H. S., GOODWIN D. A., SCHEFFEL U., WAGNER H. N. and KRAMER H. H. *Nucleonics* **25**, 62 (1967).
7. HOSAIN F., MCINTYRE P. A., POULOSE K., STERN H. S. and WAGNER H. N. *Clin. Chim. Acta* **24**, 69 (1969).
8. HAYES R. L., BYRD B. L. and CARLTON J. E. *J. nucl. Med.* **9**, 323 (1968) Abstract.
9. MISHKIN F. S. and REESE I. C. *J. nucl. Med.* **9**, 462 (1968).
10. STERN H. S., GOODWIN D. A., WAGNER H. N. and KRAMER H. H. *Nucleonics* **24**, 57 (1966).
11. COLOMBETTI L. G., GOODWIN D. A. and HINKLEY R. L. *J. nucl. Med.* **10**, 633 (1969).
12. GEMMILL W. J., SCOTT D. M. and KRAMER H. H. *J. nucl. Med.* **9**, 170 (1968).
13. ADATEPE M. H., WELCH M., ARCHER E., STUDER R. and POTCHEN E. J. *J. nucl. Med.* **9**, 426 (1968).
14. TROW R. S., BROWN D. W., AHRENS C. R., CLEAVELAND J. D. and LEE J. I. *Radiology* **93**, 611 (1969).
15. RODRIGUEZ J., MACDONALD N. S. and TAPLIN G. V. *J. nucl. Med.* **10**, 368 (1969) Abstract.
16. CHAPMAN C., LONSDALE M. D. and HAYTER C. J. *Int. J. appl. Radiat. Isotopes* **21**, 679 (1970).
17. ERJAVES M., JENKO M. and RAVNIK V. *Radiol. Diagn.* **11**, 791 (1970).
18. BUCHANAN J. W., RHODES B. A. and WAGNER H. N. *J. nucl. Med.* **10**, 487 (1969).
19. GOODWIN D. A., STERN H. S., WAGNER H. N. and KRAMER H. H. *Nucleonics* **24**, 65 (1966).
20. HISADA K. and TSUTOMU M. *Radioisotopes* (Japan) **16**, 377 (1967).
21. SZYMENDERA J., KAMINSKA J., RADWAN M. and ZOLTOWSKI T. *Br. J. Radiol.* **42**, 944 (1969).
22. COLOMBETTI L. G., GOODWIN D. A. and HERMANSON R. *J. nucl. Med.* **10**, 597 (1969).
23. FRENCH R. J. *Br. J. Radiol.* **42**, 68 (1969).
24. RHODES B. A., BOURDOISEAU M., BUCHANAN J. W., COOPER J. F. and WAGNER H. N. *J. nucl. Med.* **11**, 356 (1970) Abstract.
25. SZYNENDERA J., ZOLTOWSKI T., RADWAN M., and KAMINSKA J. *J. nucl. Med.* **12**, 212 (1971).
26. SUBRAMANIAN G. and MCAFEE J. G. *J. nucl. Med.* **11**, 365 (1970) Abstract.
27. TOUYA E., TOUYA J. J., BEKERMAN C., PERILLO W., GUELFI A. G., OSORIO A. and FERRARI M. *J. nucl. Med.* **10**, 376 (1969) Abstract.
28. HILL T., WELCH M. J., ADATEPE M. and POTCHEN E. J. *J. nucl. Med.* **11**, 28 (1970).
29. O'MARA R. E., SUBRAMANIAN G., MCAFEE J. G. and BURGER C. L. *J. nucl. Med.* **10**, 18 (1969).
30. HISADA K., MISHIMA T., HIRAKI T., TONAMI N., MATSUDAIRA M. and KORANISHI M. *Radiology* **91**, 1204 (1968).
31. BURDINE J. A., WALTZ T. A., MATSEN F. A. and RAPP F. *J. nucl. Med.* **9**, 306 (1968) Abstract.
32. FERRARI M., PAEZ A., LOPEZ SOTO C., TOUYA E., BURGOS R. and TOUYA J. J. *Thorax* **24**, 695 (1969).
33. POE N. D., DORE E. K., WILSON A. F. and TAPLIN G. V. *J. nucl. Med.* **10**, 427 (1969) Abstract.
34. ALVAREZ J. In *Radiopharmaceuticals* (Edited by SUBRAMANIAN G. *et al.*), p. 102. Soc. Nucl. Med. New York (1975).
35. PENKOSKE P., POTCHEN E. J. and WELCH M. J. *J. nucl. Med.* **10**, 646 (1969).
36. WELCH M. J. and WELCH T. J. In *Radiopharmaceuticals* (Edited by SUBRAMANIAN G. *et al.*), p. 73. Soc. Nucl. Med., New York (1975).
37. MARDELL R. *J. nucl. Med.* **11**, 470 (1970).
38. ADATEPE M. H., WELCH M., EVENS R. G. and POTCHEN E. J. *Am. J. Roentgenol., Rad. Therapy and Nuclear Med.* **112**, 701 (1971).
39. MANI R. S. and NARASIMHAN D. V. S. *Radiopharmaceuticals and Labeled Compounds*, p. 135. IAEA-SM-171 Vol. 1.
40. BESNARD M., GAUCHER C. H. and MERLIN L. *J. Radioanal. Chem.* **9**, 73 (1971).
41. JAISWAL D. K., CHANDER J., MALIK A. and SINGH B. *J. Chrom.* **84**, 417 (1973).
42. MATIN P. and GOODWIN D. A. *J. nucl. Med.* **12**, 668 (1971).

43. Goodwin D. A., Song C. H., Finston R. and Matin P. *Radiology* **108**, 91 (1973).
44. Goodwin D. A., Imbornone C. J. and Song C. H. *J. nucl. Med.* **12**, 434 (1973) Abstract.
45. Merrick M. V., Nunn A. D., Thakur M. L. and Gunasekera S. W. *Int. J. nucl. Med. Biol.* **2**, 45 (1975).
46. Goodwin D. A., Meares C. F. and Song C. H. *Radiology* **105**, 699 (1972).
47. Grove R. B., Eckelman W. C. and Reba R. C. *J. nucl. Med.* **14**, 917 (1973).
48. Robbins P. J., Silberstein E. B. and Fortman D. L. *J. nucl. Med.* **15**, 273 (1974).
49. Prichard J. H., Tubis M., Blahd W. H., Endow J. S. and McDonald N. S. *J. nucl. Med.* **15**, 526 (1974) Abstract.
50. Sundberg M. W., Meares C. F., Goodwin D. A. and Diamanti C. I. *Nature, Lond.* **250**, 587 (1974).
51. Subramanian G., McAfee J. G., Rosenstreich M. and Coco M. *J. nucl. Med.* **16**, 1080 (1975).
52. Jones A. G., Davis M. A. and Dewanjee M. K. *Radiology* **117**, 727 (1975).
53. Dewanjee M. K. and Kahn P. C. *Radiology* **117**, 725 (1975).
54. Adatepe M. H., Penkoske P., Van Amberg A., Wharton T., Evens R. G. and Potchen E. J. *Int. J. appl. Radiat. Isotopes* **22**, 498 (1971).
55. Dugan M. A., Kozar J. J., Ganse G. and Quap C. *J. nucl. Med.* **13**, 782 (1972) Abstract.
56. Buraggi G. L., Ringhini R., Rodari A. and Scaiano D. *J. Biol. Nucl. Med.* **15**, 23 (1971).
57. Sinn H., Georgi P., Clorius J. and Maier-Borst W. *Nucl. Med.* (*Stuttg.*) **13**, 180 (1974).
58. Adatepe M. H., Studer R., Welch M. and Potchen E. J. *Int. J. appl. Radiat. Isotopes* **22**, 220 (1971).
59. Goodwin D. A. *J. nucl. Med.* **12**, 580 (1971).
60. Barker S. L. and Gusmano E. A. *J. nucl. Med.* **12**, 581 (1971).
61. Thomas J. and Wiener S. N. *J. nucl. Med.* **12**, 467 (1971) Abstract.
62. Allen D. R., Hartnett D. E., Nelp W. B. and Hamilton G. W. *J. nucl. Med.* **15**, 821 (1974).
63. Scott J. R., Shysh A., Noujaim A. A., Wiebe L. I. and Ediss C. *Int. J. appl. Radiat. Isotopes* **25**, 139 (1974).
64. Tubis M., Cohen M. B., Wagner M. S. and Gilliam C. D. *Medical Radioisotope Scintigraphy*, p. 395. IAEA-SM-164 Vol. II (1973).
65. Caro R. A., Ciscato V. A., Ihlo J. E., Nicolini J. O., Palcos M. C. and Radicella R. *Int. J. appl. Radiat. Isotopes* **25**, 501 (1974).
66. Buchanan J. W., Rhodes B. A. and Wagner H. N. *J. nucl. Med.* **12**, 616 (1971).
67. Raban P., Gregora V., Sindelar J. and Alvarez-Cervera J. *J. nucl. Med.* **14**, 344 (1973).
68. Scheffel U., Rhodes B. A. and Wagner H. N. *J. nucl. Med.* **12**, 462 (1971) Abstract.
69. Isawa T., Hayes M. and Taplin G. V. *J. nucl. Med.* **12**, 138 (1971).
70. Lin M. S., Burke G. and Silverstein G. E. *Radiology* **107**, 449 (1973).
71. Cook D. J. and Lander H. *J. nucl. Med.* **12**, 765 (1971).
72. Gleason G. I. *Int. J. appl. Radiat. Isotopes* **8**, 90 (1960).
73. Greene M. W. and Tucker W. D. *Int. J. appl. Radiat. Isotopes* **12**, 62 (1961).
74. Kopecky P. and Mudrova B. *Int. J. appl. Radiat. Isotopes* **25**, 263 (1974).
75. Yano Y. and Anger H. O. *J. nucl. Med.* **5**, 485 (1964).
76. Hayes R. L., Carlton J. E. and Byrd B. L. *J. nucl. Med.* **6**, 605 (1965).
77. Weber D. A., Greenberg E. J., Dimich A., Kenny P. J., Rothschild E. D., Myers W. P. L. and Laughlin J. S. *J. nucl. Med.* **10**, 8 (1969).
78. Hayes R. L. *Radioactive Pharmaceuticals* (Edited by Andrews *et al.*), p. 603. USAEC CONF-651111 (1966).
79. Colombetti L. G. and Goodwin D. A. *J. nucl. Med.* **10**, 396 (1969) Abstract.
80. Carlton J. E. and Hayes R. L. *Int. J. appl. Radiat. Isotopes* **22**, 44 (1971).
81. Goulet R. T., Shysh A., Noujaim A. A. and Lentle B. C. *Int. J. appl. Radiat. Isotopes* **26**, 195 (1975).
82. Hnatowich D. J. *J. nucl. Med.* **16**, 764 (1975).
83. Waxman A. D., Kawada T., Wolf W. and Siemsen J. K. *Radiology* **117**, 647 (1975).
84. Yano Y. *Radiopharmaceuticals from Generator-Produced Radionuclides*, p. 117. IAEA-PL-392 (1971).
85. Stemple R. M. *Technetium-99m and gallium in medicine—a supplementary bibliography*. ORNL-TM-2091 #2 (1971).
86. Edwards C. L. and Hayes R. L. *JAMA* **212**, 1182 (1970).
87. Hartman R. E. and Hayes R. L. *J. Pharmacol. Exp. Ther.* **168**, 193 (1969).
88. Shealy C. N., Aronow S. and Brownell G. L. *J. nucl. Med.* **5**, 161 (1964).
89. Anghileri L. J. *Int. J. appl. Radiat. Isotopes* **19**, 421 (1968).
90. Konikowski T., Glenn H. J. and Haynie T. P. *J. nucl. Med.* **14**, 164 (1973).
91. Dewanjee M. K., Kahn P. C., Carmel A. and Dewanjee U. *J. nucl. Med.* **15**, 487 (1975) Abstract.
92. Burleson R. L., Johnson M. C. and Head H. *J. nucl. Med.* **15**, 98 (1974).
93. Hill J. H., Merz T. and Wagner H. N. *J. nucl.*

Med. **16,** 1183 (1975).
94. REBA R. C., ECKELMAN W. C. and REBA R. C. *J. nucl. Med.* **14,** 627 (1973) Abstract.
95. HAYES R. L., CARLTON J. E. and RAFTER J. J. *J. nucl. Med.* **7,** 335 (1966) Abstract.
96. ANGHILERI L. J. and PRPIC B. *Int. J. appl. Radiat. Isotopes* **18,** 734 (1967).
97. COLOMBETTI L. G., GOODWIN D. A. and TOGAMI E. *J. nucl. Med.* **11,** 704 (1970).
98. COLOMBETTI L. G., HUBER M. and RAVASINI R. *J. nucl Biol. Med.* **16,** 16 (1972).
99. CHESLER D. A., HALES C., HNATOWICH D. J. and HOOP B. *J. nucl. Med.* **16,** 80 (1975).
100. HNATOWICH D. J. *J. nucl. Med.* **17,** 57 (1976).
101. BROWNELL G. L., BURNHAM C. A., HOOP B., CHESLER D., CONNOR H., KAZEMI H. and BUNNELL J. *J. nucl. Med.* **13,** 417 (1972) Abstract.
102. JONES T. and CLARK J. C. *Br. J. Radiol.* **42,** 237 (1969).
103. YANO Y., MCRAE J. and ANGER H. O. *J. nucl. Med.* **10,** 384 (1969) Abstract.
104. YANO Y., MCRAE J. and ANGER H. O. *J. nucl. Med.* **11,** 674 (1970).
105. MAYRON L. W., KAPLAN E., COLOMBETTI L. G., FRIEDMAN A. M. and GINDLER J.E. *Int. J. nucl. Med. Biol.* **2,** 40 (1975).
106. JONES T., PETTIT J. E., RHODES C. G. and WATERS S. L. *J. nucl. Med.* **14,** 414 (1973) Abstract.
107. YANO Y. and ANGER H. O. *J. nucl. Med.* **9,** 412 (1968).
108. GRANT P. M., ERDAL B. R. and O'Brien H. A. *J. nucl. Med.* **16,** 300 (1975).
109. ALLEN J. F. and PINAJIAN J. J. *Int. J. appl. Radiat. Isotopes* **16,** 319 (1965).
110. MECKELNBURG R. L. *J. nucl. Med.* **5,** 929 (1964).
111. HILLMAN M., GREENE M. W., BISHOP W. N. and RICHARDS P. *Int. J. appl. Radiat. Isotopes* **17,** 9 (1966).
112. TROTT N. G. and POPHAM M. G. In *Radiopharmaceuticals from Generator-Produced Radionuclides,* p. 105. IAEA-PL-392 (1971).
113. CHAUDHURI T. K., CHAUDHURI T. K. and CHRISTIE J. H. *Int. J. appl. Radiat. Isotopes* **23,** 204 (1972).
114. CHAUDHURI T. K., CHAUDHURI T. K. and CHRISTIE J. H. *J. nucl. Med.* **14,** 346 (1973).
115. MYERS W. G. and HUNTER W. W. *J. nucl. Med.* **9,** 337 (1968) Abstract.
116. GANATRA R. D. *Radiopharmaceuticals from Generator-Produced Radionuclides,* p. 111. IAEA-PL-392 (1971).
117. BLAU M., ZIELINSKI R. R. and BENDER M. A. *Nucleonics* **24,** 60 (1966).
118. CASTRONOVO F. P., REBA R. C. and WAGNER H. N. *J. nucl. Med.* **10,** 242 (1969).
119. CASTRONOVO F. P. *J. nucl. Med.* **14,** 341 (1973).
120. NAGAI T. and WATARI K. *J. nucl. Med.* **9,** 608 (1968).
121. YANO Y. and ANGER H. O. *J. nucl. Med.* **9,** 2 (1968).

International Journal of Applied Radiation and Isotopes, 1977, Vol. 28, pp. 183–201. Pergamon Press. Printed in Northern Ireland

Gallium-67 and Indium-111 Radiopharmaceuticals

M. L. THAKUR*

Medical Research Council, Cyclotron Unit, Hammersmith Hospital, Ducane Road, London W12 0HS, U.K.

(*Received* 4 *April* 1976)

There are no compounds known to exist naturally in the human body which involve gallium or indium metal cofactors. In spite of this numerous compounds which contain ^{67}Ga and ^{111}In as their integral part have been prepared mainly for localization studies. One of the important considerations in these preparations has been the physical characteristics of ^{67}Ga and ^{111}In which are summarized together with the methods of their preparation.

Factors governing the chelation of organometallic radiopharmaceuticals, their *in vivo* stability and usefulness have been discussed. The compounds prepared have been reviewed and summarized together with the basic concept in their preparation. The results of their *in vivo* evaluations have been outlined and the merits for potential future applications of some of the compounds have been mentioned.

INTRODUCTION

THIS article reviews the data available on pharmaceuticals labeled with ^{67}Ga and ^{111}In. Prior to the general availability of these radionuclides to a radiopharmaceutical chemist, generator produced ^{68}Ga ($T_{1/2} = 68$ min) and ^{113m}In ($T_{1/2} = 100$ min) had been widely used. Since the pharmaceuticals labeled with these latter two radionuclides are reviewed elsewhere in this issue, no attempt has been made to refer to those radiopharmaceuticals.

Most of the radiopharmaceuticals used in nuclear medicine may be classified into two groups: first, those pharmaceuticals prepared for the dynamic functional studies of certain organs in the body, such as radioiodinated hippuran for renal function studies and radioiodinated sodium iodide for thyroid function studies, and second, those radiopharmaceuticals designed to localize abnormalities in the body, for example, the compounds prepared for the localization of tumors, detection of infarcts and visualization of certain other abnormal organs. Most of the compounds prepared with ^{67}Ga and ^{111}In fall in the second group of radiopharmaceuticals. The success of these radiopharmaceuticals is related to several factors which are discussed together with the physical characteristics of the radionuclides. The methods of production of the radionuclides have been reviewed and the methods of preparation of their labeled compounds are reported with data on the evaluation of their clinical applicability.

ROLE OF ORGANOMETALLICS

The human body needs both organic and inorganic compounds for biochemical synthesis. Approximately 12% of all known enzymes need a metal co-factor.[1] The importance of several metal ions, their *in vivo* incorporation into organometallics, and the relation of their concentration in the body to health have been recently described by PIER[2].

Iron, for example is an essential component of cytoplasm, catalase and myoglobin, and is required for the synthesis of vitamin B. Since iron is also an essential component of hemoglobin, isotopes of iron, e.g. ^{59}Fe and ^{52}Fe, have been used in nuclear medicine for the investigation of hematological disorders. Cobalt is a

* The article was written while on sabbatical at Mallinckrodt Institute of Radiology, Washington University School of Medicine, 510 South Kingshighway, St. Louis, MO 63110, U.S.A.

constituent of cyanocobalmin (vitamin B_{12}) and its radionuclides, namely, ^{56}Co, ^{57}Co, ^{58}Co are used to evaluate disorders of vitamin B_{12} metabolism. The radionuclides of iodine, ^{131}I, ^{125}I and ^{123}I, are used widely to investigate the function of thyroid glands.

The main source of these elements in the body is food. Plants and animals extract these elements from the Earth's crust which is rich with several of these metals. The elements gallium and indium, however, have very poor abundance in the Earth's crust. Their intake in the human body is, therefore, so small that it has not been investigated. Since data on the utilization of these two elements by the body are not available, the radiopharmaceutical chemist is unable to prepare any compound of gallium or indium that may be naturally useful for the human body. The choice of an organic compound in the preparation of the gallium-67 and indium-111 pharmaceuticals is therefore based on the biochemical properties of the compound. For example, indium-111 bleomycin was prepared as a tumor-localizing agent[3] with the concept that the cytotoxic bleomycin would carry the radionuclide to the target tissue.

The organometallics prepared on the basis of the biochemical properties of the compound may have two basic disadvantages. First, the biological properties of the organic compound may be altered or destroyed due to the complex formation. Second, upon administration of the organometallic into the body, the radioactive metal ion may be detached from the compound due to its unsuitable thermodynamic stability.

CHEMICAL AND PHYSICAL CHARACTERISTICS OF GALLIUM AND INDIUM

To form an organometallic compound of a useful thermodynamic stability, the metal ion should have certain characteristics. Monovalent metal ions like Na^+ and K^+ have less avidity for any chelating agent and form less stable compounds than bivalent Cu^{2+} or trivalent Fe^{3+} ion. Both gallium and indium are trivalent elements and belong to group IIIA in the periodic classification of elements. Since they possess only one valency state, 3+, in most of their compounds (under very exceptional circumstances, gallium and indium exhibit an oxidation state 2+ and form stable compounds like GaI_2 and InI_2), under normal circumstances no oxidation or reduction of the metal ions has been necessary before the complex formation. Gallium and indium both have a six coordination number and generally form octahedral complex structures.

Gallium is more electronegative than indium and forms complexes of equal or slightly higher stability than the stability of corresponding indium complexes. Both elements form insoluble hydroxides at approximately pH 6 which are soluble only in strong acid and base. They are highly metallic in their properties but melt at relatively low temperature; Ga at 29.8°C and In at 156°C. Their ionic (3+) radii are 0.62 Å and 0.81 Å respectively.

PRODUCTION AND DECAY CHARACTERISTICS OF ^{67}Ga AND ^{111}In

Gallium-67 and ^{111}In are accelerator-produced radionuclides. Not until 1953 was ^{67}Ga first produced for *in vivo* use.[4] Sixteen years later ^{111}In was employed for tumor localization[5] and for lymphatic scintigraphy[6] in animals. Several methods[7–19] of producing ^{67}Ga and ^{111}In are reported in the literature which are summarized in Tables 1 and 2. A comparative study of ^{67}Ga production methods has been carried out by Helus and Maier-Borst[20].

Both ^{67}Ga and ^{111}In decay by electron capture and have no β emission. Their physical half lives of 78 hr and 67 hr respectively are well suited for several *in vivo* applications. These half lives allow *in vivo* studies to be carried out for several days without having to administer excessively large quantities of radioactivity.

Considering the gamma photon emission of ^{67}Ga and ^{111}In, indium-111 is the nuclide of choice. The gamma energies of the radionuclide, 173 keV and 247 keV, are in the optimum range of detectability for the commercially available detecting devices and the abundance of the gamma emissions (89 and 94% respectively) provide 183 photons in every 100 disintegrations. Although the gamma energies of ^{67}Ga (Table 1) are in a suitable range for detection, they have poor abundance and give only 93 photons in every 100 disintegrations. Therefore, three times

TABLE 1. Gallium-67: decay characteristics and methods of production
$t_{1/2}$ = 78 hr; EC, no β; Zn-X rays—93 keV (40%); γ—184 keV (24%); 296 (22%); 388 (7%) Decays to stable zinc

Nuclear reaction	Incident beam energy	Thick target yield at E.O.B. (μCi/μhr)	Method of purification	Radiochemical impurities	Reference
$^{67}Zn(p, n)$ $^{68}Zn(p, 2n)$	25	?	Solvent extraction	?	4
$^{67}Zn(p, n)$ $^{68}Zn(p, 2n)$	21	330–400	Solvent extraction	^{66}Ga, ^{65}Zn	7
$^{65}Cu(\alpha, 2n)$	30	160	Ion exchange	^{66}Ga, ^{65}Zn	8
$^{66}Zn(d, n)$ $^{67}Zn(d, 2n)$	16	352	Ion exchange	^{66}Ga	9
$^{66}Zn(d, n)$ $^{67}Zn(d, 2n)$	16	340	Ion exchange	?	10
$^{66}Zn(d, n)$ (90% enriched)	16	946	Ion exchange	^{66}Ga	11
$^{67}Zn(p, n)$ $^{68}Zn(p, 2n)$	21	~1500	Solvent extraction	^{66}Ga	12
$^{66}Zn(d, n)$	15	58 ± 20	$Fe(OH)_3$ Co-precipitation	^{66}Ga	13
$^{67}Zn(d, 2n)$	8	30	Solvent extraction		14

TABLE 2. Indium-111: decay characteristics and methods of production
$t_{1/2}$ = 67 hr; EC, no β; Cd-X-rays; γ—173 (89%); 247 (94%) Decays to stable cadmium

Nuclear reaction	Incident beam energy	Thick target yield at E.O.B. (μCi/μhr)	Method of purification	Radiochemical impurities	Reference
$^{110}Cd(d, n)$	15	18 ± 12	$Fe(OH)_3$ Co-precipitation	^{109}In, ^{115}In	13
$^{109}Ag(\alpha, 2n)$	?	?	Ion exchange	?	15
$^{109}Ag(\alpha, 2n)$	?	?	$Fe(OH)_3$ Co-precipitation	?	16
$^{109}Ag(\alpha, 2n)$	30	200	Ion exchange	^{109}In	17
$^{110}Cd(d, n)$	?	?	In $(OH)_3$ precipitation	^{114}In, ^{115}In	18
$^{111}Cd(p, n)$	16	515	Solvent extraction	^{114m}In	19
$^{111}Cd(p, n)$	15	140	Solvent extraction	^{114m}In	14

as much ^{67}Ga as ^{111}In is necessary to obtain equal count rate at fixed geometry.[21] In spite of these disadvantages, ^{67}Ga has been used more extensively than ^{111}In in clinical applications.

CHELATING AGENTS

The chelating agents in the organometallics designed for *in vivo* use are expected to perform several roles. First, they should have a suitable functional group or groups in their molecule which would "chele" or "claw" the desired metal ion. Second, they should keep their biological activity unaltered even after the complex formation. Third, they should carry a substantial portion of the injected radioactivity to the target tissue. Fourth, the rest of the radiopharmaceutical should clear rapidly from the circulating blood and should ideally leave the body through excretion.

The chelating agents are generally classified on their functional group or ligand involved. A ligand supplies electron density for the covalent bond formed to the metal ion. Some typical

TABLE 3. Some typical ligands

Donor atom	Ligand
C	CN^-
N	NO_2^-, N_3^-, NCS^-, NH_2^-
O	OH^-, O_2^-, O_2^{2-}, CO_3^{2-}, SO_4^{2-}, $S_2O_3^{2-}$ RCO_2^-, $C_2O_4^{2-}$
P	PR_2^-, PO_4^{3-}, $P_2O_7^{4-}$
S	SCN^-, $S_2O_3^{2-}$

ligands commonly used in the formation of organometallic compounds are listed in Table 3.

When the chelating agent has one such ligand it is called a monodentate ligand. Some compounds have two or more ligands in their molecules. These compounds are referred to as bidentate or polydentate ligands. Some of these compounds used in the preparation of ^{67}Ga and ^{111}In pharmaceuticals are given in the text showing the possible sites of their attachment to a metal ion M.

The biological activity of a compound is also a direct function of such a ligand. When these ligands are used for the formation of a covalent bond to the metal ion, the organometallic formed is chemically and biologically a new compound. Since all the compounds prepared with ^{67}Ga and ^{111}In for *in vivo* use are not naturally available in the body, they would be expected to have different *in vivo* behavior than the chelating agent.

FACTORS AFFECTING DETECTABILITY

Since detection of radioactivity in a target tissue is an aim in nuclear medicine, factors affecting the accumulation play major roles in the clinical usefulness of a radiopharmaceutical. Assuming a radionuclide with ideal gamma photon is chosen, at least six factors can affect the detectability. These factors and their relation to each other are schematically shown in Fig. 1 and discussed as follows:

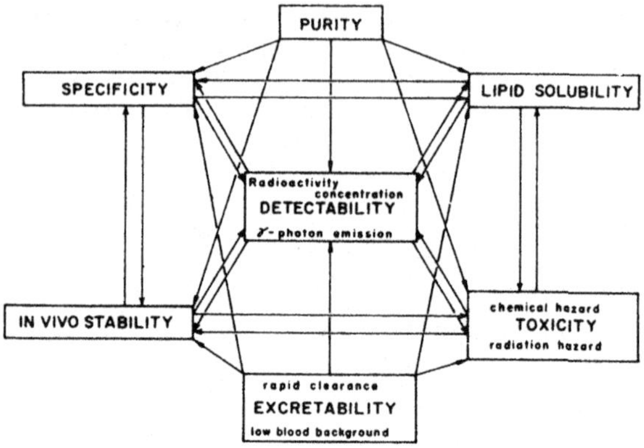

FIG. 1. Factors affecting the detectability and their relation to each other.

1. In vivo *stability*

The stability of a complex has, unfortunately, acquired several loose definitions. In this discussion stability refers to the thermodynamic stability which is related to the metal-ligand bond energies. When the organometallic compound LM* enters the body it is met with (1) several other competing ligands Li and (2) several metal ions Mi. If these have higher affinity for the radionuclide and for the ligand, the thermodynamic stability of the radiopharmaceutical will be challenged. This would result in the formation of new compounds LiM* and LMi, *in vivo* according to the following equation

$$LM^* + Li + Mi \rightarrow LiM^* + LMi$$

Most of the compounds prepared with ^{67}Ga and ^{111}In have poor thermodynamic stability. This poor stability results in translocation of the radioactivity to plasma transferrin within a short time after the radiopharmaceutical has been administered and in loss of the specificity of the original compound.

Unfortunately, little data on the stability constants of the gallium and indium compounds are available. The stability constants of only 3 gallium and 6 indium compounds have been reported, compared to those of more than one hundred compounds prepared with iron.[22] WELCH and WELCH, however, have recently derived the following (corrected) equation:[23]

$$\log K_{In} = -0.5854 + 1.0108 \log K_{Fe}$$

where $\log K_{In}$ and $\log K_{Fe}$ are the equilibrium constants for the indium and ferric ions respectively. Such an equation could also be derived for gallium compounds, and stability constants for several gallium and indium compounds may be estimated.

2. *Lipid solubility*

Lipid solubility is another factor which may influence the detectability, particularly when the organometallic is designed for penetration into the cells. Lipid solubility is related to the functional groups in a compound. Aliphatic hydrocarbon chains, aryl alkyl groups and polycyclic hydrocarbons enhance the lipid solubility of compounds of which they are part. These lipophilic agents have been recently discussed by McAfee[24]. The lipid solubility of a compound is determined by the oil/water partition coefficient. Only a few lipid soluble ^{67}Ga and ^{111}In compounds have been investigated.

3. *Purity*

The purity of a radiopharmaceutical is often ignored. A radiopharmaceutical free of any undesirable radionuclidic, chemical and radiochemical impurity is considered pure. Gallium-67 and ^{111}In have reasonably long half-lives so that essential chromatographic and spectroscopic tests can be performed to evaluate their purity. A small amount of impurities in a radiopharmaceutical can make it toxic, non-specific, unstable and lose its desired characteristics. A longer-lived radionuclide which may be present as an impurity may restrict the amount of activity that may be administered and may influence the detectability adversely.

4. *Specificity*

Specificity is an important factor that can influence accumulation and detection. The specificity of a pharmaceutical is related to the amount of the administered dose that concentrates in the target tissue. Although most of the compounds prepared with ^{67}Ga and ^{111}In are chosen to carry out a specific role in the body, none of them has been found to be specific. The purity, lipid solubility, *in vivo* stability and excretability may also affect the specificity favorably or unfavorably as the case may be.

5. *Toxicity*

Toxicity in this discussion is referred to the damage that can be caused by chemicals and radioactivity administered with a radiopharmaceutical. In most cases, the quantity of a ^{67}Ga and ^{111}In compound administered is small and, generally, safe. For example 150 μg of 8-hydroxyquinoline administered to a 70 kg man is 30 thousand fold less than its LD_{50} dose.[25] These radiopharmaceuticals are usually administered only once or at most, a few times, and thus remain non-toxic. However, they should be free of any toxic impurity.

The radiation damage that may be related to the radioactivity administered has been carefully considered by several workers.[26–28] The radiation dose, any toxic chemical and longer lived radionuclidic impurity restrict the quantity of radioactivity that can be safely administered and therefore, affect the detectability of the radiopharmaceutical.

6. *Excretion*

Excretion of the radionuclide affects the detectability by two methods. If the unutilised radioactivity is rapidly excreted, the detectability can be enhanced by (1) having less radioactive background in the body and (2) allowing more radioactivity to be administered safely. However, when the radioactivity in the non-target organs is not cleared from the body, the effect is reversed, thus restricting the amount of radioactivity that can be safely administered and giving higher blood and body background. The blood clearance and excretion of the majority of the compounds prepared with ^{67}Ga and ^{111}In have been studied and are discussed later.

COMPOUNDS PREPARED WITH GALLIUM-67

Interest in the physiological significance of gallium arose as early as 1949 when Dudley[29] tried to determine the presence of gallium in biological materials following the discovery of gallium radionuclides in fission products. Dudley injected 100 mg gallium lactate per kg body weight subcutaneously and 33 mg intravenously to rats and rabbits. Lactic acid (I) (see over) was chosen as a biodegradable compound which kept gallium in solution. Tissue distribution of the compound was estimated chemically over 5 hr to 60 days post injection.

From this work DUDLEY concluded that the bones act as "storage depots" for gallium, yielding it slowly, and that appreciable quantities still persist for as long as 60 days after a single injection of gallium lactate. In his next two publications, DUDLEY reported the LD_{50} dose of gallium lactate (47 and 43 mg/kg body weight of rats and rabbits respectively)[30] and observed[31] that gallium is transported by the plasma proteins and largely clears from the plasma within 24 hr. He also noted that gallium is cleared from the body by the kidneys and that small but appreciable quantities are "fixed" in the liver and kidneys and retained in part for more than 30 days. In the same publication he suggested that gallium may prove to be a new agent for bone metabolism studies.

DUDLEY and MADDOX[32] in the same year injected gallium-72 ($T_{1/2} = 14.1$ hr lactate to dogs and demonstrated, autoradiographically, the deposition of the gallium activity in the bones.

In 1951, LANG[33] injected low specific activity, reactor produced (350–400 μCi) gallium-72 lactate to humans with advanced bony metastases. Counting of the deposited radioactivity was performed with end window Geiger–Müller counters. The authors, with 12 patient studies, reported complete agreement with the previous results of the animal studies and found 20 fold greater concentration of the radioactive gallium in human bone lesions than in normal bone.

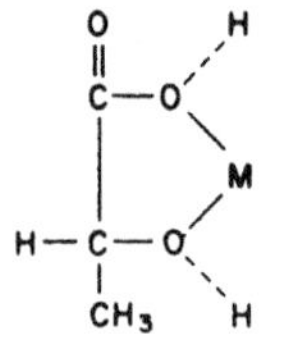

(I)

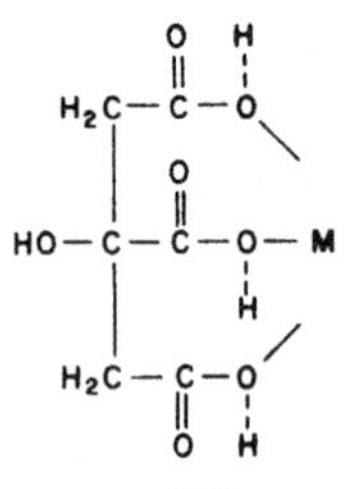

(II)

Later DUDLEY[34] suggested that gallium-72 citrate (II) was easier to prepare than any other injectable compound of gallium including gallium lactate. BRUNER *et al.* in 1953[35] noted two disadvantages of gallium-72. First, this reactor-produced radionuclide has a low specific activity, and, second, it has a relatively short half life which does not permit its use in studies to be carried beyond a few hours after administration. They therefore prepared carrier-free, 78 hr half life gallium-67 by 25 MeV proton bombardment of high purity zinc in a cyclotron.

The animal experiments with carrier-free gallium-67 citrate showed less bone uptake of the activity than the previously reported data obtained with the use of stable gallium and low specific activity ^{72}Ga. These observations were later confirmed by VAN DER WERLF[36] in the clinical trials with carrier-free ^{66}Ga and ^{67}Ga citrate.

HARTMAN and HAYES[37] in their efforts to improve gallium-67 citrate as a bone scanning agent studied the influence of gallium carrier. They concluded from their work that the deposition of gallium activity in bone was enhanced only when the plasma was saturated with gallium. When carrier-free gallium citrate was administered it was sequestered by the plasma and no gallium ions were available for bone deposition. Binding of the gallium activity to plasma depended upon the citrate concentration and pH of the solution. At pH 5.5 only 43% of the gallium activity was associated with plasma which increased to more than 90% at pH 6.5 and above. Thus, the use of carrier containing gallium citrate as a bone scanning agent was discontinued due to its toxicity and to its lack of sufficient thermodynamic stability and specificity.

While working with carrier-free gallium-67 citrate the same group[38] reported in 1969 an observation which came to them as "an unexpected bonus." They observed that the radiogallium concentrates in soft tissue tumors when administered as citrate. This accidental discovery created a new era in the story of ^{67}Ga. Gallium-67 citrate since then has become one of the main subjects of investigation in nuclear medicine and has gained reputation as a soft tissue tumor localizing agent.[39–101]

Not much is known about the mechanism by which the radiogallium concentrates in tumors. DALRYMPLE *et al.*[39] suggested passive diffusion through cell membrane as a mechanism. Considering the possible influence of lipid solubility, CLAUSEN *et al.*[102] tried to extract the gallium-67 activity associated with tumor tissue into *n*-butanol. Only 26.6% of the activity was extracted. The rest of the activity, they confirmed, was firmly bound to haptoglobin and transferrin,[103] which has 14 binding sites for gallium when gallium is in a colloidal state.[102] CLAUSEN *et al.* found that the tumor to blood and tumor to surrounding soft tissue ratios depended upon the type of tumors. The ratios were highest (316–388) for reticulum cell sarcoma and lowest (1.2–0.3) for the malignant melanoma. Since proteins transport the activity, the possible influence of vascular permeability on the accumulation of radiogallium in the soft tissue tumors must be considered. A great deal of discussion on the mechanism of radiogallium uptake in soft tissue tumors is given in recent articles.[104–106] However, nothing has been found as yet which would help a radiopharmaceutical chemist to improve the specificity of the tumor localizing agent.

Very little has been reported concerning the radiochemical purity of privately prepared or commercially available gallium-67 citrate. Since gallium forms a 1:1 complex of a moderate stability with citrate ions and since there is a high enough concentration of citrate ions to sequester carrier free gallium ions, no free gallium activity should be expected provided there is a negligible concentration of other metal ions in the solution. Furthermore, if there was a small percentage of free gallium ions they would soon form gallium transferrin.[107] However, difference in the gallium-67 citrate preparations available from different commercial suppliers have been observed by WAXMAN *et al.*[108] Upon analyzing the products chromatographically using 85% methanol as a solvent, different *Rf* values were obtained with a compound supplied by one of three manufacturers. The *Rf* value shifted when citrate ions were added to one of the other two products. No conclusion can be reached from these observations and further work is required to elucidate this phenomenon.

Not many other compounds of ^{67}Ga have been thoroughly evaluated. KONIKOWSKI *et al.*[107] compared the kinetics of ^{67}Ga-chloride, lactate, citrate and ^{67}Ga-iron DTPA (Diethylenetriaminepentaacetic acid) in mice brain sarcomas and kidneys. A known quantity of iron was mixed with ^{67}Ga chloride solution before a required (1:1) quantity of DTPA was added. The pH of the solution was adjusted to 7–7.5. The hexadentate DTPA forms a much stronger complex with gallium than the other three compounds. The DTPA bound gallium-67 activity transfers to plasma transferrin much more slowly than the other three compounds, clears much faster from the blood and has a smaller degree of concentration in the tumors than the other three compounds. The initial concentration of the radioactivity in the liver was inversely proportional to the thermodynamic stability of the complex ions. The initial liver concentration was much higher with ionic ^{67}Ga-chloride than with ^{67}Ga-lactate and citrate. The stronger ^{67}Ga-DTPA species showed appreciably smaller liver uptake than the other three compounds.

Bleomycin, a cytotoxic multidentate antitumor agent, has been labeled with ^{67}Ga by THAKUR *et al.*[109] and has been compared with ^{111}In-labeled and ^{99m}Tc-labeled bleomycin as a tumor localizing agent in animals. Later, GROVE *et al.*[110] compared the compound with ^{111}In-labeled and ^{57}Co-labeled bleomycin in patients. Adequate radiochemical purity of the compound was reported by both groups. No advantages of ^{67}Ga bleomycin as a tumor localizing agent have been demonstrated compared to ^{111}In bleomycin and no further work has been carried out.

DEWANJEE *et al.*[111] have prepared ^{67}Ga phytate and evaluated it as lymph node scanning agent in animals. Phytic acid (1,2,3,4,5,6-cyclohexanehexolphosphoric acid) is a multidentate complexing agent often used for the removal of traces of heavy metal ions. Sodium phytate has also been used clinically in hypercalciuria. The use of ^{67}Ga phytate as a lymph node scanning agent has been compared with ^{99m}Tc phytate, ^{99m}Tc-sulphur colloid and ^{67}Ga-hydroxide colloid. Promising results have been reported. However, no further reports have been available in the literature.

BURLESON *et al.* have labeled rabbit[112] and human[113] leukocytes with ^{67}Ga citrate for the localization of abscesses. Approximately 250 ml human blood was incubated with 6 mCi gallium-67 citrate at room temperature for 30 min. Plasma was removed by centrifugation and activity associated with cells was administered back to the patients. Only about 6% of the initial activity was bound to leukocytes. Since abscesses are a focal accumulation of leukocytes, this agent should have been very specific for the localization of abscesses. However, the poor labeling efficiency has restricted its *in vivo* applicability.

COMPOUNDS PREPARED WITH INDIUM-111

Most of the earlier work on the preparation of radioactive indium compounds for medical applications was carried out with generator-produced ^{113m}In ($T_{1/2} = 100$ min). The disadvantages of its short half-life for certain applications were soon apparent and necessitated a search for a better replacement. Indium-111 has a suitable physical half life (67 hr) for several applications and has desirable gamma photon energies (173 keV–89% and 247 keV–94%) for external detection with commercially available detecting devices. Although the first reference to an ^{111}In compound for *in vivo* use was not recorded until 1969, its attractive physical characteristics have initiated preparation of a variety of useful compounds which are more numerous than the compounds prepared with ^{67}Ga.

1. *Indium*-111 *chloride*

Indium-111 was first administered to tumor-bearing rats and dogs in the simple ionic form indium chloride. Twenty-four to 48 hr after injection "nearly all"[(5)] of the tracer was accumulated in the tumors irrespective of the cell type.

It was soon apparent from clinical trials[(114)] that the indium activity administered as chloride concentrated in soft tissue tumors as well as in bones. The uptake of the radioindium has been attributed to the formation of ^{111}In transferrin. A number of investigators[(115–119)] have further studied ionic ^{111}In as a potential tumor localizing and bone marrow imaging agent. The *in vivo* conversion of ionic indium to indium transferrin is so well established that GOODWIN *et al.*[(115)] and FARRER *et al.*[(118–119)] have entitled their reports as "^{111}In-labeled transferrin . . ." although only ^{111}In chloride was administered. Recently, indium-111 citrate has been used in patients for bone marrow imaging.[(120)] Since the thermodynamic stability of the ^{111}In citrate complex is weak compared with that of indium transferrin, indium citrate is essentially the same compound as ^{111}In chloride from a chemist's point of view.

The mechanism by which the radioindium accumulates in the bone marrow is not well understood. In addition to transferrin bound indium, LILIEN *et al.*[(116)] have suggested that some of the radioactive indium ions may be replacing iron in porphyrin. BEAMESH and BROWN[(121)] could not confirm these results and GLAUBITT *et al.*[(120)] using ^{111}In citrate suggested that the bone marrow uptake of the indium activity depended upon several factors, including (1) bone blood flow, (2) the compounds derived from the indium citrate in the body and (3) the avidity of phagocytic or other cells for ^{111}In citrate.

2. *Indium*-111 *labeled ferric hydroxide*

Indium hydroxide is very toxic;[(122)] however, indium does coprecipitate with ferric hydroxide which is far less toxic. Therefore, carrier-free ^{111}In-labeled colloidal ferric hydroxide[(6,123)] has been used for lymph node scanning. More than 20% of the administered dose has been accumulated per gram of an excised lymph node.[(123)]

When the iron concentration and pH of the solution is raised, bigger particles, macro-aggregates, are formed which are stabilized with gelatin and which have been used for lung scintigraphy.[(124)]

3. *Indium*-111 *DTPA*

The multidentate ligand diethylenetriamine-pentacetic acid (III) forms a strong 1 : 1 complex with indium at pH 6–7. Since indium is carrier free, a very small (10^{-3}M or less) quantity of

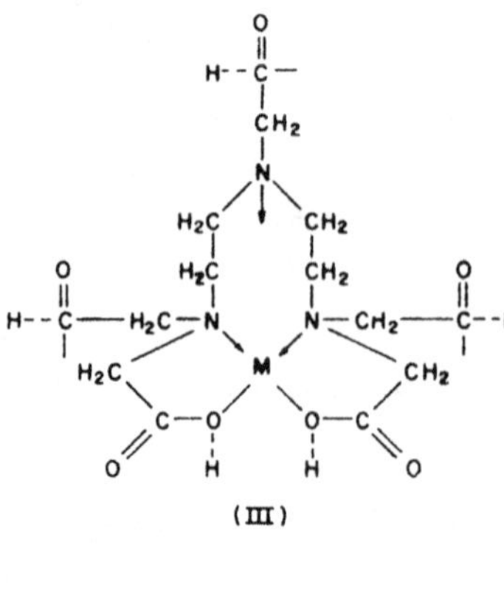

(III)

DTPA solution is used. The pH and the isotonicity of the solution are adjusted if necessary. The radiochemical purity of the compound is checked by paper or thin layer chromatography (solvent-HCl: acetone, *Rf* values $^{111}In^{3+}$ 0.0 ^{111}In-DTPA 0.75) and the solution is sterilized.

The complex has been mainly used for cisternography.[125–129] For this purpose the solution is administered intrathecally and the dynamics of cerebrospinal fluid are studied. The activity absorbed from the cerebrospinal fluid is excreted by the kidneys, thus decreasing the blood background. Indium-111 DTPA has been compared with other radiopharmaceuticals used for cisternography, and has been found to be the best.[126] However, very rarely, aseptic meningitis has been reported as an adverse reaction in children following the intrathecal administration of the agent.[127] The whole body radiation dose received from ^{111}In DTPA has been estimated by GOODWIN *et al.*[126]

4. *Indium*-111 *labeled bleomycin*

Since the use of ^{111}In chloride as a tumor localizing agent was limited, the search for a better compound continued. The physical characteristics of the radionuclide were accepted for this application. However, an effective carrier was needed which could carry the radionuclide to the target tissue.

Bleomycin, a mixture of closely related antibiotic substances, has been shown to have cytostatic properties on malignant cells in culture and therapeutic effects on artificially induced tumors in animals.[130] Intravenously administered bleomycin clears rapidly from the circulating blood and about 40% is excreted in the urine within 24 hr. These chracteristics of the compound make it a highly suitable agent to localize tumors. However, two more important criteria must be satisfied. First, bleomycin should form a complex with a radionuclide of such a thermodynamic stability that it would be able to carry the radioactivity to the target. Second, the newly formed compound should retain the biological properties of the basic organic molecule.

Bleomycins have one major structure in common but they differ in their terminal amino groups. The major structure has several chelating groups in its molecule and has been shown to have affinity for bivalent cations such as Cu^{2+}.[131] The complex forms swiftly and simply upon the addition of cupric chloride to a bleomycin solution.

It has been shown that the chelation with trivalent indium can be accomplished both in acidic and neutral media.[3] However, the rate of chelation at room temperature was slow and elevation of the temperature (40°C) of the reaction mixture was necessary for the completion (> 99%) of the complex formation within 30 min. At room temperature (22°C) after a 30 min incubation period, 15–20% of the ^{111}In remained unbound.[132] Since the pH of the solution into which the mixtures were incubated had no influence on the rate of complex formation and since bleomycin may hydrolyze in acidic media,[133] the use of isotonic saline solution was preferred. Since the carrier-free indium does not hydrolyze at pH 6–6.5, the use of isotonic saline for an injectable preparation appeared attractive. Some workers[26,134,135] have chosen to add acidic solution of ^{111}In to an aqueous solution of bleomycin and to adjust the pH to neutral before or after incubation. One thing common to all these methods of preparation of ^{111}In bleomycin has been the method of the determination of labeling efficiency. The complex has been analyzed by thin layer chromatography using a mixed solvent of 10% ammonium chloride added to an equal volume of methanol.

THAKUR[136] reported that 1 mg of bleomycin chelates as much as 80 μg of In^{3+}. If the molecular weight of bleomycin is assumed to be about 1400, the results suggest that a 1 : 1 complex is formed. This information, however, does not provide any clue as to the binding site for the metal.

The thermodynamic stability of the indium bleomycin complex is also unknown. THAKUR *et al.*[109] have reported that 4 hr after administration of ^{111}In bleomycin to a patient, almost 95% of the circulating activity is associated with the plasma transferrin, but that at 5 min after administration greater than 95% of the circulating activity remains bleomycin bound. The blood clearance and fecal[109] and urinary excretion[109,137] of ^{111}In activity administered as labeled bleomycin to laboratory animals are similar to data obtained using tritium (^{3}H) labeled bleomycin.[138] However, not enough data are available to claim that the biological properties of bleomycin chelated with ^{111}In are unaltered.

The tissue distribution of ^{111}In-labeled and ^{67}Ga-labeled bleomycin has been compared in tumor-bearing animals by THAKUR *et al.*[109] and by GROVE *et al.*[134] Both groups reported a higher tumor to blood radioactivity ratio with ^{111}In. Since the detection of less than microgram quantities of bleomycin in tissue samples still remains a problem,[139] the actual amount of bleomycin accumulated in the tumor could not be estimated. In attempts to increase the uptake of ^{111}In activity and specificity, separated A_2 and B_2 moieties of bleomycin labeled with indium have been used by THAKUR *et al.*[140] and by ECKELMAN *et al.*[141] No advantage of the separated moieties has been reported.

In spite of the encouraging results with ^{111}In bleomycin in animals, it has not been widely used as a tumor detecting agent. The blood clearance of the agent measured as ^{111}In in humans is slower[27] than in the animals, and the accumulation of the radioactivity in the liver and bone marrow has been constantly reported to cause difficulty in detecting small lesions. GROVE *et al.*[142] and PATERSON *et al.*[143] have compared the ^{111}In-labeled bleomycin with ^{67}Ga citrate in patients and have concluded that ^{111}In-labeled bleomycin did not offer any advantage over ^{67}Ga citrate. However, several other groups[144–149] have reported encouraging results in clinical trials in adults and children.[150]

The mechanism of the ^{111}In-labeled bleomycin uptake in tumors has not been studied. The initial blood clearance of the ^{111}In activity administered as ^{111}In-labeled bleomycin is faster than the blood clearance of ^{111}In chloride. This again suggests that the indium activity is not transferred to the plasma transferrin immediately. In the determination of the role of bleomycin, it would be very useful to know whether the activity accumulated by the tumors is carried to the cells as labeled transferrin, labeled bleomycin, or both. However, lack of a suitable analytical method for the detection of a small quantity of stable bleomycin in tissue sample has prevented work in this direction.

Attempts have been made by GREGORIDIAS and NEERUNJUN[151] to improve the specificity of ^{111}In-labeled bleomycin by entrapping it in liposome. (Liposomes are minute phospholipid vesicles composed of a closed system of one or more bilayers alternating with aqueous compartments.)[152] Liposome entrapped agents, when injected, do not come in contact with blood. Their clearance from plasma and tissue distribution therefore is controlled by the entrapped agent. Also, the liposome with the entrapped agent is taken up by the cell by endocytosis, and provides a convenient mechanism for "homing" the agent in the target tissue.

Furthermore, the uptake of ^{111}In activity by cells exposed to liposomes containing the radiolabeled agent was most pronounced when the liposomes were associated with an antibody corresponding to the cells studied *in vitro*. Superficially, it appears to be a very attractive and promising technique, but its fate as a tumor localizing agent cannot be evaluated until further studies have been performed.

5. *Miscellaneous compounds*

In order to evaluate the influence of chelating agents on the tissue distribution of ^{111}In and to evaluate chelating agents as potential tumor localizing agents, a number of ^{111}In complexes have been investigated by MERRICK *et al.*[153] and GOODWIN *et al.*[154] The compounds investigated were chloride, citrate, lactate, acetate, hydroxyethylethylenediamine-triaceticacid (HEDTA), tetraphenyl porphyrin (TPP) (IV), bleomycin, 8-hydroxyquinoline and mercapto quinoline. MERRICK *et al.* have described the methods of preparation and the

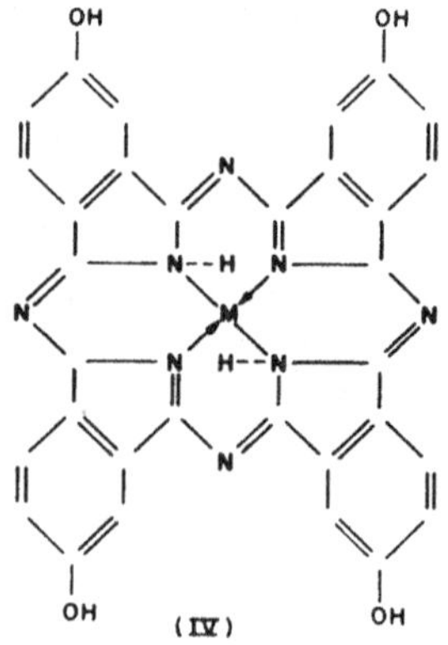

(IV)

quality control of the compounds they have prepared.

Of the compounds evaluated, chloride is ionic and citrate, lactate and acetate are weak complexes. After administering these agents labeled with ^{111}In, the activity distributes similarly to ^{111}In transferrin. Indium-111 HEDTA is a strong complex and has no specific tumor-localizing properties. However, it clears from the blood rapidly and this rapid clearance is expected to enhance the chances of tumor localization. The TPP, 8-hydroxyquinoline and mercapto quinoline complexes were chosen since they dissolve in lipids and penetrate through cell membrane.

These ^{111}In-labeled agents were compared in laboratory animals, but unfortunately the results were disappointing. The ionic form was cleared from the blood the least, the weak complexes intermediate and HEDTA and bleomycin most. The TPP demonstrated highest activity in the liver. The 8-hydroxyquinoline and mercapto quinoline gave very poor tumor to blood ratios ($\sim$0.1 and 4 respectively) at 24 hr. Most of the ^{111}In 8-hydroxyquinoline retained in the blood and lungs may be due to undissolved particles. Most of the ^{111}In activity administered as mercapto quinoline remained in the liver similar to that of ^{111}In TPP.

McAfee and Thakur[155] have investigated ^{111}In labeled nitrilotriacetic acid (V), acetyl acetone (VI) and tetracycline for labeling isolated leukocytes. Less then 1% of the activity was accumulated by the leukocytes.

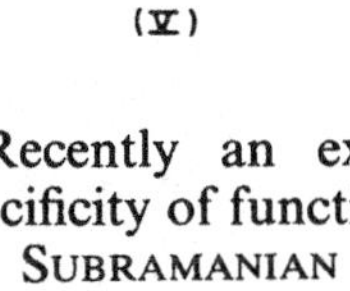

(V)

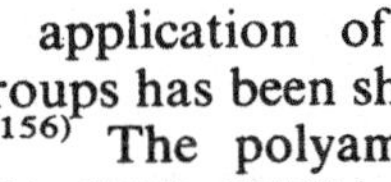

(VI)

Recently an excellent application of the specificity of functional groups has been shown by Subramanian *et al.*[156] The polyaminoacetic acids, namely EDTA (VII), DTPA (III) and nitrilotriacetic acid (V), have been modified by substituting methylene phosphoric acid for

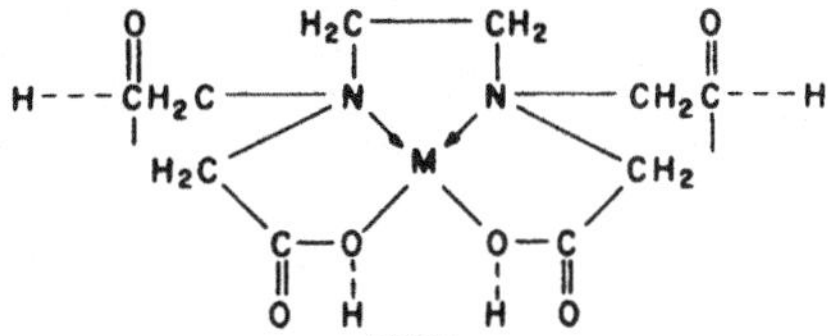

(VII)

acetic acid. The new compounds have retained their strong chelating properties and have gained additional bone seeking properties due to the newly added phosphate groups. The new compounds, ethylenediaminetetra (methylene phosphoric) acid (EDTMP), hexamethylenediaminetetra (methylene phosphoric) acid (HMDTP), dimethylenetriaminepenta (methylene phosphoric) acid (DTPMP) and nitrilotris (methylene) phosphoric acid (NTMP) have been obtained as their sodium, potassium or ammonium salts in aqueous solution. The required amount of ^{111}In chloride solution was mixed with 25–50 mg of the desired phosphonates in 4–5 ml solution and pH of the mixture was adjusted to 7–8. The amount of free indium was determined by electrophoresis, the solutions were sterilized by Millipore filtration, and administered to rabbits. Excellent gamma camera scintigraphs of rabbit skeletons have been obtained with ^{111}In-labeled EDTMP (VIII), HMDTP and DTPMP. The poor results obtained with ^{111}In NTMP have been

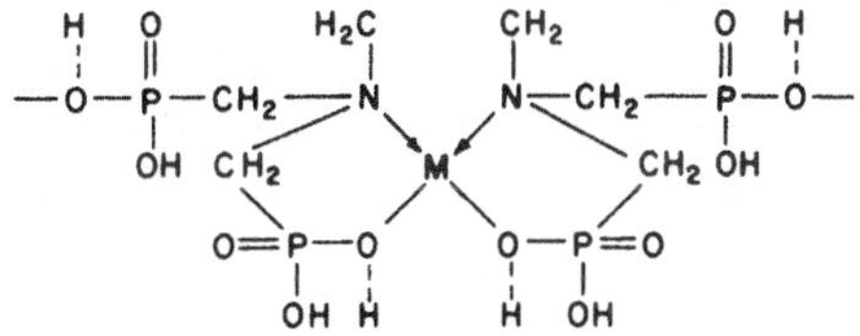

(VIII)

attributed to the poor thermodynamic stability of the complex. These compounds (^{111}In HMDTP, ^{111}In EDTMP and ^{111}In DTPMP) exhibit the best combination of purity, *in vivo* stability, clearance, specificity and detectability.

6. *Indium-111 labeled bifunctional chelate*

Another approach has been used to increase the thermodynamic stability and, possibly, the specificity of ^{111}In-labeled compounds. Goodwin *et al.*[157,26] have modified the EDTA molecule so that the molecule retains its initial metal binding capability and gains additional ability to link to another desired compound, which may have tumor or any other tissue specificity. The EDTA molecule has been synthetically attached to a benzenediazonium group which has been coupled to a protein by an overnight reaction at 4°C in an aqueous buffer solution of pH 8.1. The unbound reagents and buffer ions were then eliminated by dialysis against citrate buffer at pH 6. The ^{111}In chloride was then added to the conjugated protein solution for chelation with the multidentate modified EDTA molecule. The newly formed ^{111}In compound has been shown to have excellent *in vitro* and *in vivo* stability demonstrated by electrophoresis of rabbit plasma obtained up to 6 days after administration of the compound.

Bovine fibrinogen and human serum albumin conjugated ^{111}In-labeled benzenediazonium EDTA compounds have been administered to tumor-bearing mice. Twenty-four hours later 6.69% and 4.63% respectively of the administered dose of the two compounds were retained per gram of blood. The kidneys retained the highest activity (18.13% and 11.37% per gram respectively) of all the tissues, and tumor to blood ratios were 1.23 and 2.12 respectively. Although these ratios are unsatisfactory at the moment, the technique has brilliantly improved the thermodynamic stability of the ^{111}In compounds. Now an improvement in its specificity is needed so that the compound can be used as a tumor-detecting agent.

7. *Indium*-111-*labeled* 8-*hydroxyquinoline*

Recently a new radiopharmaceutical of ^{111}In has been evaluated in nuclear medicine. McAfee and Thakur[155,158] have surveyed several soluble and particulate radioactive agents for labeling phagocytic leukocytes. Indium-111-labeled 8-hydroxyquinoline (oxine) gave the most promising results.

Oxine is a bidentate ligand. Three molecules of oxine (IX) chelate one atom of indium at pH 5 to 6. The complex is lipid soluble and has a moderate thermodynamic stability. Thakur *et al.*[21] have shown that the ^{111}In oxine added to whole blood forms ^{111}In transferrin. However, they have further demonstrated that when

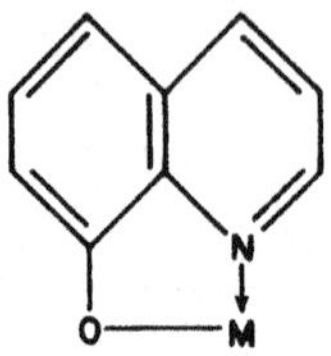

IX

the complex is added to isolated leukocytes suspended in saline greater than 95% of the activity is taken up by the cells by passive diffusion. An autoradiograph of a typical labeled leukocyte is shown in Fig. 2.

When the labeled leukocytes are administered to dogs, the activity remains associated with the cells up to 27 hr after injection, and clears from the circulating blood with the half time of approximately 7 hr. The ^{111}In-labeled leukocytes were administered to animals with experimentally induced sterile[21] and non-sterile[159] abscesses. Twenty-four hours later the radioactivity in the abscesses was 70 ± 23 times higher than the activity in an equal weight of blood. The ^{67}Ga citrate administered simultaneously to the same animals gave abscess to blood ratios of only 4.5 ± 1.6.[21] A typical whole body scan of a dog is shown in Fig. 3. This animal had induced abscesses in the left chest wall and both forearms, and was injected with ^{111}In-labeled leukocytes and scanned 24 hr later.

The 150 μg oxine in the preparation is 3 × 10^4 below its LD_{50} dose. The estimated radiation dose received by a 70 kg man from 1 mCi of ^{111}In-labeled leukocytes is shown in Table 4.[160]

The preparation is thus nontoxic, stable and specific. The ^{111}In-labeled leukocytes have potential applications in addition to detecting abscesses, including the determination of leukocyte kinetics and the detection of infarcts,[161] areas of inflammation and certain types of tumors.[162]

Table 4. Estimated absorbed radiation dose in rads for one mCi.

	Ga-67 citrate	In-111 leukocytes
Whole body	0.26	0.15
Liver	0.46	1.0
Spleen	0.53	0.46

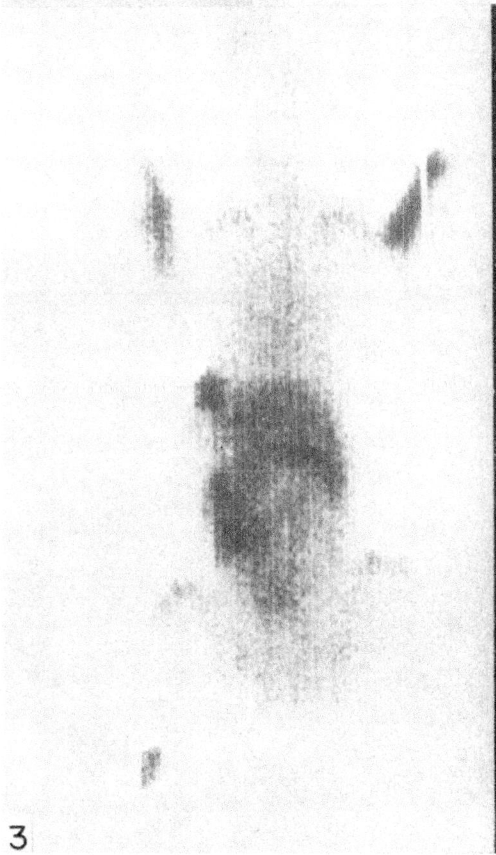

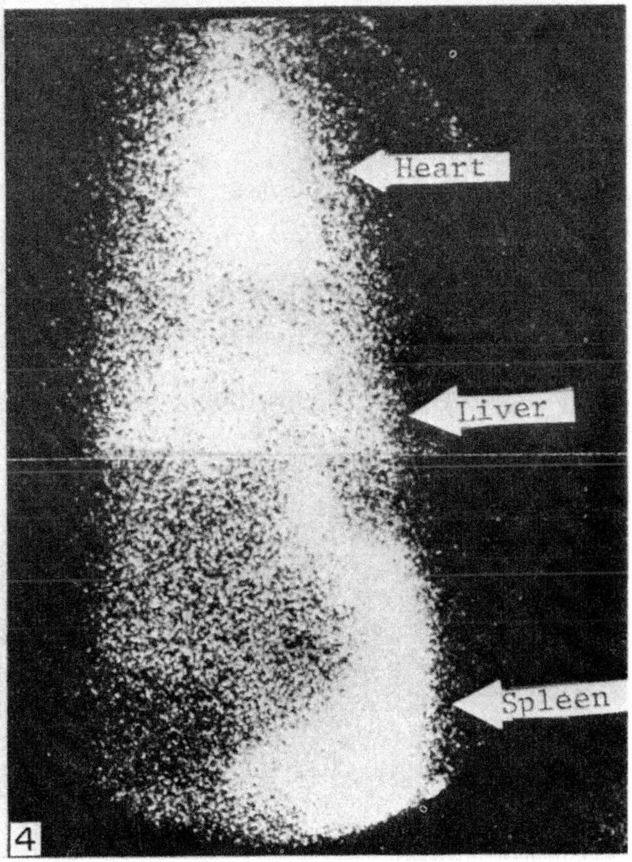

FIG. 2. Autoradiograph of indium-111 labeled canine leukocyte (× 1000).

FIG. 3. Anterior whole body scan obtained 24 hr after administration of indium-111 labeled autologus leukocytes in a dog bearing abscesses in fore arms and the left chest wall.

FIG. 4. Anterior gamma camera images obtained 24 hr after administration of indium-111 labeled red cells in a dog. The heart and the spleen are clearly outlined by the blood pool activity. Practically no change in the quality of the images had been observed between a few minutes and 24 hr after injection of the labeled red cells.

Indium-111-labeled oxine has also been used by THAKUR *et al.* to label red cells.[163] The cells were isolated from the plasma, leukocytes and platelets by the methyl cellulose technique described previously.[21] They were washed twice with isotonic saline and centrifuged each time for 5 min at 1850*g*. Finally, they were suspended in sterile physiological saline of approximately the same volume of blood (~6 ml) from which the cells were separated. The ^{111}In oxine complex prepared in the manner described previously[21] was added to the cells and incubated for about 20 min at room temperature. A small aliquot was then centrifuged and the activity associated with the cells and in the supernatant was counted. Approximately 95% labeling efficiency was achieved.

The labeled cells were administered intravenously, to a dog, without further centrifugation and serial gamma camera images were obtained. Figure 4 shows the cardiac liver and spleen blood pool activity 24 hr after administration of ^{111}In-labeled red cells.

The ^{111}In oxine complex has been used further to label platelets.[164,165] Fresh venous blood was obtained in a 50 ml syringe containing 7 ml sterile acid citrate dextrose solution. The platelet rich plasma was obtained by a 15 min centrifugation at 180g. More than 99% of the red cells and 90% of the leukocytes are eliminated by this step. The platelet rich plasma was separated in conical sterile plastic tubes with siliconized pasteur pipets and further centrifuged at 2000g for 15 min. The platelets were washed twice and finally resuspended in 10 ml isotonic saline pH 6.5.

The ^{111}In oxine complex, dissolved in 50 μl of ethanol and diluted to 250 μl with sterile physiological saline, was added dropwise to the platelet suspension and the mixture was incubated at 37°C for about 20 min. More than 95% of the activity is taken up by the platelets. The function of the labeled platelets has been studied by their ability to aggregate on the addition of low concentration of adenosine diphosphate and collagen as stimulating agents. No ill effects have been observed.

The labeled canine platelets have been administered to dogs with venous thrombi induced by an alteration of the intima with an electric current. Three hours after administration of the labeled platelets the induced venous thrombosis in the femoral vein has been detectable. The results indicate that the ^{111}In-labeled platelets have a potential ability to evaluate vascular damage and detect thrombosis.

Thus, the lipid solubility of the ^{111}In oxine has been utilized to label cellular blood components, and the labeled cells have been shown to have specificity for the detection of a variety of disorders.[166] The specificity combined with the good imaging characteristics of the ^{111}In should enable ^{111}In oxine to be a useful compound.

CONCLUSION

There is no organometallic compound of gallium or indium known to exist naturally in the human body. However, a number of compounds prepared with ^{67}Ga and ^{111}In have been evaluated for localization studies. The detectability by external monitoring of these lesions depends upon several factors. Low specific activity gallium-72 ($T_{1/2} = 14.1$ hr) lactate was first reported to be a bone seeking agent. The disadvantages of its toxicity and unsuitable half life were soon apparent. The longer-lived, carrier-free, cyclotron-produced gallium-67 ($T_{1/2} = 78$ hr) citrate was therefore introduced. It failed to localize in bones but was accidentally found to accumulate in soft tissue tumors. Since then ^{67}Ga citrate has been used as a radiopharmaceutical for studying soft tissue tumors in spite of its poor detectability and lack of specificity.

Efforts have been made to introduce an organometallic compound with better detectability than ^{67}Ga citrate. Indium-111 has better decay characteristics and similar chemical properties to those of gallium since both elements are in the same group of periodic classification of elements. However, ^{111}In citrate has not been as effective for tumor localization as ^{67}Ga citrate. Both radionuclides administered in ionic or weakly bound chelated form such as citrate are known to translocate to plasma transferrin. The radioactivity clears from the circulating blood slowly, accumulates in bone marrow and to a certain extent, in tumors. However, better results have been achieved with ^{67}Ga citrate. The reasons for the superiority are difficult to understand and further studies are required.

Due to the desirable characteristics of indium-111, compounds labeled with this radionuclide which have different thermodynamic stability and different specificity has been investigated. Indium-111-labeled bleomycin, an antibiotic and antitumor cytostatic agent, has been of some clinical interest. However, its thermodynamic stability and specificity are still inadequate. Attempts have been made to increase the thermodynamic stability and to retain or increase the specificity of other organometallics. The use of benzenediazonium-EDTA chelated with ^{111}In and conjugated with proteins is shown to have the desired thermodynamic stability. It remains to be seen if the molecule can be made tumor specific by conjugating it with bleomycin or its derivaties or other compounds known to have cytostatic properties.

The ^{111}In complexes of polyaminocarboxylic acids which are known to have higher thermodynamic stability have been successfully modified for use as bone seeking agents. Since the bone to blood activity ratios are optimum at approximately one hour after administration of the radiopharmaceutical, the ^{113m}In ($T_{1/2}$ = 100 min) may prove to be a better label for bone imaging.

A lipid soluble complex ^{111}In oxine has been shown to label cellular blood components such as leukocytes, red blood cells and platelets without loss of viability. Pathologically, it has been shown that a number of disorders in the body result in the focal accumulation of leukocytes. These disorders include abscesses, certain types of tumors, infarcts, inflammatory processes and thrombosis. Platelets are also known to accumulate in thrombosis, in an area of vascular damage, as well as in transplanted kidneys at the time of rejection. Cells labeled with ^{111}In remain viable with their function unaltered. In theory, therefore, these labeled cells should prove valuable for the localization of several abnormal lesions in the human body. In addition to this, the labeled cells may also be useful in the determination of leukocytes and platelet kinetics. These recent advances with ^{111}In-oxine labeled leukocytes and platelets potentially offer new applications for this radionuclide in nuclear medicine.

Acknowledgements—I am indebted to Dr. R. E. Coleman and Dr. M. J. Welch for their critical comments and helpful suggestions; to Dr. M. M. Ter-Pogossian for his interest and encouragement. I am also grateful to Julius Hecht for the preparation of the drawings, to my wife Lalita for arranging some of the references and to Diane Zaltsman for her continued efforts in converting the handwritten manuscript into a readable one.

REFERENCES

1. Albert A. *Selective Toxicity*, 5th Edition, p. 345. Chapman and Hall, New York (1973).
2. Pier S. M. *Texas Reports on Biology and Medicine* **33,** 85 (1975).
3. Thakur M. L. *Int. J. appl. Radiat. Isotopes* **23,** 357 (1973).
4. Bruner H. D., Hayes R. L. and Perkinson J. D. X. *Radiology* **61,** 602 (1953).
5. Hunter W. W., Jr. and DeCock H. W. *J. nucl. Med.* **10,** 343 (1969) Abstract.
6. Goodwin D. A., Finston R. A., Colombetti L. G., Beaver J. E. and Hupf H. B. *J. nucl. Med.* **10,** 337 (1969) Abstract.
7. Hupf H. B. and Beaver J. E. *Int. J. appl. Radiat. Isotopes* **21,** 75 (1970).
8. Silvester D. J. and Thakur M. L. *Int. J. appl. Radiat. Isotopes* **21,** 630 (1970).
9. Steyn J. and Meyer B. R. *Int. J. appl. Radiat. Isotopes* **24,** 365 (1963).
10. Neirinckx R. D. and Merwe M. J. *Radiochem. Radioanal. Lett.* **7,** 31 (1970).
11. Neirinckx R. D. *Int. J. appl. Radiat. Isotopes* **27,** 1 (1976).
12. Brown L. C., Callahan A. P., Skidmore M. R. and Wilson T. B. *Int. J. appl. Radiat. Isotopes* **24,** 665 (1973).
13. Gruverman I. J. and Kruger P. *Int. J. appl. Radiat. Isotopes* **5,** 21 (1959).
14. Dahl J. R. and Tilbury R. S. *Int. J. appl. Radiat. Isotopes* **23,** 431 (1972).
15. DePasquali G., Von Goeler E. and Peacock R. N. *J. Inorg. Nucl. Chem.* **11,** 257 (1959).
16. Neirinckx R. D. *Radiochem. Radioanal. Lett.* **4,** 152 (1970).
17. Thakur M. L. and Nunn A. D. *Int. J. appl. Radiat.* **23,** 139 (1972).
18. Drehmann U. and Puerschel U. *J. Prakt. Chem.* **14,** 122 (1961).
19. Brown L. C. and Beets A. L. *Int. J. appl. Radiat. Isotopes* **23,** 57 (1972).
20. Helus F. and Maier-Borst W. *Radiopharmaceuticals and Labelled Compounds* Vol. 1, p. 317. IAEA, Vienna (1973).

21. THAKUR M. L. unpublished data.
22. *Stability Constants* (Compiled by BJERRUM J., SCHWARZENBACH G. and SELLEN L. G.) Part I, The Chem Soc. London (1957).
23. WELCH M. J. and WELCH T. J. *Radiopharmaceuticals* (Edited by SUBRAMANIAN G., RHODES B. A., COOPER J. F. and SODD V. J.), p. 73. Society Nucl. Med., New York (1975).
24. MCAFEE J. G. *Radiopharmaceuticals* (Edited by SUBRAMANIAN G., RHODES B. A., COOPER J. F. and SODD V. J.), p. 3. Soc. Nucl. Med., New York (1975).
25. THAKUR M. L., COLEMAN R. E. and WELCH M. J. *J. Lab. Clin. Med.* in press.
26. GOODWIN D. A., SUNDBERG M. W., DIAMANTI C. I. and MEARES C. F. *Radiopharmaceuticals* (Edited by SUBRAMANIAN G., RHODES B. A., COOPER J. R. and SODD V. J.), p. 80. Soc. Nucl. Med., New York (1975).
27. WILLIAMS E. D., MERRICK M. V. and LAVENDER J. P. *Br. J. Radiol.* **48,** 275 (1975).
28. MIRD/Dosimetry Report No. 2. *J. nucl. Med.* **14,** 755 (1973).
29. DUDLEY H. C. *J. Pharm. Exp. Ther.* **95,** 482 (1949).
30. DUDLEY H. C. and LEVINE M. D. *J. Pharm. Exp. Ther.* **95,** 487 (1949).
31. DUDLEY H. C., MADDOX G. E. and LARUE H. C. *J. Pharm. Exp. Ther.* **96,** 135 (1949).
32. DUDLEY H. C. and MADDOX G. E. *J. Pharm. Exp. Ther.* **96,** 224 (1949).
33. LANG F. *Ann. Int. Med.* **35,** 1237 (1951).
34. DUDLEY H. C. *J. Am. chem. Soc.* **72,** 3822 (1950).
35. BRUNER H. D., HAYES R. L. and PARKINSON J. D. *Radiology* **61,** 602 (1953).
36. VAN DER WERLF J. TH. *Acta Radiol.* **41,** 343 (1954).
37. HARTMAN R. E. and HAYES R. L. *J. Pharm. Exp. Ther.* **168,** 193 (1969).
38. EDWARD C. L. and HAYES R. L. *J. nucl. Med.* **10,** 103 (1969).
39. DALRYMPLE G. V., BOYD C. M., MOSS A. J. and SANDERS J. L. *J. nucl. Med.* **12,** 349 (1971) Abstract.
40. EDWARD C. L. and HAYES R. L. *J. nucl. Med.* **10,** 103 (1969).
41. EDWARD C. L. and HAYES R. L. *JAMA* **212,** 1182 (1970).
42. HIGASI T., HISADA T. and NAKAYAMA T. *et al. Radioisotopes* **19,** 311 (1970).
43. VAIDYA S. G., CHAUDHRI M. A. and MORRISON R. *et al. Lancet* **2,** 911 (1970).
44. WINCHELL H. S., SANCHEZ P. D. WATANABE C. K. *et al. J. nucl. Med.* **11,** 459 (1970).
45. ITO Y., OKAYAMA S. and AWAMO T. *et al. Radiology* **101,** 355 (1971).
46. LARSON S. M. and SCHALL G. L. *JAMA* **218,** 257 (1971).
47. LAVENDER J. P., BARKER J. and CHAUDHRI M. A. *Br. J. Radiol.* **44,** 361 (1971).
48. SUZUKI T., HONJO I. and HAMAMOTO K. *et al. Am. J. Roentgol.* **113,** 92 (1971).
49. SCHWARTZENDRUBER D. C., NELSON B. and HAYES R. L. *J. Nat. Cancer Inst.* **46,** 941 (1971).
50. EDWARDS C. L., HAYES R. L. and NELSON B. *J. nucl. Med.* **13,** 428 (1972).
51. HIGASI T., NAKAYAMA Y. and MURATA A. *et al. J. nucl. Med.* **13,** 196 (1972).
52. JONES A. E., KOSLOW M., JONSTON G. S. and OMMAYA A. K. *Radiology* **105,** 693 (1972).
53. LANGHAMMER H., GLAUBITT G. and GREBE S. F. *et al. J. nucl. Med.* **13,** 25 (1972).
54. LARSON S. M., MILDER M. S. and TRETNER I. *et al. J. nucl. Med.* **13,** 448 (1972).
55. MANFREDI J. Q., QUINONES J. and BARTAK S. *J. nucl. Med.* **13,** 452 (1972).
56. NELSON B., HAYES R. L. and EDWARDS C. L. *et al. J. nucl. Med.* **13,** 92 (1972).
57. VAN DER SCHOOT J. B. *Radiol. Clin. Biol.* **41,** 371 (1972).
58. BAKSHI S. and BENDER M. A. *J. Surg. Oncol.* **5,** 539 (1973).
59. FRATKIN M. J. and SHARPE A. R. *J. nucl. Med.* **14,** 499 (1973).
60. JOHNSTON G. S. and JONES A. E. *et al. J. Surg. Oncol.* **5,** 529 (1973).
61. LOMAS F., DIDOS P. E. and WAGNER H. N. *Radiology* **106,** 246 (1973).
61a. LITTENBERG R. L., TAKETA R. M. *et al. Ann. Int. Med.* **79,** 403 (1973).
62. OKUYAMA S., ITO Y., SUZUK M. and SATO T. *Sci. Rep. Res. Inst., Tohoku, Univ.* **20,** 1 (1973).
63. WAXMAN A. D., LEE G., WOLFSTEIN R. and SIEMSEN J. K. *J. nucl. Med.* **14,** 903 (1973).
64. CALDAROLA L., DESILIPPI P. G., POLLONI R. and BADELLINO F. *J. nucl. Med.* **15,** 480 (1974).
65. EDELING C. J., HENRIKSEN O. and FOGH J. *J. nucl. Med.* **13,** 144 (1974).
66. FRANKEL R. S., JOHNSON K. W., MABRY J. J. and JOHNSTON G. S. *Radiology* **111,** 365 (1974).
67. GELRUD L. G. and ARSENÉAU J. G. *et al. J. Lab. Clin. Med.* **83,** 489 (1974).
68. HEIDENREICH P., KRIEGEL H. and HOT G. *Int. J. appl. Radiat. Isotopes* **25,** 557 (1974).
69. HENKIN R. E., POLCYN R. E. and QUINN J. L. *Radiology* **110,** 151 (1974).
70. JAMES O., WOOD E. J. and SHERLOCK S. *Gut* **15,** 404 (1974).
71. KINOSHITA F., USHIO T., MAEKAWA A., ARIWA R. and KUBO A. *J. nucl. Med.* **15,** 227 (1974).
72. QUE L., MCCARTHEY W., HANKIN A. and HOFFER P. *Am. J. Digs. Dis.* **19,** 271 (1974).

73. SILBERSTEIN E. B., KORNBLUT A., SHUMVICH D. A. and SAENGER E. L. *Radiology* **110,** 605 (1974).
74. WEINSTEIN M. D. and MIALE A. *S. Med. J.* **67,** 1433 (1974).
75. WALLNER R. J., CROLL M. N. and BRADY L. W. *J. nucl. Med.* **15,** 308 (1974).
76. ADLER S., PARTHASARATHY K. L., BAKSHI S. P. and STUTZMAN L. *J. nucl. Med.* **16,** 255 (1975).
77. ANDREWS G. A. and EDWARDS C. L. *JAMA* **233,** 1100 (1975).
78. BELANGER M. A., BEAUCHAMP J. M. and NEITZSCHMAN *J. nucl. Med.* **16,** 470 (1975).
79. DEROO M. H. K. *J. nucl. Med.* **16,** 250 (1975).
80. ERJAVEC M., SNAJDER J. and TURNER T. *Int. J. nucl. Med. Biol.* **2,** 86 (1975).
81. ERJAVEC M., MOVBIN T. and SNAJDER J. *J. nucl. Biol. Med.* **19,** 53 (1975).
82. ERICA G. A., CODD J. E., NEWTON W. T. and DONATI R. M. *Radiology* **117,** 731 (1975).
83. FRANKEL R. S., RICHMAN S. D., LEVENSON S. M. and JOHNSTON G. S. *Radiology* **114,** 393 (1975).
84. FELMAN A. H. and STANFORD T. S. *Radiology* **117,** 649 (1975).
85. HARVEY W. C., PODOLOFF D. A. and KOPP D. T. *J. nucl. Med.* **16,** 2 (1975).
86. HURWITZ S. R., HAGAN P. L. and ALAZRAKI N. P. *J. nucl. Med.* **16,** 280 (1975).
87. HEIDENDAL G. A., ROOS P., THIJS L. G. and WIENER J. D. *J. nucl. Med.* **16,** 793 (1975).
88. HABIBIAN M. R., STAAB E. V. and MATTHEWS H. A. *JAMA* **233,** 1073 (1975).
89. HENKIN R. E. CRC Critical Reviews *Clin. Rad. Nucl. Med.* **7,** 65 (1975).
90. HOPKINS G. B. and MENDE C. W. *J. nucl. Med.* **16,** 609 (1975).
91. KUMAR B., COLEMAN R. E. and ALDERSON P. O. *Arch. Surg.* **110,** 1237 (1975).
92. LEVIN J. and KEW M. C. *J. nucl. Med.* **16,** 949 (1975).
93. LENTLE B. C., CASTOR W. R., KHALIG A. and DIERICH H. *J. nucl. Med.* **16,** 374 (1975).
94. OSTER M. W., GELRUD L. G., LOTZ M. J., HERZIG G. P. and JOHNSTON G. S. *JAMA* **232,** 377 (1975).
95. TEATES C. D. and HUNTER J. G. *Radiology* **116,** 383 (1975).
96. WAXMAN A. D., SIEMSEN J. K., LEE G. C., WOLFSTEIN R. S. and MOSER L. *Radiology* **116,** 675 (1975).
97. LEPANTO P. B., ROSENSTOCK J., LITTMAN P., ALAVI A., DONALDSON M. and KUHL D. E. *Am. J. Roentgenol* **126,** 179 (1976).
98. PATERSON A. H. G., YOXALL A., SMITH I., TAYLOR D. M. and MCCREADY V. R. *Cancer Res.* **36,** 452 (1976).
99. SIEMSEN J. K., SIEGFRIED F. G., SARGENT E. N. and WENTZ D. *Radiology* **118,** 375 (1976).
100. SILBERSTEIN E. B. *Am. J. Med.* **60,** 226 (1976).
101. TEDESCO F. J., COLEMAN R. E. and SIEGEL B. A. *JAMA* **235,** 59 (1976).
102. CLAUSEN J., EDELING C. J. and FOGH J. *Cancer Res.* **34,** 1931 (1974).
103. GUNASEKERA S. W., KING L. J. and LAVENDER P. J. *Clin. Chem. Acta* **39,** 401 (1972).
104. SWARTZENDRUBER D. C., NELSON B. and HAYES R. L. *J. Nat. Cancer Inst.* **46,** 941 (1971).
105. RICHMAN S. D., LEVENSON S. M., JONES A. E. and JOHNSTON G. S. *Semin. Nucl. Med.* **5,** 103 (1975).
106. PATERSON A. G. H. and MCCREADY V. R. *Br. J. Radiol.* **48,** 520 (1975).
107. KONIKOWSKI T., GLEEN H. J. and HAYNIE T. P. *J. nucl. Med.* **14,** 164 (1973).
108. WAXMAN A. D., KAWADA D., WOLF W. and SIEMSEN J. K. *Radiology* **117,** 647 (1975).
109. THAKUR M. L., MERRICK M. V. and GUNASEKERA S. W., *Radiopharmaceuticals and Labelled Compounds*, Vol. 2, p. 183. IAEA (1973).
110. GROVE R. B., ECKELMAN W. C. and REBA R. C. *J. nucl. Med.* **14,** 627 (1973) Abstract.
111. DEWANJEE M. K., KAHN P. C., CARMEL A. and DEWANJEE U. *J. nucl. Med.* **15,** 487 (1974) Abstract.
112. BURLESON R. L., JOHNSON M. C. and HEAD H. *J. nucl. Med.* **15,** 98 (1974).
113. BURLESON R. L., HOLMAN B. R. and TOW D. E. *Surgery Gynecol. Obstet.* **141,** 379 (1975).
114. HUNTER W. W., JR. and ROCCOBONO X. J. *J. nucl. Med.* **11,** 328 (1970) Abstract.
115. GOODWIN D. A., GOODE R., BROWN L. and IMBOVNONE C. J. *Radiology* **100,** 175 (1971).
116. LILIEN D. L., BERGER H. G., ANDERSON D. P. and BENNETT L. R. *J. nucl. Med.* **14,** 184 (1973).
117. RAYUDU V. S., SHIRAZI P. H., FRIEDMAN A. and FORDHAM E. W. *J. nucl. Med.* **14,** 397 (1973).
118. FRARRER P. A., SAHA G. B. and SHIBATA H. N. *J. nucl. Med.* **13,** 429 (1972) Abstract.
119. FRARRER P. A., SAHA G. B. and KATZ M. *J. nucl. Med.* **14,** 394 (1973) Abstract.
120. GLAUBITT D. H. M., SCHLÜTER I. H. and HOBERLAND K. U. R. *J. nucl. Med.* **16,** 769 (1975).
121. BEAMESH M. R. and BROWN E. B. *Blood* **43,** 703 (1974).
122. CASTONOVO F. and WAGNER H. *J. nucl. Med.* **11,** 307 (1970).
123. GOODWIN D. A., FISTON R. H., COLLEMBETTI L. G., BEAVER J. E. and H' PF H. *Radiology* **94,** 175 (1970).
124. GOODWIN D. A. *J. nucl. Med.* **12,** 580 (1971).
125. HOSAIN F. and SOM P. *Br. J. Radiology* **45,** 677 (1972).

126. GOODWIN D. A., SONG C. H., FISTON R. and MARTIN P. *Radiology* **108,** 91 (1973).
127. ALDERSON P. O. and SIEGEL B. A. *J. nucl. Med.* **14,** 609 (1973).
128. GILDAY D. L. and KELLAM J. *J. nucl. Med.* **14,** 399 (1973) Abstract.
129. WAGNER M. S., MORTON R. J., GRAHAM L. S., GREENFIELD L. and BENNETT L. R. *J. nucl. Med.* **14,** 462 (1973) Abstract.
130. UMEZAWA H. *Proc. Euchem Conf. antibiotics* Arhus, Denmark, September (1972).
131. Umezawa H., SUHARA Y., TAKIDA T. and MAEDA K. *J. Antibiotics*, series A (Tokyo) **19,** 210 (1966).
132. THAKUR M. L., unpublished data.
133. TAKITA T., MURAOKA T., MAEDA K. and UMEZAWA H. *J. Antibiotics* series A (Tokyo) **21,** 79 (1968).
134. GROVE R. B., ECKELMAN W. C. and REBA R. C. *J. nucl. Med.* **14,** 917 (1973).
135. ROBINS P. J., SILBERSTEIN E. B. and FORTMAN D. L. *J. nucl. Med.* **15,** 273 (1974).
136. THAKUR M. L. *Radiopharmaceutical and Labelled Compounds*, Vol. 2, p. 203. IAEA, Vienna (1973).
137. KONIKOWSKI T., HAYNIE T. P. and GLENN H. J. *J. nucl. Med.* **16,** 738 (1975).
138. Bleomycin, Applications for approval to carry out clinical trials. Vol. I, Lundbeck Ltd., 6902.
139. THOMPSON M. and WILLIAMS C. R. *Anal. Lett.* **9,** 179 (1976).
140. THAKUR M. L., GUNSASEKERA S. W. and MERRICK M. V. Unpublished data.
141. ECKELMAN W. C., RZESZOTARSKI W. J., SIEGEL B. A., KUBOTA H., CHELLIAH M., STEVENSON J. and REBA R. C. *J. nucl. Med.* **16,** 1033 (1975).
142. GROVE R. B., REBA R. C., ECKELMAN W. C. and GOODYEAR M. *J. nucl. Med.* **15,** 386 (1974).
143. PATERSON A. H. G., TAYLOR D. M. and MCCREADY V. R. *Br. J. Radiology* **48,** 832 (1975).
144. MERRICK M. V., LAVENDER J. P., POOLE G. W., STRADLING P., THAKUR M. L. and WALTER L. H. *Br. J. Radiol.* **48,** 279 (1975).
145. VERMA R. C., SILVERSTEIN M. J., GREENFIELD L. D. and BENNETT L. R. *J. nucl. Med.* **16,** 578 (1975) Abstract.
146. HORN N. L., VERMA R. C. and BENNETT L. R. *J. nucl. Med.* **16,** 537 (1975) Abstract.
147. LILLIEN D. L., JONES S. E., O'MARA R. E., SALMON S. E. and DURIE B. G. M. *Cancer (Philadelphia)* **35,** 1036 (1975).
148. JONES S. E., LILLIEN D. L. and O'MARA R. E. *et al. Med. Ped. Oncology* **1,** 11 (1975).
149. GOODWIN D. A., SUNDBERG M. W., DIAMANTI C. I. and MEARS C. F. *Proc. 18th Clinical Conf. Cancer, 1973*, p. 57. Yearbook Medical Publishers, Inc., New York (1975).
150. O'MARA R. E., HANDMAKER H., CORRIGAN T. and LILLIEN D. L. Presented to the 62nd Annual Meeting of RSNA, Chicago, November (1975).
151. GREGORIDIAS G. and NEERUNJUN E. D. *Biochem. Biophys. Res. Comm.* **65,** 537 (1975).
152. SEGAL A. W., GREGORIDIAS G. and BLACK C. D. V. *Clin. Sci. Molec. Med.* **49,** 99 (1975).
153. MERRICK M. V., NUNN A. D., THAKUR M. L. and GUNASEKERA S. W. *Int. J. nucl. Med. and Biology* **2,** 45 (1975).
154. GOODWIN D. A., IMBORNONE C. J. and SONG C. H. *J. nucl. Med.* **12,** 434 (1971); and private communication.
155. MCAFEE J. G. and THAKUR M. L. *J. nucl. Med.,* **17,** 480 (1976).
156. SUBRAMANIAN G., MCAFEE J. G., ROSENSTREICH M. and COCO M. *J. nucl. Med.* **16,** 1080 (1975).
157. GOODWIN D. A., MEARES C. F., DIAMANTI C. I. and SUNDBERG M. W. 1st World Cong. Nucl. Med., Tokyo, Japan (1974).
158. MCAFEE J. G. and THAKUR M. L. *nucl. Med.* **17,** 488 (1976).
159. THAKUR M. L., COLEMAN R. E. and WELCH M. J. *Radiology* **119,** 731 (1976).
160. THAKUR M. L., EICHLING J. O., WELCH M. J. and COLEMAN R. E. unpublished data.
161. MALLORY G. K., WHITE P. D. and SALCEDELO-SALGAR J. *Am. Heart J.* **18,** 647 (1939).
162. WILLIAMS W. J., BENTLER E., ERSLER A. J. and RUNDLES R. W. *Hematology*, p. 657. McGraw-Hill, New York (1972).
163. THAKUR M. L., COLEMAN R. E. and WELCH M. J. unpublished data.
164. THAKUR M. L., WELCH M. J., JOIST J. H. and COLEMAN R. E. *J. nucl. Med.* abstract, **17,** 580 (1976).
165. THAKUR M. L., WELCH M. J., JOIST J. H. and COLEMAN R. E. *Thrombosis Res.* **9,** 345 (1976).
166. SEGAL A. W., THAKUR, M. L., ARNOT R. N. and LAVENDER J. P. *Lancet,* 1056 (1976).

International Journal of Applied Radiation and Isotopes, 1977, Vol. 28, pp. 203–206. Pergamon Press. Printed in Northern Ireland

The Affinity of Radiolabeled Bone-Seeking Compounds for Injured Myocardium in the Rat

ALUN G. JONES* and MICHAEL A. DAVIS
Department of Radiology, Harvard Medical School and Peter Bent Brigham Hospital, Boston, Massachusetts, U.S.A.

(*Received* 4 *April* 1976)

The *in vivo* behavior of a number of bone-seeking radiolabeled compounds has been studied in a rat model of myocardial infarction. The compounds include six phosphonates and four inorganic phosphates, including pyrophosphate, labeled with ^{99m}Tc, and two phosphonates labeled with ^{113m}In. The uptake of radioactivity in damaged myocardial muscle has been compared to that in normal myocardium, bone, muscle and blood, and the resulting contrast ratios to those obtained with ^{99m}Tc-pyrophosphate, the current clinical agent of choice for external visualization of infarcted myocardium.

INTRODUCTION

CONSIDERABLE interest has been generated in the nuclear medicine community following the discovery that technetium-labeled radiopharmaceuticals originally designed for other purposes localize in infarcted myocardial tissue.[1,2] Of the compounds tested, ^{99m}Tc-Sn pyrophosphate has been the most successful both in animal models and in man,[3] and is now the agent of choice in those institutions performing such studies. This work reports on an investigation of the *in vivo* behavior of several labeled compounds, six organic phosphonates and four inorganic phosphates, using a rat model of damaged myocardium. The model has been described by ADLER *et al.*[4] who have demonstrated the validity of the technique as an initial screening procedure for potential infarct imaging agents.

MATERIALS AND METHODS

The six organic phosphonates tested in this study were the sodium salts of DTPMP, diethylenetriamine penta(methylene phosphonic acid), an analog of DTPA; EDTMP (or ENTMP), ethylenediamine tetra(methylene phosphonic acid), an analog of EDTA; MDP, methylene diphosphonate,[5] the agent of choice for routine bone scanning in our clinics; HEDP, 1-hydroxyethylidene-1,1-diphosphonic acid; TEDTMP, triethylenediamine tetra(methylene phosphonic acid), also known as HMDTP, hexamethylenediamine tetra(methylene phosphonic acid); NTMP, nitrilo tri(methylene phosphonic acid), which possesses an ammonia-like structure. The four inorganic phosphates tested were: the linear molecules sodium pyrophosphate (PP_i) and tripolyphosphate (PP_3), and the cyclic sodium trimetaphosphate (TRIMETA) and tetrametaphosphate (TETRAMETA). For comparative purposes the radiopharmaceutical ^{99m}Tc-Sn-DTPA, which does not display the properties of a skeletal imaging agent, was also studied.

Labeling procedure

The four multidentate phosphonates[6] were labeled with ^{99m}Tc by the following method: solutions were made up in sterile distilled water and titrated to pH 6.5 with sodium hydroxide to a final concentration of 40 mg/ml of the free acid. One mg of stannous chloride was dissolved in 0.5 ml of this solution and the required ^{99m}Tc activity added in 4 ml isotonic saline. The concentration required for the animal studies was normally 1 mCi/ml.

HEDP was used as Osteoscan™ (Procter and

*For reprints contact: Alun G. Jones, Harvard Medical School, 50 Binney Street, Boston, MA 02115, U.S.A.

Gamble) and prepared according to the manufacturer's instructions. MDP kits prepared in this laboratory for routine clinical use, were reconstituted by simple addition of $^{99}TcO_4^-$ in saline. Each kit contains 10 mg of the disodium salt of MDP and 1 mg $SnCl_2$.

DTPMP and EDTMP were also labeled with ^{113m}In derived from a ^{113}Sn–^{113m}In generator (New England Nuclear). The required activity of carrier-free $^{113m}InCl_3$ in 1 ml 0.05 M HCl was added to 0.5 ml of the stock sodium salt solution (pH 6.5), followed by 1 ml of 0.05 M $NaHCO_3$ buffer solution (pH 8.9). The final pH was 7.5.[7]

The inorganic phosphates were labeled with ^{99m}Tc, again by stannous chloride reduction. The solutions contained 10 mg of the phosphate in isotonic saline to which had been added 1 mg $SnCl_2$. The final volume following addition of pertechnetate was 4–5 ml. Pyrophosphate kits (Mallinckrodt Nuclear) were prepared according to the manufacturer's instructions. The PP_3, TRIMETA and TETRAMETA (Sigma Chemical Co.) were assayed for hydrolysis products to ensure their purity.

Kits of stannous-DTPA (CIS Radiopharmaceuticals) were also prepared according to the recommended procedure.

Chromatographic analysis

^{99m}Tc labeling efficiency was determined by paper chromatography on Whatman #1 or #3MM paper using freshly prepared 85% methanol as the solvent. ^{113m}In labeling was assayed by thin-layer chromatography (Gelman ITLC™) using 0.1 N NH_4OH as eluant. Chromatograms were developed at the time of administration of the agents.

Biologic distribution

The myocardial injury was inflicted as described by Adler *et al.*[4] Male Sprague–Dawley rats in the weight range 190–240 g were used. In each case, the radioactive agent was administered three hours after the creation of the injury. The animals were anesthetized with ethyl ether, a small incision made in the medial aspect of the thigh to expose the saphenous vein and the test material injected under visual observation using a 27 gauge needle.

The animals were all sacrificed at one hour after injection; no time course was studied, although agents are known to show time dependent uptake in damaged myocardium.[8] Sacrifice was by a combination of ether anesthesia and cardiac exsanguination. The heart, liver, spleen, stomach, kidneys, intestines and bladder were excised from the carcass. Samples of muscle (abdominal and skeletal) and bone (femur and tibia) and blood were also taken. The myocardial lesion was carefully isolated from the heart, and a sample of adjacent (or border) tissue and two samples of uninjured myocardium taken. Only those lesions which appeared white in color were considered acceptable. Alternative appearances were the less severe hemorrhagic injuries or the more severe cases of puncture of the myocardial wall.

All tissue samples were weighed as rapidly as possible after separation and later counted in a NaI (Tl) scintillation detector. Values were expressed as % injected dose/g tissue (% I.D./g) and % I.D./total organ. In the case of blood, bone and muscle, the total tissue was assumed to be 6, 10 and 50%, respectively, of the total body weight.

RESULTS AND DISCUSSION

Table 1 shows the uptake (% I.D./g tissue) of all the agents tested using the animal model described previously. The number of animals in each case was a minimum of five, except in the case of ^{99m}Tc-DTPA, where there were three.

Table 2 shows the target to non-target ratios considered to be the most relevant in assessing the effectiveness of a skeletal imaging agent for the purpose of imaging damaged myocardial tissue. It can be seen that the MI/normal ratio (injured to uninjured myocardium) varies considerably in this group of materials. The complex phosphonates TEDTMP and MDP show the highest values, followed closely by $^{99m}Tc\text{-}PP_i$ and ^{99m}Tc-HEDP. Reference to column 1 in Table 1, however, shows that none of the phosphonates have uptake in the damaged tissue as high as $^{99m}TcPP_i$, i.e. 2.23% I.D./g, which is twice that of the nearest phosphonate, TEDTMP. Furthermore, the data in Table 2 demonstrate a more favorable MI/blood ratio for pyrophosphate and, perhaps more importantly, a greater contrast ratio between injured myocardium and bone.

TABLE 1. Organ distribution (% I.D./g) of radiolabeled phosphates and phosphonates in rat infarct model one hour after administration (mean ± S.D.)

Agent	Damaged myocardium	Normal myocardium	Bone	Muscle	Blood
^{99m}Tc-TEDTMP	1.22 ± 0.26	0.04 ± 0.01	2.35 ± 0.41	0.02 ± 0.01	0.13 ± 0.01
^{99m}Tc-MDP	0.86 ± 0.33	0.03 ± 0.01	2.44 ± 0.55	0.01 ± 0.01	0.08 ± 0.02
^{99m}Tc-HEDP	0.79 ± 0.25	0.03 ± 0.01	2.31 ± 0.45	0.02 ± 0.01	0.08 ± 0.02
^{99m}Tc-EDTMP	0.98 ± 0.17	0.05 ± 0.01	2.28 ± 0.34	0.02 ± 0.01	0.10 ± 0.04
^{99m}Tc-DTPMP	0.29 ± 0.12	0.02 ± 0.01	2.34 ± 0.18	0.01 ± 0.01	0.04 ± 0.01
^{99m}Tc-NTMP	1.06 ± 0.04	0.17 ± 0.06	0.92 ± 0.43	0.05 ± 0.01	0.73 ± 0.22
^{99m}Tc-PP$_i$	2.23 ± 0.41	0.09 ± 0.02	2.30 ± 0.41	0.03 ± 0.01	0.15 ± 0.04
^{99m}Tc-TETRAMETA	0.94 ± 0.18	0.10 ± 0.02	1.73 ± 0.33	0.06 ± 0.01	0.11 ± 0.03
^{99m}Tc-PP$_3$	1.30 ± 0.78	0.28 ± 0.11	1.62 ± 0.05	0.08 ± 0.02	0.49 ± 0.07
^{99m}Tc-TRIMETA	0.53 ± 0.19	0.17 ± 0.06	0.21 ± 0.13	0.02 ± 0.01	0.09 ± 0.03
^{99m}Tc-DTPA	0.69 ± 0.19	0.06 ± 0.01	—	0.04 ± 0.01	0.19 ± 0.04
^{113m}In-EDTMP	0.81 ± 0.23	0.04 ± 0.01	3.36 ± 0.4	0.02 ± 0.01	0.11 ± 0.1
^{113m}In-DTPMP	0.22 ± 0.01	0.02 ± 0.01	2.98 ± 0.2	0.01 ± 0.01	0.08 ± 0.01

TABLE 2. Contrast ratios between damaged myocardium and other tissues at one hour (mean ± S.D.)

Agent	MI ratio	MI/bone	MI/muscle	MI/blood
^{99m}Tc-TEDTMP	30.5 ± 10.0	0.52 ± 0.17	61.0 ± 19.5	9.4 ± 2.0
^{99m}Tc-MDP	28.7 ± 12.3	0.35 ± 0.12	66.8 ± 11.8	12.8 ± 2.1
^{99m}Tc-HEDP	26.7 ± 8.9	0.34 ± 0.11	61.9 ± 4.17	9.5 ± 1.4
^{99m}Tc-EDTMP	19.6 ± 5.1	0.43 ± 0.10	42.7 ± 10.1	9.6 ± 2.2
^{99m}Tc-DTPMP	14.5 ± 7.1	0.12 ± 0.05	23.8 ± 11.6	7.9 ± 3.4
^{99m}Tc-NTMP	7.3 ± 0.9	1.15 ± 0.43	23.7 ± 4.7	1.5 ± 0.5
^{99m}Tc-PP$_i$	24.8 ± 4.6	0.97 ± 0.18	67.0 ± 11.8	14.9 ± 3.0
^{99m}Tc-TETRAMETA	9.8 ± 1.9	0.54 ± 0.15	15.6 ± 2.1	8.5 ± 1.9
^{99m}Tc-PP$_3$	4.6 ± 2.1	0.80 ± 0.48	16.3 ± 7.8	2.7 ± 1.2
^{99m}Tc-TRIMETA	3.1 ± 1.4	2.52 ± 1.80	26.5 ± 13.3	5.9 ± 3.7
^{99m}Tc-DTPA	11.5 ± 4.2	—	17.3 ± 4.8	3.6 ± 1.0
^{113m}In-EDTMP	20.2 ± 5.7	0.24 ± 0.09	40.5 ± 11.5	7.4 ± 2.1
^{113m}In-DTPMP	12.0 ± 0.9	0.07 ± 0.01	22.0 ± 2.3	2.8 ± 0.3

The results therefore bear out the clinical observation that pyrophosphate is the most successful of the current commercial agents for myocardial infarct imaging. They also suggest that trials are warranted with ^{99m}Tc-MDP, once this material becomes more widely available.

Fig. 1 shows a plot of % I.D./g in bone vs the MI/Normal ratio. According to this graph. there appears to be a region with a linear relationship between these two parameters and above which a saturation effect occurs. This finding may be fortuitous, however, because when the % I.D./g in damaged myocardium vs % I.D./g in bone is plotted for each of the technetium-labeled agents (Fig. 2), little correlation is evident. Of note is the way in which pyrophosphate stands apart from the other agents when the data is treated in this fashion in contrast to the presentation in Fig. 1. This merely serves to underline the importance of considering all the relevant uptake values and not only the observed myocardial contrast ratios.

The ^{113m}In-labeled bone scanning agents[7,9] show comparable properties to their technetium analogs in this animal model, but result in inferior contrast ratios to those observed with

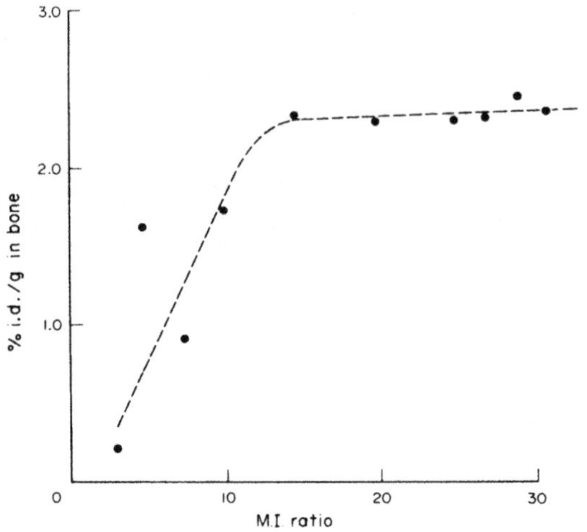

FIG. 1. The % injected dose (I.D.) per gram of bone plotted against the MI ratio (uptake in injured vs. uninjured myocardium) for the ^{99m}Tc-labeled phosphorus-containing compounds. Data taken from Tables 1 and 2.

pyrophosphate. These agents have been broached as a possible means of sequential scanning of myocardial infarcts[10] at time intervals less than that allowed by a ^{99m}Tc-labeled compound alone. It is interesting to note that when a mixture of ^{113m}In- and ^{99m}Tc-DTPMP is injected simultaneously into normal rats at the same dose level as was used in this study, the overall uptake expressed as % I.D. in

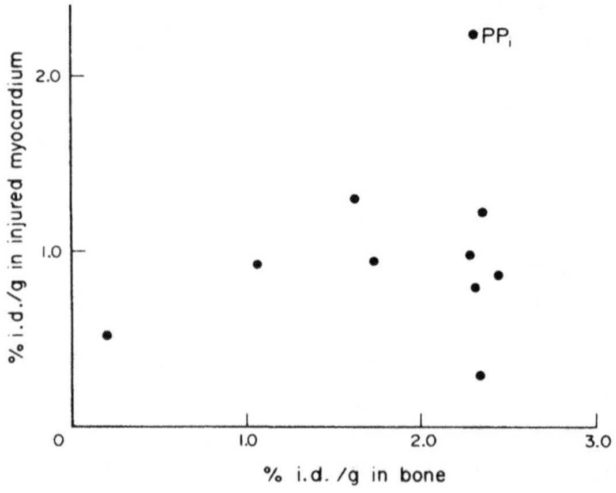

FIG. 2. The % I.D./g of injured myocardial tissue plotted against % I.D./g of bone for the ^{99m}Tc-labeled phosphorus-containing compounds. Data taken from Table 1. Note the unique position of ^{99m}Tc-PP$_i$.

the whole skeleton remains the same, but the individual contribution of each isotope is depressed. For example injection of either ^{113m}In-DTPMP or ^{99m}Tc-DTPMP shows an uptake of 2.98 or 2.34 % I.D./g, respectively, in normal Sprague–Dawley rats of the weight range specified in this work, giving approximately 50–55% in the total skeleton. Simultaneous injection of both materials shows $25.0 \pm 2.5\%$ ^{99m}Tc and $33.3 \pm 4.0\%$ ^{113m}In in the skeleton, with the corresponding % I.D./g being depressed to 1.51 ± 0.29 and 1.99 ± 0.23, respectively.[11] This may possibly indicate a saturation of available sites as well as competition between the same carrier molecules labeled with different radionuclides. It is not inconceivable that attempts to scan sequentially within short time intervals may also suffer similar interference. Thus, it may be that different carrier molecules acting by different mechanisms may be required to transport the different radionuclides.

Acknowledgements—The authors wish to acknowledge ALICE D. CARMEL for expert technical assistance and REBEKAH A. TAUBE for reviewing the manuscript. The guidance and encouragement of Dr. S. JAMES ADELSTEIN is deeply appreciated.

This work was partially supported by USPHS Grant GM 1 674 and by Grant from the American Heart Association (75-796) and its N.E. Massachusetts Chapter (1279).

REFERENCES

1. HOLMAN B. L., DEWANJEE M. K., IDOINE J. *et al. J. nucl. Med.* **14,** 595 (1973).
2. ROSSMAN D. J., STRAUSS W. H., SIEGEL M. E. *et al. J. nucl. Med.* **16,** 875 (1975).
3. PARKEY R. W., BONTE F. J., MEYER S. L. *et al. Circulation* **50,** 540 (1974).
4. ADLER N., CAMIN L. L. and SHULKIN P. *J. nucl. Med.* **17,** 203 (1976).
5. SUBRAMANIAN G., MCAFEE J. G., BLAIR R. J. *et al J. nucl. Med.* **16,** 744 (1975).
6. JONES A. G. and DAVIS M. A. *J. nucl. Med.* **16,** 540 (1975).
7. JONES A. G., DAVIS M. A. and DEWANJEE M. K. *Radiology* **117,** 727 (1975).
8. ZWEIMAN F. G., HOLMAN B. L., O'KEEFE A. *et al. J. nucl. Med.* **16,** 975 (1975).
9. SUBRAMANIAN G., MCAFEE J. G. and ROSENSTREICH M. *et al. J. nucl. Med.* **16,** 1080 (1975).
10. HOLMAN B. L., JONES A. G., DAVIS M. A. *et al. J. nucl. Med.* **17,** 508 (1976).
11. JONES A. G. (unpublished results).

International Journal of Applied Radiation and Isotopes, 1977, Vol. 28, pp. 207–211. Pergamon Press. Printed in Northern Ireland

Cyclotron Production of Rhodium-101*m* through its Precursor Palladium-101

KENNETH L. SCHOLZ,* VINCENT J. SODD* and JAMES W. BLUE†

(*Received* 4 *March* 1976)

The preparation of high-purity ^{101m}Rh through its natural precursor, ^{101}Pd was investigated. The $^{103}Rh(p,3n)^{101}Pd$ and natural $Pd(p,xn)^{101}Ag \rightarrow {}^{101}Pd$; $Pd(p,pxn)^{101}Pd$ reactions were studied. An anion exchange method was used to isolate the ^{101}Pd produced and to then isolate ^{101m}Rh after ingrowth. Maximum yield of ^{101m}Rh was 4 mCi/μA-hr achieved when rhodium metal was irradiated with 37-MeV protons. Rhodium-101*m* was also incorporated in rhodium II butyrate, a compound used in chemotherapy.

1. INTRODUCTION

PLATINUM-GROUP metals incorporated into coordination complexes have been shown to be potent antitumor agents. The first, discovered by ROSENBERG,[1] was *cis*-dichlorodiammineplatinum II. Much research has been performed to determine biochemical effects and mechanisms for this and related platinum-containing compounds, including the use of ^{193m}Pt and ^{195m}Pt as radioactive tracers.[2]

Rhodium was the second platinum-group metal to be studied.[3] Work was begun much more recently and few compounds have been investigated. To date, rhodium II carboxylates (acetate, propionate, butyrate)[3] have received the most attention. However, no radioactive isotopes of rhodium have been developed to use as labels for these compounds nor has rhodium been used in any type of nuclear medicine procedure. We, therefore, decided to investigate rhodium isotopes that could be used in humans as possible tumor-localizing radiopharmaceuticals. The isotopes of rhodium having half-lives greater than one hour are listed in Table 1. An inspection of their decay properties shows that ^{101m}Rh is probably the isotope of choice for use in low-radiation dose radiopharmaceuticals.

Rhodium-101, ^{102}Rh and ^{102m}Rh are undesirable because of long half-lives; and, although ^{99}Rh, ^{99m}Rh, ^{100}Rh and ^{106m}Rh have shorter half-lives, they emit high-energy gamma rays unsuitable for imaging. Rhodium-105 has a 36-hr half-life and may possibly be useful, but it decays by β^- emission and usable gamma-ray probability is only 25% per disintegration. Rhodium-103 is the only stable isotope of rhodium.

A simplified decay scheme[6] for ^{101m}Rh is shown in Fig. 1. Rhodium-101*m* decays with a 4.3 d half-life by electron capture (93%) to stable ^{101}Ru and emits 307 (88%) and 545-keV (4%) gamma-rays; it also decays by isomeric transition (7%) to ^{101}Rh. However, because of the 3.2 yr ^{101}Rh half-life and the low branching percentage to ^{101}Rh, the ^{101}Rh activity resulting from ^{101m}Rh decay is just 0.03% of the original activity, e.g. the decay of 1 mCi of ^{101m}Rh results in 0.3 μCi of ^{101}Rh. Rhodium-101 decays by electron capture to ^{101}Ru and emits mainly 127 and 198-keV gamma-rays.

A previous method for producing ^{101m}Rh was designed to give sources for decay scheme determinations. Ruthenium-101 enriched to 91% was irradiated with 6-MeV protons to produce ^{101m}Rh.[7] This direct method will also yield ^{101}Rh as well as other undesirable rhodium isotopes from isotopic impurities in the

*Nuclear Medicine Laboratory, FDA, Cincinnati General Hospital, Cincinnati, OH 45267, U.S.A.

†Lewis Research Center, NASA, Cleveland, OH 44135, U.S.A.

TABLE 1. Rhodium isotopes with half-lives longer than one hour

Rhodium isotope	Half-life[4]	Decay mode[5]		Major gamma-ray energy keV and abundance (%)[4]
^{99}Rh	15 days	β^+ EC		89 (100) 322 (22) 618 (13)
^{99m}Rh	4.7 hr	EC β^+	(90) (10)	341–1556 (unknown)
^{100}Rh	21 hr	EC β^+	(93) (7)	540 (88) 820 (25) 2380 (39)
^{101}Rh*	3.2 yr	EC		127 (81) 198 (69) 326 (12)
^{101m}Rh*	4.3 days	EC IT	(92.8) (7.2)	307 (88) 545 (4)
^{102}Rh	206 days	EC β^+ β^-	(84) (16)	475 (46)
^{102m}Rh	2.9 yr	EC		475 (95) 631 (56) 697 (44) 767 (34) 1047 (44)
^{103}Rh	stable			
^{105}Rh	36 hr	β^-		306 (5) 319 (20)
^{106m}Rh	2.2 hr	β^-		451 (25) 512 (86) 616 (20) $>$700 (197)

*Rhodium-101 and ^{101m}Rh decay data from Ref. 6.

target. Because of this problem with directly produced ^{101m}Rh, we investigated the possibility of producing ^{101m}Rh through its natural precursor ^{101}Pd. Palladium-101 has an 8.3-hr half-life and decays solely to ^{101m}Rh. A check of radioactive palladium isotopes that decay to rhodium shows that only the decay of ^{100}Pd ($T_{1/2}$, 4 d) to ^{100}Rh ($T_{1/2}$, 21 hr) could cause interference. However, it appeared that this could be minimized by the proper choice of irradiation energies and times for chemical separation.

Palladium-101 had been previously produced[8] for decay scheme measurements by irradiating thick rhodium targets with 30 and 42-MeV protons, but no yields were given.

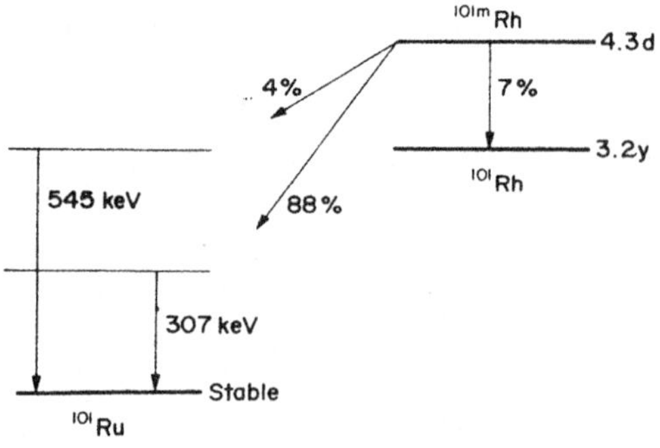

FIG. 1. Simplified decay scheme of rhodium-101*m*.

The present work deals with the production of high-purity ^{101m}Rh through ^{101}Pd. The excitation function for the ^{103}Pd$(p,3n)^{101}$Pd reaction was determined. Palladium-101 was also produced by irradiating natural palladium with 41-MeV protons; palladium has 6 stable isotopes ranging in mass from 102 to 110. Previously reported ion-exchange procedures[9] were adopted to isolate palladium from both target materials and to then recover ^{101m}Rh after ingrowth.

2. EXPERIMENTAL

2.1 *Irradiations*

Irradiations were performed at the Lewis Research Center (NASA) isochronous cyclotron which currently accelerates protons to 46 MeV; beam energies were known to within ± 0.5 MeV. Beam current measurements were made by integrating the charge collected in a Faraday cup. Target materials included palladium foils and rhodium in the form of foil, powder and rhodium chloride trihydrate. Yields for nuclear reactions studied were determined using palladium and rhodium foils approximately 315 mg/cm^2 thick. They were counted without chemical processing.

2.2 *Analysis of gamma-ray spectra and data treatment*

Identification and assay of gamma-ray emitting radionuclides were carried out using a spectrometer consisting of a Ge(Li) crystal 6 cm^2 in area and 0.7-cm thick with a 4096-channel pulse-height analyzer. The resolution (FWHM) of the spectrometer for the 661.6-keV gamma ray of ^{137}Cs was 3.2 keV. Yield data were determined using an 88% probability per decay for the 307-keV emission of ^{101m}Rh and an 18% probability for the 296-keV gamma-ray of ^{101}Pd.[6]

2.3 *Chemical procedures*

Chemical separations were based on the premise that palladium, including the desired ^{101}Pd, could be isolated free from all rhodium and silver nuclides. A suitable decay time was allowed for ^{101m}Rh ingrowth after which it was then collected. This indirect procedure used for both target elements was necessary to remove directly produced radioactive rhodium isotopes and to insure that the ^{101m}Rh was carrier free. This is obvious in the use of rhodium targets, but we found that the high purity palladium (99.9%) foils used as targets also contained rhodium (0.05%). The only contaminant present was a small amount of ^{100}Rh ($T_{1/2}$, 21 hr). Separations were based on a previously reported[9] anion-exchange procedure using Dowex 1X8, 50–100 mesh, utilizing the fact that both rhodium and palladium form chloride anion complexes.

Palladium metal targets were dissolved in hot aqua regia, taken to dryness and dissolved in 4 N HCl. Silver nitrate was added and the solution was filtered to remove silver isotopes produced during the irradiation. The solution containing 0.5 g palladium was loaded onto an ion-exchange column (1.5 cm dia., 12 cm high); and the rhodium was eluted with 4 N HCl and discarded. After allowing the ^{101}Pd to decay overnight, the top of the ion-exchange resin containing palladium was removed and put on a fresh bed of resin (4 cm high). The location of the palladium was determined by the characteristic red-brown color of its chloride. This was done to insure complete removal of rhodium and to prevent stable palladium breakthrough from contaminating the ^{101m}Rh. The ingrown ^{101m}Rh was then eluted with 4 N HCl. Palladium breakthrough was monitored by checking the eluent for ^{100}Pd or ^{101}Pd.

The chemistry when rhodium chloride trihydrate was used as a target was similar to the scheme above except that 20 mg of palladium carrier was added; and the silver nitrate step was not needed since silver isotopes are not produced by the proton irradiation of rhodium.

Rhodium-101*m* was incorporated into rhodium II butyrate using a previously described[10] method by refluxing rhodium chloride and sodium butyrate in absolute ethanol for one hour. The reaction proceeded to completion (90% overall yield) when 6 mg of rhodium was present. No reaction occurred when carrier was absent. Further studies are necessary to determine the minimum carrier necessary to achieve a good chemical yield. We also have started animal distribution studies using this ^{101m}Rh-rhodium II butyrate,[10] and using ^{101m}Rh as eluted from the ion-exchange column after the

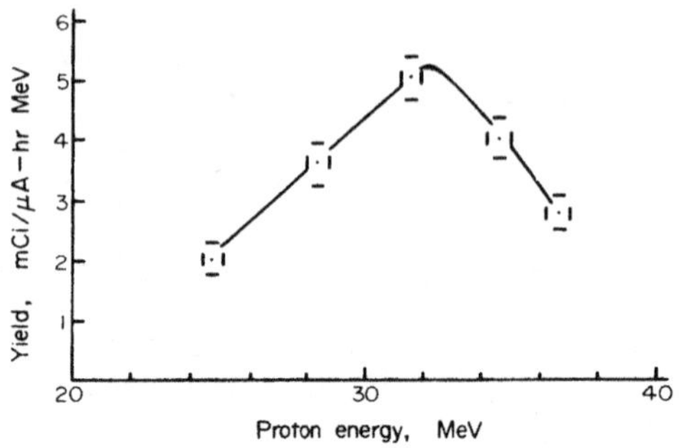

FIG. 2. Excitation function for the $^{103}Rh(p,3n)$ ^{101}Pd reaction.

removal of hydrochloric acid and redissolution in 0.01 N HCl. It is projected that because of the convenient half-life of ^{101m}Rh it can be useful as a tumor localizing agent in processes that take several days for optimal *in vivo* uptake.

3. RESULTS AND DISCUSSION

The excitation function for the $^{103}Rh(p, 3n)$ ^{101}Pd reaction is shown in Fig. 2. The indicated errors for energy are a combination of uncertainties in the incident proton energy and the target thickness; while the errors shown in the yield data result from counting errors and uncertainties in detector efficiency and photon abundance. The yields are shown in mCi/μA-hr for a one-MeV energy loss in the target. The yields shown were for short (10 min) irradiations when compared with the half-life of ^{101}Pd; therefore, no correction was made for ^{101}Pd that decayed during the irradiation. The foils used were actually 0.8–3.8-MeV thick but the yields were normalized to a one-MeV energy loss. The irradiation energies defined by the points in Fig. 2 are averages of the energies of the protons entering and exiting each foil. The peak yield is at about 32-MeV proton energy and the total integrated yield in the energy range studied is about 50 mCi/μA-hr. The theoretical maximum of ^{101m}Rh available from this yield of ^{101}Pd is 4 mCi/μA-hr; however, only 40–70% of this yield can actually be achieved because of chemical losses, decay of ^{101m}Rh during in-growth from ^{101}Pd and use of a chemical compound as a target instead of metallic rhodium. The ^{101m}Rh collected by ion exchange is initially free from ^{101}Rh. Some ^{100}Rh is present, but a few days after elution, the ^{100}Rh was less than 0.2% of the ^{101m}Rh activity. A Ge(Li) spectrum of high-purity ^{101m}Rh is shown in Fig. 3; inspection at 540 keV shows that no ^{100}Rh is present in this sample taken 3 days after separation. The small peaks in the 100–200-keV range result from very low intensity transitions of ^{101m}Rh.

A significant problem in the use of rhodium for targets is that rhodium and many of its compounds are extremely difficult to dissolve. We could not dissolve rhodium foil at all; a one hour fusion with sodium peroxide produced only surface discoloration and no discernible loss of thickness. Fusion with sodium peroxide did completely attack rhodium powder; however, after proton irradiation of one test sample, the rhodium powder did not react completely. Because of these problems with metallic targets we investigated rhodium compounds and decided that rhodium chloride trihydrate showed promise, since before irradiation it easily dissolved in water.

Rhodium chloride trihydrate (638 mg) in an aluminum holder covered with 0.127 mm tungsten foil was irradiated for one hour with a 0.7-μA beam of protons. After irradiation most of the rhodium salt (95%) dissolved easily in water. The remaining fraction was removed by filtration since it dissolved slowly and would interfere with the ion exchange preparation of carrier-free ^{101m}Rh.

Palladium metal targets dissolve quickly in aqua regia; however, the yields of ^{101m}Rh from

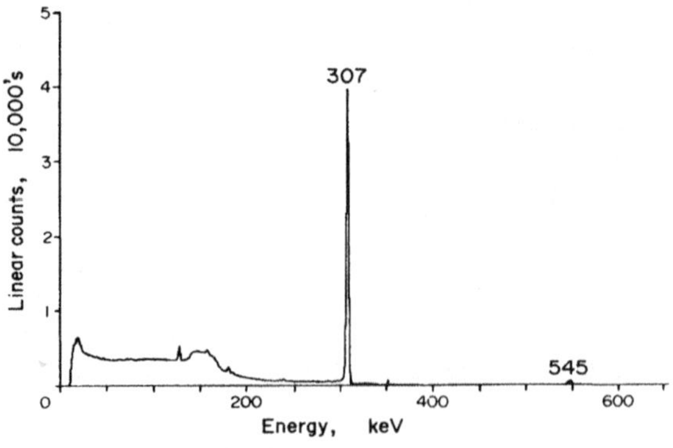

FIG. 3. Ge(Li) spectrum of high-purity Rhodium-101*m*.

palladium targets are much lower than those achieved with rhodium targets. The natural Pd$(p,xn)^{101}$Ag$\rightarrow$ ^{101}Pd and Pd$(p,pxn)^{101}$Pd reactions were studied on metallic targets which degraded the initial proton energy from 41 to 35 MeV. The typical yield of ^{101}Pd after a short irradiation was 0.5 mCi/μA-hr; the maximum yield of ^{101m}Rh available would be 40 μCi/μA-hr. It is planned to study this reaction at higher energies. It appears that a proton irradiation energy of 60–65 MeV to optimize the ^{105}Pd$\rightarrow$ ^{101}Ag + ^{101}Pd and the ^{106}Pd$\rightarrow$ ^{101}Ag + ^{101}Pd reactions would give the highest yield while keeping the target to a reasonable thickness. Palladium-105 and ^{106}Pd have natural abundances of 22.2 and 27.3%, respectively.

We conclude that high-purity ^{101m}Rh can be produced through its natural precursor ^{101}Pd; and that ^{101m}Rh can be incorporated into molecules of biological significance.

REFERENCES

1. ROSENBERG B., VAN CAMP L., TROSKO J. E. and MANSOUR V. H. *Nature, Lond.* **222,** 385 (1969).
2. LANGE R. C., SPENCER R. P. and HARDER H. C. *J. nucl. Med.* **13,** 328 (1972).
3. BEAR J. L., GRAY H. B., JR., RAINEN L., CHANG I. M., HOWARD R., SERIO G. and KIMBALL A. P. *Cancer Chemother. Rep.* Part 1 **59,** 611 (1975).
4. BOWMAN W. W. and MACMURDO K. W. *Atomic Data and Nuclear Tables* **13** (2–3), 225–8 (1974).
5. LEDERER C. M., HOLLANDER J. M. and PERLMAN I. *Tables of Isotopes*, 6th Edition. Johnson and Wiley, New York (1967).
6. *Nuclear Data Sheets* **10** (1), 59 (1973).
7. SHARMA B. L. *Nucl. Phys.* **19,** 550 (1960).
8. EVANS J. S., KASHY E., NAUMANN R. A. and PETRY R. F. *Phys. Rev.* **138,** B9 (1965).
9. BERMAN S. S. and MCBRYDE W. A. E. *Can. J. Chem.* **36,** 835 (1958).
10. KITCHENS J. and BEAR J. L. *Thermochimica Acta* **1,** 537 (1970).

International Journal of Applied Radiation and Isotopes, 1977, Vol. 28, pp. 213–227. Pergamon Press. Printed in Northern Ireland

Radiochemical Quality Control of Short-lived Radiopharmaceuticals

KENNETH A. KROHN* and ANNE-LINE JANSHOLT
Department of Radiology, School of Medicine, University of California, Davis, CA 95616, U.S.A.

(*Received* 16 *February* 1976)

Radiochemical purity, the fraction of radioactivity present in the specified chemical form, is a major factor determining the reproducibility of a nuclear medicine diagnostic procedure. Impurities may arise during preparation and storage of radiopharmaceuticals and will frequently modify organ distribution and specificity, possibly leading to an incorrect diagnosis of the patient's health. This review is a guide to radiochemical analytical principles for those who develop new short-lived radiopharmaceuticals. The applicability of some recently developed methods is contrasted with conventional techniques, and guidelines are suggested for predicting the best separatory mechanism to apply to a new radiochemical quality control problem. The selected method(s) should be sensitive, reproducible, separate all possible components without causing changes in sample composition, and preferably be convenient to perform.

INTRODUCTION

IN EVALUATING the quality of a radiopharmaceutical preparation, several properties must be tested to assure its safety and efficacy. The properties tested always include radionuclidic, radiochemical, and chemical purity; sterility and apyrogenicity become essential considerations when the radiopharmaceutical is intended for parental administration. At one time, the responsibility for quality control was generally borne by the commercial manufacturer of the radiopharmaceutical. However, with the increased preparation of radiopharmaceuticals labeled with short-lived isotopes by independent investigators, the responsibility for quality control has shifted to individual nuclear medicine laboratories where the preparations are being made. Therefore, a guide to radiochemical quality control principles for independent researchers seems appropriate.

Most drugs are administered to patients for their therapeutic effects, whereas most radiopharmaceuticals are administered for diagnostic purposes. Diagnostic amounts of high specific activity radiopharmaceuticals neither perturb the steady-state physiology of the patient nor produce pharmacologic effects or toxicologic, immunologic, or allergic reactions. However, any radiopharmaceutical preparation may contain chemical, radiochemical, and radionuclidic components other than those intended to be present. These impurities may degrade image quality, increase absorbed radiation dose, or localize in areas other than those intended, ultimately giving incomplete or incorrect information to the nuclear medicine physician. Radiochemical impurities in a radiopharmaceutical preparation would rarely produce a serious toxic reaction but may lead to a serious error in diagnosis.

Radiochemical purity refers to the fraction of a specific radioisotope that is present in the desired chemical form; whereas radionuclidic purity refers to the fraction of the total radioactivity that is present as the specified radioisotope. Radionuclidic impurities originate from the target materials during bombardment in the reactor or cyclotron and cannot always be removed by chemical purification methods. For radionuclidic and radiochemical quality, one

*Address all reprint requests to: Kenneth A. Krohn, School of Medicine, 4301 X Street, Sacramento, CA 95817, U.S.A.

does not usually require absolute purity but asks whether the preparation is sufficiently pure for its intended purpose. Purity requirements will be more stringent for some radiopharmaceuticals than for others and may vary for each use to which a given radiopharmaceutical is applied. Establishing criteria for the minimum acceptable radiochemical purity of a radiopharmaceutical is a function of both chemical and biologic properties and will be discussed later. Sterility and apyrogenicity are absolute quality requirements for radiopharmaceuticals intended for parenteral administration.

Satisfactory radionuclidic and radiochemical purity does not guarantee the chemical purity of a radiopharmaceutical preparation. Chemical purity is an additional variable requiring analysis by radiopharmaceutical scientists; however, it is primarily determined by the quality of the different chemicals used. When analytical grade chemicals are used in the compounding of radiopharmaceuticals, chemical purity is more consistent and predictable than is radiochemical purity.

In this paper we will present a philosophy for establishing radiochemical purity criteria and outline some techniques for developing quality control methods for new radiopharmaceuticals. We will attempt to review conventional and recently developed analytical methods applicable to radiochemical quality control and to offer the scientist developing a quality control protocol for a new radiopharmaceutical, some guidance on the potential applicability of these techniques.

Measurements of radiochemical purity are often the simplest of all radiopharmaceutical quality control tests. Specific methods for determining radiochemical impurities in commonly available radiopharmaceuticals which were recently reviewed in the literature[1–3] and several methods described in the U.S. Pharmacopeia XIX[4] will not be discussed in this paper. Recent developments in separation science have established chromatography as the most versatile and selective separation method, and when combined with the sensitivity of radiation detectors, it provides a rapid and sensitive method for measuring radiochemical impurities. Although not infallible, chromatographic and electrophoretic techniques have become the methods of choice for radiochemical quality control, due not only to their versatility but also to the speed and ease with which results can be obtained.

SOURCES OF RADIOCHEMICAL IMPURITIES

Cyclotrons and reactors produce the radionuclidic impurities which contaminate radiopharmaceuticals, but what makes radiochemical impurities? Table 1 illustrates these impurities originating in all production steps, from isotope production to synthesis and purification of compounds, as well as from various degradative processes occurring during radiopharmaceutical storage. It is important to understand the relative significance of each of these processes, when they are likely to occur, and how impurities may be tested for and controlled or eliminated.

Methods for isolating radionuclides from irradiated targets include distillation, precipitation, solvent extraction, and ion exchange chromatography. The method is selected which yields the highest radionuclidic purity, but this selection is often at the expense of chemical or radiochemical purity. For labeling reactions, chemical and radiochemical purity are critically important and, along with carrier concentration, should be measured prior to initiating labeling procedures. For example a radiochemical impurity in ^{18}F-fluoride interfered with amino acid labeling,[5] and the sporadic presence of an impurity in ^{123}I-iodide prepared by Crocker Nuclear Laboratory correlated with a reduction in our protein iodination yields. Commercial radioiodine preparations have also been criticized for variable quality which affected the yield of lactoperoxidase catalyzed iodination.[6]

TABLE 1

Sources of radiochemical impurities
Radioisotope production and isolation
Impurities in radiopharmaceutical synthesis
Synthesis side-reactions leading to labeled derivatives
Incomplete preparative separation
Breakdown during storage
Radiation-induced decomposition
Chemical instability—hydrolysis and oxidation
Reaction with chemicals in the carrier medium

Radiopharmaceutical scientists strive for a quantitative labeling yield when preparing radiopharmaceuticals; however, this goal is often unattainable, leaving unbound isotope as a radiochemical impurity in the reaction product. A preparative separation designed to remove impurities is, therefore, a common part of many radiopharmaceutical preparations. Ideally, the purification step will separate the labeled from the unlabeled compound, thereby removing all carrier and improving the specific activity of the radiopharmaceutical. This happens when radioiodinated amino acids are purified chromatographically; however, it does not occur when radioiodinated proteins are purified, and it is not necessary for radiochemical quality control tests.

A more subtle radiochemical impurity arises from synthetic side-reactions which occur at the time of radiopharmaceutical synthesis, leading to labeled derivatives of the desired radiopharmaceutical. It is necessary to test the biologic behavior of such derivatives by techniques described in the next section to determine if these radiochemical impurities will be detrimental to the *in vivo* use of the radiopharmaceutical. Especially when labeling large biologic molecules with isotopes of heterologous atoms, can labeled derivatives be produced with biologic modifications inconsistent with the intended diagnostic use of the radiopharmaceutical.

A radiopharmaceutical completely in the desired chemical form following preparative separation may contain radiochemical impurities when used at some later time. Radiation-induced decomposition is the most commonly mentioned cause of chemical breakdown during storage of radiopharmaceuticals. Each nuclear decay event deposits many thousand keV of energy into a small volume of radiopharmaceutical. This energy may rupture chemical bonds within the molecule where the radioactive decay took place (primary radiolysis) or it may rupture chemical bonds in neighboring molecules (secondary radiolysis). Primary radiolytic decomposition occurs with nearly every radioactive decay event and is the reason for avoiding "doubly-labeled" molecules in radiopharmaceutical preparations. Many authors have blamed radiolytic effects for the appearance of inorganic iodine in solutions of iodinated organic molecules.[1,7] The magnitude of radiolytic effects varies with the energy absorbed by the radiopharmaceutical solution as well as with the specific activity of the preparation. Its effect can be reduced by decreasing specific activity, but it can never be totally eliminated. Radiation chemists have discovered a number of chemical additives (e.g. electron scavengers, such as ethanol) which inhibit secondary radiolysis reactions. The rigorous exclusion of oxygen improves the stability of some radiochemicals. The addition of compounds with a chemical identity similar to the labeled radiopharmaceutical has also been recommended as an effective means of protecting radiopharmaceuticals from radiolytic decomposition.[7] Added carrier molecules must be considered chemical impurities in the preparation and should be kept to a minimum quantity and be used only when absolutely necessary, when proven effective and harmless to the patient.

While radiation inevitably causes decomposition of radiopharmaceuticals, chemical decomposition may occur independent of radioactivity. Chemical decomposition does not usually result in complete destruction of the labeled molecule but involves primarily the separation of the radioisotope from the compound to which it was bound (Table 2). It frequently includes the substitution of either a proton or hydroxyl group from a water molecule for the radioisotope and

TABLE 2. Chemical decomposition pathways

*I-Mol	+	H_2O	→	H-Mol	+	*I	Hydrolysis
*I-Mol	+	Mol′	→	*I-Mol′	+	Mol	Chemical reaction
*I-Mol	+	I	→	I-Mol	+	*I	Reaction with carrier
*I-Lig	+	Lig′	→	*I-Lig′	+	Lig	Competitive chelation
*I-Lig	+	I	→	I-Lig	+	*I	Equilibration with carrier

*I: Any radioisotope Mol, Mol′: Different molecules
I: A stable nuclide of the same element Lig, Lig′: Different ligands

is termed "hydrolysis." These reactions are frequently pH dependent and always depend upon the chemical composition of the medium in which the radiopharmaceutical is dissolved. Decomposition can also be catalyzed by reactive hydroxyl and siloxyl groups on the surface of glass. Chemical decomposition can most easily be controlled by storing a radiopharmaceutical in the proper buffer and container. A common example of hydrolysis is the substitution of a proton for an iodine atom in radioiodinated proteins. The resulting free radioiodide atom is a radiochemical impurity in the radiopharmaceutical, and it should be quantitated before use. A more complex example of hydrolysis is the water/oxygen induced oxidation of Tc(IV) to TcO_4^-. Some transition metal chelates break down in solution because of competitive reactions with other ligands. A chelation compound exists in equilibrium with both free ligand and metal and, hence, is susceptible to chemical modifications by competing ligands and metal ions of the same charge. Such reactions can be eliminated by carefully controlling the chemical composition of the radiopharmaceutical solution.

Endothermic chemical reactions depend on temperature, and reaction rates decrease as temperature decreases. The rate of hydrolysis and chelate breakdown can be reduced by lowering the temperature at which samples are stored. Temperature should not be too low, as freezing can be detrimental to certain chemicals. For example the tertiary structures of large biologic molecules are often destroyed by the crystallization that occurs when aqueous salt solutions are frozen. Some organic molecules (e.g. DMSA) become insoluble at low temperatures and should not be refrigerated.

When adding chemicals to a radiopharmaceutical preparation (e.g. radiolysis scavengers, bacteriostatic agents, or enzyme inhibitors), their possible reactivity with the radiopharmaceutical must be considered. The tracer concentration of many radiopharmaceuticals makes them unstable with respect to chemical combination with other molecules in solution. Such reactions are variable and unpredictable and have been particularly troublesome for carrier-free solutions of ^{131}I and ^{32}P.[6,7] Phenol, a common bacteriostatic agent, can react with free radioiodine, leading to a radiochemical impurity.[7] Biological preparations may contain proteolytic enzymes which slowly degrade the molecule. We have found that crude ammonium sulfate precipitated fibrinogen preparations may contain enough fibrinolytic enzyme activity to completely degrade the fibrinogen to lower molecular weight split products within 24 hr at room temperature.

Chemical reactions with the radiopharmaceutical storage container must also be considered as a potential mechanism leading to radiochemical impurities. The very low molar concentration of radioactive chemicals makes them unusually susceptible to impurities present on the surface of various containers. A well cleaned glass surface can be an active site for the catalysis of chemical reactions, and a microgram of a metal syringe needle dissolved by buffer salts can contribute to the alteration of a radiopharmaceutical.

There are many possible sources of radiochemical impurities in radiopharmaceutical preparations. Some of these can be controlled or eliminated but others cannot. However, every effort expended to protect the quality of a radiopharmaceutical will optimize the efficacy with which it can be used.

ESTABLISHING PURITY LIMITS

Everyone agrees that a radiopharmaceutical must be pure. But what does "pure" mean? The word has no simple, generally accepted meaning. Pure water might mean deionized water to the scientist, water for injection to the pharmacist, or simply drinking water to the layman. An acceptable standard for purity should emphasize sufficient purity for the intended use. In some instances, this will require a high degree of absolute purity, whereas in others it will require only the absence of certain specific impurities.

Cohen has argued that radiopharmaceutical specifications should "not be systematically severe but should take into account the real biological inconveniences that might result from the presence of radiochemical impurities."[7] The biodistribution of potential radiochemical impurities may be detected by labeling them and studying their distribution in appropriate animal models. Although not a useful technique for routine analysis of short-lived radiopharmaceuticals, it can be useful while developing a new product, to decide which labeled side products or degradation fractions are acceptable contaminants in the final product and which must be removed before use. Radiochemical impurities whose biologic pathways differ from that of the radiopharmaceutical may concentrate in organs in a pattern that interferes with interpretation of the study.

Only by understanding the extent to which chemical or radiochemical impurities can affect the biological distribution of a new radiopharmaceutical can the nuclear scientist define safe and reasonable limits of quality for a new product. For example radiochemical quality control tests were able to separate rose bengal into many fractions;[8] however, these fractions all had similar biological properties, making this radiochemical quality control test unnecessary in defining the practical or usable quality of the ^{123}I-Rose Bengal preparation. On the other hand, improvements in radiochemical quality control have revealed "radiochemical impurities" that were chemically similar but demonstrated better biologic properties than the intended radiopharmaceutical. An example is the observation of Kojima *et al.* that a radiochemical impurity in *I-19-Iodocholesterol, *I-6-Iodomethylnorcholesterol, actually proved to be the better adrenal scanning agent in rats.[9] Recent reports on thin-layer chromatography of ^{67}Ga-citrate in 85% methanol/water showed that preparations from different manufacturers migrated differently under constant chromatographic conditions.[10] This chemical difference may correlate with differences in biologic behavior.

As a general criteria, 95% radiochemical purity is a reasonable minimum standard that may be revised upward or downward for special cases. Important factors to consider are the blood clearance rates and excretion rates of both the desired radiopharmaceutical and the possible radiochemical impurities. A level of 5% free radioiodine would normally be acceptable for radioiodinated plasma proteins, because the radioiodide is cleared from the blood by thyroid trapping or excreted in the urine at a much faster rate than that by which labeled protein is cleared from the blood. On the other hand, one is reluctant to accept even 1% free radioiodide in ^{123}I-hippuran because hippuran is cleared from the blood much faster than iodide. If radiochemical impurities clear more slowly than the desired radiopharmaceutical, the *in vivo* radiochemical purity actually decreases with time. If a radiopharmaceutical is designed so that the biological $T_{1/2}$ is shorter than the physical $T_{1/2}$, a very high initial purity is required because after a few biological half-lives the radiochemical impurity may become the predominating radiopharmaceutical circulating in the blood stream. The consequence may be excessive body background which could camouflage disease otherwise detected by the radiopharmaceutical.

SPECIFIC SEPARATORY TECHNIQUES APPLICABLE TO SHORT-LIVED RADIOPHARMACEUTICALS

Before selecting specific methods for a chemical separation, the general mechanisms and methods of separatory systems should be considered. The methods by which chemical compounds are separated can be broadly divided into equilibrium processes and rate processes, but neither is capable of completely separating a mixture into the proper number of totally pure fractions. Distillation, extraction, and precipitation are common examples of equilibrium processes and generally result in one very pure fraction, with the other fraction containing molecules having a wide range of properties. The distillate and precipitate contain a pure compound, but many different chemicals can remain in the supernatant. Modern developments in separation science have emphasized rate processes, such as chromatography, electrophoresis, and ultracentrifugation. Each rate process separates a chemical mixture into a continuum of compounds, from those which exhibit a large degree of some chemical or physical property to those with total absence of the same property. Some properties which are the basis for rate separations include molecular weight and charge, acidity or basicity, volatility, or the presence of various chemical groups. Understanding the chemical or physical differences which might exist between a radiopharmaceutical and its potential radiochemical impurities, therefore, allows one to predict the basis of selectivity that may be effectively exploited in developing a separatory system, thus reducing the number of separatory techniques to be investigated before selecting a radiochemical quality control protocol. Having determined the chemical property by which a mixture may be separated, it is worthwhile to review some advantages and disadvantages of the more well-established radiochemical quality control techniques.

Precipitation is one of the oldest separatory techniques known to chemists. It allows for only a two phase separation, but in some instances that may be sufficient. For example the U.S. Pharmacopeia XIX recommends a precipitation test for the radiochemical purity of ^{51}Cr-chromate.[4] The chemical properties of the material that precipitates are always more uniquely defined than are those that stay in solution. Consequently, in any precipitation test for radiochemical purity, the desired radiopharmaceutical should be precipitated while the radiochemical impurities remain in the supernatant. We have used coprecipitation of ^{123}I-*o*-iodohippurate with hippuran as a preparative separation and as a radiochemical purity test. Crystallization is a more selective precipitation process which has been used for separating optical isomers of amino acids and sugars. It requires carrier amounts of the compound that is precipitated; however, it will separate molecular isomers that can be separated by no other known method.

Chromatography is a process by which a mixture of compounds is separated while being swept over a stationary phase adsorbent by a mobile solvent for which the components in the mixture have varying affinities. Solute molecules are carried in the direction of solvent flow, but with the proper eluent as well as adsorbent and operating conditions, each component of the sample will migrate at a different velocity, resulting in the desired separation. Chromatographic techniques vary, but common to all is a stationary phase, a solvent and solvent delivery system, a mechanism for introducing an unknown sample for analysis, and a method for detecting the distribution of separated molecules. The nomenclature for chromatography defines the physical state of the mobile and stationary phases. If the stationary phase is a liquid or solid contained in a column, the method is identified as gas or liquid column chromatography, depending on the physical state of the mobile phase. Alternatively, if the stationary phase is paper or a thin slab of adsorbent operated as an open bed, the method is termed paper or thin-layer chromatography. The principles and practice of gas,[11] liquid[12–14] and thin-layer[12,15] chromatography have been described in the literature.

Liquid column chromatography is further subdivided to describe the physical state of the stationary phase and the chemical mechanism by which it interacts with solute molecules. If the stationary phase is a solid material which reversibly adsorbs and desorbs molecules, the technique is called liquid–solid adsorption chromatography. Common examples of adsorption column packings are silica gel and alumina which tend to separate mixtures based on the number and type of polar groups. An extension of adsorption chromatography is ion-exchange chromatography in which the stationary phase has charged sites which have differing affinities for solute ions of opposite polarity.

Alternatively, the stationary phase may be a liquid adsorbed onto a porous solid support. The liquid stationary phase must be immiscible in the mobile phase and have varying chemical affinities for the compounds to be separated. This techniques is called liquid–liquid partition chromatography, and the chemical principle is analogous to liquid–liquid extraction using a separatory funnel but with greater efficiency. Generally, the stationary phase is more polar than the mobile (normal phase); however, recent applications have been described for reverse-phase partition chromatography in which the immobile phase is less polar than the mobile phase. In principle, the selectivity in partition chromatography is almost unlimited as the nature of the two liquid phases can be varied through a wide range of chemical properties; however, in practice, polarity and chain length have been the major properties exploited in achieving separation.

The eluent in column chromatography is pumped through the stationary phase particles, and a suitable system is provided for on-line injection of the mixture to be separated. In the application of column chromatography to radiochemical quality control analysis, the column eluate may be fraction collected for specific wet chemical procedures or monitored by a flow-through radioactivity detector connected to a ratemeter to give a continuous record of the radioactivity profile eluted from the column.

Several recent technological advances have significantly improved conventional liquid chromatography. Theoretically, small bore chromatographic columns with small spherical particles give the best resolution, but they require high column inlet pressures. New pumps and plumbing have been developed that make high pressure liquid chromatography (HPLC) a practical technique.

The major advance making HPLC such a versatile separatory tool has been the development of new column packings. Originally packing materials were porous with a high surface area that led to peak broadening from the slow diffusion of solute molecules into and out of deep pores containing stagnant mobile phase. Recently, "pellicular" column packings have been developed that have a spherical nonporous core with a uniformly thin (~ 30 μm) porous outer shell of siliceous or polymeric adsorbent, thus reducing the path length for diffusion and maintaining closer equilibrium between eluent and adsorbent. The practical consequence is greatly increased speed and resolution, but the small quantity of adsorbent dictates more dilute samples and buffer. Techniques are now being developed that allow partitioning agents to be chemically bound to pellicular packings, thus eliminating a major problem in conventional liquid–liquid chromatography, namely that stationary phase is slowly dissolved and eluted by mobile phase, thus constantly changing the chemical properties of the column. The bibliography lists several references which give specific details on new packing materials for HPLC. The paperback by PERRY *et al.*[12] and the reviews by KIRKLAND[16] and MAJORS[17] will initiate anyone considering applying HPLC to a specific separatory problem.

While advances in column technology have dominated the attention of liquid chromatographers, the solvent also plays a critical role in maximizing column performance; therefore, an experimentor should select the right solvent with care. In practice, the only way to choose the correct eluent is by trial experiments, but some guidelines can simplify the search. The solvent must dissolve the mixture to be separated, and it must not react with solute molecules or irreversibly alter the column packing. After preliminary screening by the above criteria, considerations of viscosity and polarity predominate. A viscous solvent (> 0.6 cP) slows down equilibration between the mobile and stationary phases, decreases resolution, and requires higher column inlet pressures. A solvent with low viscosity (< 0.2 cP) is avoided because its low boiling point may lead to bubbles in the column.

Selecting a solvent with the proper polarity is the last step in the development of a separation that gives maximum resolution per unit time. The trial-and-error process can be guided by ordering, from weak to strong, the relative polarity of potential solvents. Called an eluotropic series, the solvents are listed in Table 3 for alumina adsorption chromatography.[13] The ordering is roughly the same for all polar adsorbents and increases regularly with solvent polarity. For nonpolar adsorbents, the solvent strength is essentially the reverse of that for alumina, running from water and methanol on the weak end to aromatic hydrocarbons on the strong end. Using the appropriate eluotropic series, the experimentor starts with a solvent of intermediate strength and systematically works toward either end, depending on the results obtained, until the best solvent is determined. With some experience, this can be done efficiently, but trials can be more rapidly performed on thin-layer plates, with the results being directly applicable to columns. Selecting solvents for partition chromatography is more difficult but is also guided by an eluotropic series, complicated by a wide spectrum of stationary phase strengths. The reader is referred to PERRY[12] or KIRKLAND[13] for a detailed discussion of this subject.

Gel permeation chromatography (GPC) is another column technique that is used to separate compounds by molecular size. The gel, a porous column packing, excludes molecules of high molecular weight from penetrating into pores of the packing but allows smaller components to diffuse into the gel pores. Large molecules, completely excluded from

TABLE 3. Eluotropic series for alumina absorbents*

Solvent	Solvent strength	Viscosity cP, 20°	Boiling point, °C
n-Pentane	0.00	0.23	36
i-Octane	0.01	0.54	118
n-Heptane	0.01	0.41	98.4
Cyclopentane	0.05	0.47	49.3
Carbon disulfide	0.15	0.37	45
Carbon tetrachloride	0.18	0.97	76.7
i-Propyl ether	0.28	0.37	69
Toluene	0.29	0.59	110.6
Chlorobenzene	0.30	0.80	132
Benzene	0.32	0.65	80.1
Ethyl ether	0.38	0.23	34.6
Chloroform	0.40	0.57	61.2
Methylene chloride	0.42	0.44	41
Tetrahydrofuran	0.45	0.55	65
Methylethylketone	0.51	0.43	79.6
Acetone	0.56	0.32	56.2
Dioxane	0.56	1.54	104
Ethyl acetate	0.58	0.45	77.1
Dimethyl sulfoxide	0.62	2.24	190
Nitromethane	0.64	0.67	100.8
Acetonitrile	0.65	0.37	80.1
Pyridine	0.71	0.94	115.5
i-Propanol	0.82	2.3	82.4
Ethanol	0.88	1.20	78.5
Methanol	0.95	0.60	65.0
Ethylene glycol	1.11	19.9	198
Acetic acid	Large	1.26	118.5

*Reference[12], p. 55.

the pores of the gel, move quickly through the column and are eluted first. Smaller molecules penetrate the gel pores and are, therefore, diluted into a much larger column volume and eluted from the column last, whereas molecules of intermediate size are fractionated by the gel bed. The volume at which completely excluded molecules are eluted is the void volume (V_0), and the volume at which very small molecules are eluted is called the bed volume (V_b). Molecules eluted in a volume V_x between the void volume and bed volume are characterized by the parameter $K_d = (V_x - V_0)/(V_b - V_0)$. Ideally, a radiochemical quality control test should be designed such that the principal radiopharmaceutical is eluted with K_d between 0.2 and 0.8.

Among the materials that can be polymerized to give chromatographic gels are dextran,[18] acrylamide,[19] styrene,[20] agarose,[18] and silica glass.[21] Each of these gels can be prepared in ways that vary pore size, and a given pore size can be used to effectively separate molecules over a two order of magnitude range in molecular size. Gels that are effective in overlapping size ranges can be combined to achieve a complete separation of molecules having widely different molecular weights. The gel is relatively chemically inert but has been reported to adsorb some molecules. This effect is especially pronounced when carrier free quantities of material are being separated as well as with chemically active gels such as porous glass. These materials can be chemically treated to reduce the surface activity, but this has met with only limited success. Using a stronger solvent may also help to reduce adsorptive processes.

Gel chromatography requires a pump or

hydrostatic pressure to move the eluent through the gel. The eluent is generally an aqueous salt solution, but some gels can be used with organic solvents. The user must be careful to thoroughly degas solvents used in GPC, as bubbles are a major cause of trouble in gel separations. Gel chromatographic separations are very predictable, and procedures reported in the literature can generally be reproduced. A column can be uniquely defined by bed volume, eluent buffer and flow rate, gel porosity, and particle size. Once set up, gel radiochemical quality control tests can be done reproducibly and quickly.

Gels prepared from dextran, acrylamide, and agarose are easily compressed and are, therefore, operated at low pressure. Polystyrene and glass beads withstand high pressure and can be used with higher column flow rates to yield more rapid separations. The resolution of GPC columns, however, is never as good as for HPLC, with the latter being preferred for complex mixtures but the former being adequate for simple mixtures.

Gel chromatography has frequently been used for preparative chromatography of radiopharmaceuticals. The large molecular weight difference between labeled biologic molecules and unbound ions makes GPC an ideal technique for these separations. PERSSON has applied GPC to the radiochemical quality control of ^{99m}Tc radiopharmaceuticals.(22) Because some chemical forms of technetium are not eluted from polydextran gels, PERSSON elected to scan the gel column by a technique analogous to electrophoresis strip scanning. His technique allowed the separation of various technetium radiopharmaceuticals, pertechnetate, and reduced technetium, a separation which is not possible with paper or thin-layer chromatography in a single solvent system. The radioactivity peak profile of radiopharmaceuticals eluted from a polydextran column has also been used to evaluate the presence of radiochemical impurities in radioiodinated proteins.(23) These radiochemical impurities had different biologic distributions and were, therefore, significant contaminants in the radiopharmaceutical preparation.(24)

A recent modification to gel chromatography is the development of techniques for binding biologic molecules to a porous gel matrix. Differences in the biologic affinity of the immobilized ligand for solute molecules are exploited in a powerful new technique called

TABLE 4

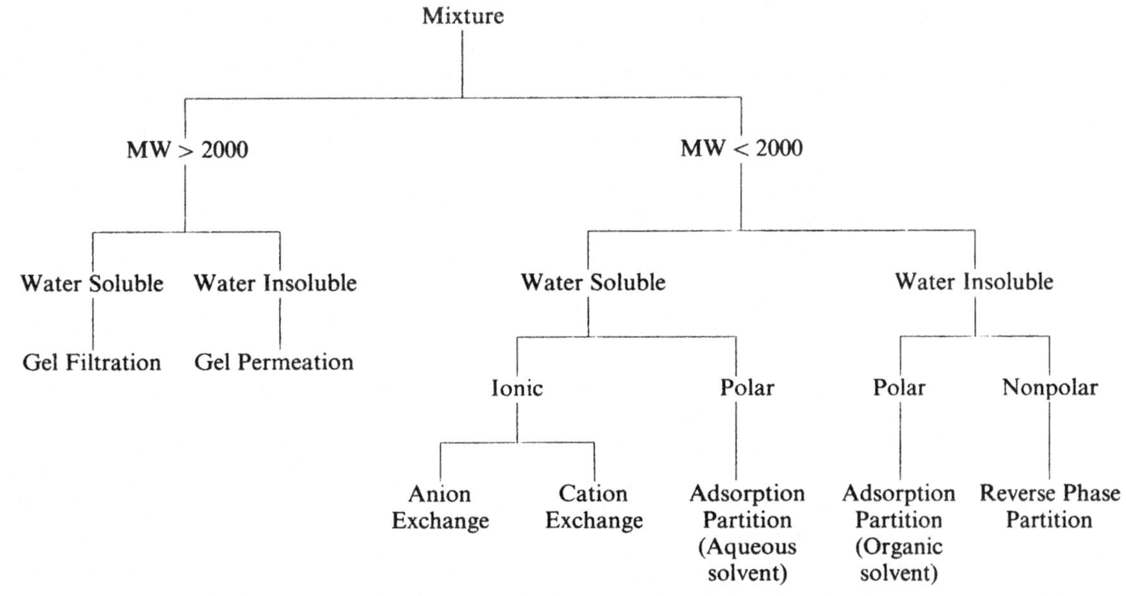

affinity chromatography.[25,26] The immobilized ligand adsorbs only substances with specific biologic properties and elutes molecules that are chemically similar but have suffered subtle biologic damage. After unbound substances are eluted, the solvent can be changed to desorb the bound material.

Of interest to the nuclear medicine researcher is that proteins, nucleic acids, steroids, antibiotics, enzymes, and cells can be chemically attached to gel substrates in ways that retain their biologic activity. Antisera to any of these molecules could be preparatively separated to give pure antibodies which can be stripped from the column and subsequently bound to a second affinity matrix. There they can serve as an adsorption column to test the purity of radioactively labeled antigens toward which they have biologic affinity. The usefulness of this separatory scheme has not been proved, but affinity columns containing immobilized antibody ligands should be useful for testing radiopharmaceuticals for the presence of labeled molecules whose biologic activity has been modified or destroyed by the labeling procedure. This is an important question in radiochemical quality control for which *in vitro* measurement techniques have been inadequate. Development and refinement of this technique should lead to better radiochemical quality testing of biologic radiopharmaceuticals which will improve the diagnostic usefulness of a large class of potentially very useful radiopharmaceuticals.

Paper and thin-layer chromatography are open-bed techniques for adsorption chromatography. The adsorbent in paper chromatography is a strip or sheet of paper across which a liquid phase migrates. Ascending, descending and horizontal solvent delivery systems can be used, with the solvent moving by capillary or gravitational force. Chromatography paper may be impregnated with chemicals to further increase chemical interaction with solute molecules.

In thin-layer chromatography (TLC), the stationary phase is a thin coating of dry adsorbent applied to aluminum, polymer, or glass. Silica gel and alumina are the most commonly used adsorbents. The solvent delivery system is the same as for paper chromatography and the chemical mechanism by which TLC separates mixtures is the same as for liquid–solid adsorption column chromatography. Considerations in selecting solvents are the same as for column adsorption chromatography.

Both paper and thin-layer chromatography are easier to perform and less expensive than column chromatography, but they give less resolution. The speed at which paper and thin-layer chromatograms can be developed varies from a few minutes to several hours or days depending upon the compounds to be separated and the degree of purity required, as well as properties of the solvent and adsorbent and the operating conditions for the system. The atmosphere in the chromatography chamber should be saturated with solvent vapor prior to development, since evaporation during development will decrease reproducibility. Evaporation is more rapid from the edges of the strip and will result in an uneven solvent front. Only a few microliters of solution should be applied to paper or TLC media. If several sample aliquots are required, the spot should dry between applications. This will make the spot as small as possible and give maximum resolution. When carrier-free compounds may be irreversibly adsorbed at the point of application, carrier amounts of the radiopharmaceutical and suspected impurities should be added. As an example, in the quality control of ^{99m}Tc-albumin, if not enough carrier albumin was present, the paper chromatogram in saline solvent showed diffuse streaking and no distinct peak.[27]

The migration of individual compounds in a sample is conveniently characterized by R_f values for both paper and thin-layer chromatography. The R_f is the distance an individual compound moves from the point of application divided by the distance the solvent front moves. R_f values are constant for each compound in a specific adsorbent/solvent system under carefully defined conditions. It is necessary to establish R_f values for both the desired compound and the probable impurities in the chromatographic system to be used for quality control by running each compound in parallel with the mixture. If an unknown radiochemical impurity is detected, it can be isolated for identification by cutting out that section of the strip, dissolving the compound in a suitable solvent, and subjecting it to further analytical tests. For radiochemical quality control procedures, the radioactivity distribution is most conveniently measured by a strip scanner which will be described later.

A recent development in TLC is Instant Thin-Layer Chromatography™ (ITLC). This important technological advance involved development of a glass microfiber cloth that could be impregnated with adsorbent. ITLC media are commercially available impregnated with silica gel or polysilicic acid, for nonpolar or polar compounds, respectively. They are supplied ready-to-use, thus eliminating the tedious step of preparing TLC adsorbent plates. ITLC supports give better resolution than conventional TLC supports, and with small volume chambers, development times can be reduced to less than 10 min with some solvents. A further advantage is that ITLC strips can be conveniently cut into sections for counting, whereas other ready-to-use TLC strips are brittle and difficult to cut. BILLINGHURST[28] has used the Sephrachrom™ ITLC system for quality control of several radiopharmaceuticals labeled with short-lived radionuclides, and a compilation of methods was reported by ICE and his associates.[29]

In addition to adsorption, partition and gel chromatography, another mechanism can be employed to separate charged molecules. The technique, *electrophoresis*, depends on the different migration rates of charged molecules in an electric field. Migration is primarily influenced by the polarity and magnitude of charge on a molecule and its size and shape, and also by the applied voltage, distance between electrodes, and duration of separation. For compounds containing acidic or basic groups, migration is primarily a function of buffer pH and ionic strength, but viscosity, temperature and sample solubility also affect separation.

The electrophoresis strip support media is generally paper or cellulose acetate, which may have chemical affinity for solute molecules but serves primarily to prevent mixing of the molecules as they migrate across it. Cellulose acetate yields more reproducible separations and better resolution than does paper. The liquid phase serves to keep the mixture in solution with a suitable charge and as a conductor of current. The support medium should be allowed to soak in the buffer before applying the sample, and excess evaporation should be avoided during sample application. A few microliters of the sample are usually applied as a band to the strip after blotting excess buffer. If the solution is very dilute, it may be necessary to apply the sample in several small portions, and sometimes it is necessary to add a carrier. Positioning of the sample relative to anode and cathode depends upon the charge sign of the molecules to be separated. If positively and negatively charged species are present, the sample is applied equidistant between anode and cathode. A sample should never be applied so that it touches the bridge supporting the electrophoresis strip.

The R_f concept is not applicable to electrophoresis, however R_m has been defined as the distance of migration of a test substance divided by the distance an arbitrary control compound migrates under the same conditions. These values are constant, but if several strips are run at the same time, the absolute migration rate can be variable from strip to strip. For radiochemical quality control procedures, radioactivity distribution after electrophoretic separation is most conveniently determined by a strip scanner which will be described later.

Paper and cellulose acetate stationary support media exhibit very little selectivity for solute molecules, however silica gel adsorbents can be used to increase electrophoretic selectivity and resolution. When conventional silica gel thin-layer strips were used, however, it was difficult to evenly and reproducibly saturate the nonconducting adsorbent with buffer. The recently introduced glass fiber support media which we described earlier for ITLC are easier to use and offer the advantages of shorter separation times, increased resolution, and easy and uniform saturation with buffer ion.[30] When an active adsorbent support is used, electrophoresis can be complemented by chromatography performed separately and in a direction perpendicular to the electrophoresis. This technique will separate very complex mixtures and has been beautifully applied to human serum components,[31] although we know of no instances in which the technique has

TM: Gelman Instrument Company, Ann Arbor, Michigan.

been applied to radiopharmaceutical quality control. Electrophoresis may also be done on permeation gels. Both slab gels and cylindrical columns take advantage of gel permeation and electrophoretic mechanisms to separate large molecules. Resolution is much better than for either electrophoresis or GPC alone, but the technique requires several hours of skilled effort. Gel electrophoresis has been useful in the quality control of proteins by detecting *in vitro* some subtle alterations in the biologic conformation of the protein.[32]

The radioactivity profile of paper or thin-layer chromatograms or electrophoresis strips is measured either by cutting the strip and counting each segment separately or by using a recording strip scanner with a radiation detector. Autoradiography has also been described as a method for obtaining a quantitative measure of radiochemical purity[7] but is not a useful technique for short-lived radiopharmaceuticals. When cutting a chromatographic or electrophoresis strip, cut the strip into many small sections of equal width rather than into two or three unequal sections where you suspect the peaks divide. Cutting the strip into many sections prevents missing small or partly overlapping peaks and detects a high general background caused by poor separation. Even if initial testing has shown that the possible compounds are completely separated, cut out a small segment between the peaks to show that it contains insignificant activity compared to the peaks.

In automatic strip scanning, the strip is carried past a collimated radiation detector at a constant speed. In commercial systems, the detector is usually a scintillator or a gas flow proportional detector connected to an analog recording device that moves the chart paper at the same speed as the strip. The recorder should be run on a setting that will give nearly full-scale deflection for the largest peak. With more sophisticated equipment, data can be obtained in a digital form; however, with analog recording devices, the individual peaks must be quantitated. Plainimetry is commonly used to integrate chart recorder peaks but is inaccurate for small peaks. Because the chart and strip move at the same speed, the strip can easily be placed next to the chart record, allowing peaks to be accurately cut and counted, and stained peaks from added carrier molecules can be used as further guides to help the experimentor cut the strip at the proper place. We have found this technique too accurate and easy to warrant the expense of adding digital recording equipment to our radiochromatogram strip scanner. A home-made radiochromatogram scanner was recently described which combines a strip chart recorder and an abandoned analog ratemeter to make an inexpensive and useful instrument.[33]

SELECTING AND EVALUATING RADIOCHEMICAL QUALITY CONTROL METHODS

Having now read more than you ever wanted to know about separation science, how do you extract from the methods available the one which could best solve an analytical problem in your laboratory? Begin by answering some questions designed to bring out the chemical differences in the sample components to be separated (Table 4):

Are there differences in molecular size of the components?

Are any components ionizable?

Are chemical groups of different polarity present?

In what solvents is the sample soluble?

If a ten per cent molecular size difference exists, gel chromatography will probably separate the mixture conveniently. The selection of gel type and solvent is readily made by referring to tables of gel properties,[12,18,19] from which the behavior of your column can be accurately predicted. If only very subtle differences exist, such as in tertiary structure of large molecules, gel electrophoresis or affinity chromatography may be necessary. These require more skill but will result in greater selectivity and resolution.

Alternatively, if there are charge differences between sample components, electrophoresis and ion-exchange chromatography should be considered first. If more subtle chemical differences must be exploited as the basis for selectivity, an adsorption technique should be tried before going on to partition separations for which solvent selection, reproducibility and chromatography operations are more difficult.

To select between the adsorption techniques of paper, thin-layer or liquid column chromatography, ask another series of questions about the procedure you wish to develop:

How many compounds must be separated?

How many samples will be run at once? How often?

How concentrated (volume and molarity) are the samples?

Must the fractions be collected after separation?

What chromatographic and radiation detection equipment is available?

If four or more compounds must be separated, the superior resolution of columns will be required, but radiochemical quality control usually requires the separation of only two or three compounds so that the resolution of open-bed techniques is adequate. Open-bed techniques require less expensive equipment and permit multiple samples to be run in parallel; however, their operation cannot be automated to the extent that HPLC can. Open-bed adsorbents will accept higher concentration samples than HPLC, but it is more difficult to retrieve the fractions after separation. The best approach for polar molecules is to try thin-layer chromatography first. A satisfactory separation for radiochemical quality control will probably be achieved, but if resolution is inadequate, the adsorbent and solvent (or one that is slightly less polar) can be conveniently translated to a HPLC technique.

The criteria applied in evaluating radiochemical quality control methods are of two types. The most important consideration is the sensitivity and reproducibility of the assay. A radiochemical quality control procedure must distinguish all possible contaminants. This may require more than one analytical system. The second criterion, especially for short-lived radiopharmaceuticals, is that the procedure should be convenient for the experimentor. Ideally, it should be simple, rapid, and inexpensive to perform. The sources of radiochemical impurities are such that the concentration of these impurities is time-dependent and the control test should, therefore, be done in temporal proximity to use of the radiopharmaceutical.

In order to make a radiochemical quality control test definitive, a procedure should be selected in which the principal radiopharmaceutical is separated into a fraction that defines its properties uniquely. Because precipitation tests separate samples into two fractions only, they are generally avoided as tests of radiochemical quality. Chromatographic and electrophoretic procedures separate chemical mixtures into a continuous spectrum of molecules having from more to less of that specific chemical or physical property which is the basis of selectivity for the separation. Procedures must be avoided that elute the principal radiopharmaceutical in the column breakthrough or bed volume or that cause it to migrate with R_f or R_m values of either 0 or 1. Many different molecules migrate with the solvent front in any one chromatographic system, and molecules with even greater differences will all stay at the origin; therefore, the extreme ends of the chromatogram are less definitive than in the middle. An analysis that does not separate the principal radiopharmaceutical from all potential impurities is worse than no radiochemical quality control test at all if it gives the user a false sense of security in the quality of the preparation. It is not ideal, but it is sometimes necessary and acceptable for radiochemical impurities to migrate with the solvent front or to stay at the origin. For example, $InCl_3$ has an R_f of 0 in all reported solvent systems.

Carrier-free aliquots of radiopharmaceuticals may behave differently from macroscopic samples. We have described irreversible adsorption at the site of injection onto some adsorbents and suggested a solution. Electrophoresis can cause oxidation of some samples or exposure to air while a small aliquot is being applied and dried on a paper, TLC, or electrophoresis strip may cause chemical modification. With some radiopharmaceuticals, it is necessary to work in an inert nitrogen atmosphere rather than in the air. Some of the simple chromatographic separations for ^{99m}Tc radiopharmaceuticals that have been described in the literature and recommended by radiopharmaceutical manufacturers give falsely high measurements of pertechnetate impurity caused simply by the analytical procedure. Having two complementary analytical procedures may help relieve the frustration of such artifacts.

When high radioactive background is detected in the continuous flow monitoring of column chromatograph effluent or in strip scanning of paper, TLC, or electrophoresis

TABLE 5. Suggested solutions for some common problems in chromatography

Problem	For ion exchange separation	For adsorption and partition separation	For gel permeation separation
No retention	Adjust pH closer to pKa (increase ionization) Use stronger resin	Reduce solvent polarity Increase adsorbent polarity Consider reverse phase	Use higher exclusion limit gel
Excessive retention	Adjust pH to suppress ionization Use weaker resin Increase temperature	Increase solvent polarity Reduce adsorbent polarity	Use lower exclusion limit gel If elution after one column volume, change solvent
Poor resolution	Use longer column Reduce concentration of competing mobile phase ions Reduce flow rate Use different resin (consider non-ionic mode)	Use longer column Reduce solvent polarity Reduce flow rate Use adsorbent (with same polarity but different functional groups)	Use larger column bed volume Reduce flow rate Use gel with narrower effective fractionation range

strips, it may indicate poor resolution or impurities being formed during the procedure. The best way to solve this problem is to select a better solvent. Table 5 lists difficulties frequently encountered by novice chromatographers and suggests some solutions. Manufacturers of chromatographic and electrophoretic equipment and column packings are generally willing and able to objectively advise the researcher on solutions to specific separatory problems, and several suppliers of chromatographic equipment conduct traveling seminars with hands-on laboratories that are helpful in getting started in chromatography.

In summary, several analytical methods must be compared before choosing one with no adverse effects on the radiopharmaceutical being tested, which distinguishes all possible contaminant radiochemicals, and which gives reproducible results. A definitive radiochemical quality control protocol may take time to develop, but it will give the preparer confidence in the radiochemical composition of a radiopharmaceutical. The nuclear medicine physician, in turn, can expect less variability in the biological distribution of the radiopharmaceutical in healthy patients and hence, the diagnostic sensitivity will be the greatest possible for that agent.

REFERENCES

1. COHEN Y. and BESNARD M. In *Radiopharmaceuticals* (Edited by SUBRAMANIAN G., RHODES B. A., COOPER J. F. and SODD V. J.), pp. 207–227. Soc. Nucl. Med., New York (1975).
2. *Analytical Control of Radiopharmaceuticals.* IAEA, STI/Pub/253, Vienna (1970).
3. STELMACH H. A. and QUINN J. L., III *Semin. nucl. Med.* **4,** 295 (1974).
4. *The United States Pharmacopeia*, 19th revision, Mack Printing Company, Easton, PA. (1974).
5. WELCH M. J., STRAATMANN M. and KROHN K. A. unpublished observation (1975).
6. DAVID G. S. and THOMAS R. J. *Science* **184.** 1381 (1974).
7. COHEN Y. In *Radioactive Pharmaceuticals* (Edited by ANDREWS G. A., KNISELEY R. M., WAGNER H. N. and ANDERSON E. B.), pp. 67–91. U.S. Atomic Energy Commission, Oak Ridge, Tennessee (1966).

8. CHRISTY B., KING G. and SMOAK W. M. *J. nucl. Med.* **15,** 484 (1974).
9. KOJIMA M., MAEDA M., OGAWA H., NITTA K. and ITO T. *J. nucl. Med.* **16,** 666 (1975).
10. WAXMAN A. D., KAWADA T., WOLF W. and SIEMSEN J. K. *Radiology* **117,** 647 (1975).
11. SCHUPP O. E. *Gas Chromatography*. Wiley, New York (1968).
12. PERRY S. G., AMOS R. and BREWER P. I. *Practical Liquid Chromatography*. Plenum, New York (1973).
13. KIRKLAND J. J. (Editor). *Modern Practice of Liquid Chromatography*. Wiley, New York (1971).
14. BROWN P. R. *High Pressure Liquid Chromatography: Biochemical and Biomedical Applications*. Academic Press, New York (1973).
15. STAHL E. *Thin-Layer Chromatography: A Laboratory Handbook*, 2nd Edition. Springer, New York (1969).
16. KIRKLAND J. J. *Anal. Chem.* **43,** 36A (1971).
17. MAJORS R. E. *American Laboratory*, p. 13 (October, 1975).
18. FISCHER L. *An Introduction to Gel Chromatography*. American Elsevier, New York (1971).
19. DETERMANN H. *Gel Chromatography*. Springer, New York (1968).
20. GRIESER M. D. and PIETRZYK D. J. *Anal. Chem.* **45,** 1348 (1973).
21. HALLER W. *Nature, Lond.* **206,** 693 (1965).
22. PERSSON B. R. R. In *Radiopharmaceuticals* (Edited by SUBRAMANIAN G., RHODES B. A., COOPER J. F. and SODD V. J.), pp. 228–235. Soc. Nucl. Med., New York (1975).
23. KROHN K. A., SHERMAN L. and WELCH M. *Biochim. Biophys. Acta* **285,** 404 (1972).
24. METZGER J., SECKER-WALKER R., KROHN K., WELCH M. and POTCHEN E. J. *J. Lab. Clin. Med.* **82,** 267 (1973).
25. LOWE C. R. and DEAN P. D. G. *Affinity Chromatography*. Wiley, New York (1974).
26. WEETALL H. H. *Anal. Chem.* **46,** 602A (1974).
27. LIN M. S., KRUSE S. L., GOODWIN D. A. and KRISS J. P. *J. nucl. Med.* **15,** 1018 (1974).
28. BILLINGHURST M. W. *J. nucl. Med.* **14,** 793 (1973).
29. SHEN V., HETZEL K. R. and ICE R. D. *Radiochemical Purity of Radiopharmaceuticals Using Gelman Seprachrom ITLC Chromatography*. Procedure Manual, Tech. Bulletin #32. Gelman Instrument Company, Ann Arbor, Michigan (Feb. 1975).
30. HAER F. *An Introduction to Chromatography on Impregnated Glass Fiber*. Ann Arbor–Humphrey Science, Ann Arbor, Michigan (1969).
31. SCHULTZE H. E. and HEREMANS J. F. *Molecular Biology of Human Proteins*, Vol. 1. Elsevier, Amsterdam (1966).
32. MOSESSON M. W., ALKJAERSIG N., SWEET B. and SHERRY S. *Biochemistry* **6,** 3279 (1967).
33. SUNDERLAND M. L. *J. nucl. Med.* **16,** 225 (1975).

International Journal of Applied Radiation and Isotopes, 1977, Vol. 28, pp. 229–233. Pergamon Press. Printed in Northern Ireland

A Method for Determining the pH Stability Range of Gallium Radiopharmaceuticals

S. KULPRATHIPANJA and D. J. HNATOWICH
Massachusetts Institute of Technology and Massachusetts General Hospital, Boston, MA, U.S.A.

(*Received* 6 *March* 1976)

A method employing paper chromatography has been used to assess the stability of several Ga-68 labeled radiopharmaceuticals in basic, neutral and acidic solutions. Ascending paper chromatography with a pyridine, ethanol, water solvent was shown to be capable of resolving the decomposition products of gallium complexes, namely gallium hydroxide and gallate, from that of the complexes themselves. The stability of gallium-labeled citrate, pyrophosphate, EDTA, and ethylenediaminetetramethylene phosphonate (EDTMP) were each investigated by paper chromatographic analysis in which the solvent was adjusted to seven pH values in the range 2.5–9.6. The acidity of the solvent was therefore relied upon to alter the chemical environment of the compound under study. At the concentration of complexing agents used in this study, no decomposition products for Ga-EDTA and Ga-EDTMP were observed throughout the entire pH range. Gallium citrate was found to be stable in the pH range from about 2–7. As the pH is raised above 7, the degree of Ga-citrate dissociation increases with the formation of increasing concentrations of gallium hydroxide and gallate. Gallium pyrophosphate was found to be stable only in acid solutions of pH less than about 5. This latter compound was found in our laboratory to be unsuitable as a bone imaging agent due to poor bone localization accompanied by excessive liver accumulation. This unfavorable *in vivo* result may be expected since the complex has now been shown to be unstable at the pH of blood.

INTRODUCTION

IMPROVEMENTS in positron scintigraphy, particularly in connection with tomographic reconstruction techniques, has led to a re-awakening of interest in positron-emitting radiopharmaceuticals. Among the available positron-emitting radionuclides, one of the most attractive for medical applications is ^{68}Ga. This nuclide has a 68 min half-life and is available as a generator product by decay of its 287 day parent ^{68}Ge.

Our laboratory is currently involved in the development of new radiopharmaceuticals labeled with ^{68}Ga. Several paper chromatographic systems have been developed to assay for radiocontaminants in preparations of these radiopharmaceuticals.[1] Our recent studies using these chromatographic techniques have shown that certain complexes of gallium dissociate during analysis when the mobile-phase solvent is either too acidic or basic. The solvent is therefore capable of altering the chemical environment of the complex deposited on the paper such that the complex may dissociate. In this case the presence of dissociation products will be apparent in the resulting chromatogram.

This report describes a study in which several gallium-labeled compounds were analyzed by an ascending paper chromatographic method in which the mobile-phase solvent was adjusted to several pH values. The method was used to establish the stability of gallium-labeled compounds in acidic, neutral, and basic solutions.

MATERIALS AND METHODS

Preparation of labeled compounds

The compounds which have been investigated are ^{68}Ga-labeled citrate, pyrophosphate, ethylenediaminetetraacetate (EDTA) and ethylenediaminetetramethylene phosphonate (EDTMP). In addition, gallium hydroxide ($Ga(OH)_3$) and gallate ($Ga(OH)_4^-$) were prepared by neutralizing, with NaOH, acid solutions of gallium chloride ($GaCl_3$) in the absence of any complexing agent.

Gallium-68 was obtained by eluting a ^{68}Ge–^{68}Ga generator (New England Nuclear Corp., Billerica, Mass.) with 0.005 M Na-EDTA at pH 7. The generator eluant contained about 1.5 mCi/ml of ^{68}Ga as Ga-EDTA and was used directly in the study of gallium in this chemical form. The preparation of other chemical forms require that gallium be separated from EDTA. This is accomplished in our laboratory according to a procedure described elsewhere[1] in which the generator eluant is made strongly acidic with HCl and the resulting chloride complex of gallium, $GaCl_4^-$, is adsorbed on an anion exchange resin column. All traces of EDTA are then washed from the resin and distilled water is used to remove gallium as the neutral trichloride, $GaCl_3$. The eluant from the exchange column is acidic and contained about 11 mCi/ml of ^{68}Ga. Gallium complexes are prepared by adding the desired complexing agent to the eluant and neutralizing with NaOH. In the case of Ga-citrate, 0.2 ml (8 mg) of Na-citrate was added; in the case of Ga-pyrophosphate, 0.2 ml (8 mg) of Na-pyrophosphate was added; and in the case of Ga-EDTMP, 0.1 ml (8 mg) of Na-EDTMP (Monsanto Chemical Company)[2] was added. The three stock solutions were adjusted to pH 8 prior to use.

Paper chromatography

The solvent used in this study was a 1:2:4 pyridine, ethanol, water mixture; Whatman No. 1 chromatographic paper ($5\frac{1}{2} \times 1$ in. strips) was used as supporting medium.[1] Gallium preparations were spotted $\frac{5}{8}$ in. from the bottom of the paper and the papers were developed until the solvent front had travelled $4\frac{3}{8}$ in. The development time was about 40 min. After drying, the papers were analyzed using a radiochromatogram scanner (Packard, Model 7201) equipped with a NaI scintillation detector.

Preparations of each compound under study were adjusted with dilute NaOH or HCl to each of the following pH values: 2.5, 4.8, 6.0, 7.2, 8.4 and 9.6. Volumes of solvent were also adjusted to the six pH values. A pH meter equipped with a glass combination electrode was used for pH measurements. Each of the six preparations were analyzed in each of the six solvents, so that 36 analyses were performed, in duplicate, for each compound.

RESULTS AND DISCUSSION

Although gallium exists in three valency states (+1, +2 and +3), only the +3 oxidation state is stable in aqueous solution. Gallium undergoes the following hydrolysis reactions in solution.[3]

$$Ga^{+3} + OH^- \leftrightarrows Ga(OH^{+2} \quad K_1 = 2.5 \times 10^{11}$$

$$Ga(OH)^{+2} + OH \leftrightarrows Ga(OH)_2^{+1} \quad K_2 = 6.2 \times 10^9.$$

It may be calculated that at pH about 2.5, gallium exists primarily as $Ga(OH)^{+2}$ whereas at slightly higher pH values, gallium exists primarily as $Ga(OH)_2^{+1}$. At yet higher values, gallium forms the insoluble hydroxide $Ga(OH)_3$.[4] Since gallium is an amphoteric element, $Ga(OH)_3$ is soluble in basic solutions and will dissolve to form gallate, $Ga(OH)_4^{-1}$. In the presence of chelating or complexing agents, the production of these species may be suppressed due to competition between hydroxyl ions and complexing species for gallium ions. However, if a gallium complex is placed in a chemical environment in which it dissociates, the formation of $Ga(OH)_3$ and $Ga(OH)_4^{-1}$ must be expected.

Figures 1–3 show that radiochromatograms obtained for $GaCl_3$, Ga-citrate, and Ga-pyrophosphate, respectively, at each of the six solvent pH values. Table 1 lists the *Rf* values obtained in this study.

It was found for each of the compounds studied, that the results of the analyses were identical regardless of preparation pH but, depending on the compound, would vary with solvent pH. Thus, altering the pH of the preparation, at least in the pH range from 2.5 to 9.6, made no difference to the radiochromatograms. This can only be the case if the solvent, when in contact with the complex on the paper, can alter the chemical environment so as to either dissociate a complex or, alternatively, cause association of a previously dissociated complex. (Apparently because of the many-fold excess of the complexing species over gallium ions on the paper, a complex will reform if the pH of the solvent falls within the range in which the complex is stable. It had been observed earlier[1] that this may not occur if the prepar-

TABLE 1. Rf values

Solvent pH:	2.5	4.8	6.0	7.2	8.4	9.6
Compound						
Ga-citrate	0.95	0.95	0.91	0.91	0.85	0.85
Ga-pyrophosphate	0.93	0.76	0.77	0.77	0.77	0.77
Ga-EDTA	0.94	0.88	0.89	0.90	0.88	0.90
Ga-EDTMP	0.90	0.89	0.83	0.81	0.80	0.83

ation is adjusted to more extreme values of pH. In these cases, the solvent may not be able to overcome the effects of excessive hydrogen or hydroxyl ion present on the paper.) Since gallium is a labile element,[5] i.e. will reach chemical equilibrium rapidly, the reactions on the paper will be completed almost immediately after contact with the solvent front.

The radiochromatograms obtained with gallium chloride are shown in Fig. 1. Three peaks are evident in the chromatograms obtained at solvent pH of 2.5 and 4.8 (Figs. 1a–b). In these acidic solutions, we expect the presence of $Ga(OH)^{+2}$, $Ga(OH)_2^{+1}$ and $Ga(OH)_3$. It had been determined earlier[1] that $Ga(OH)_3$ remains at the origin (*Rf* 0.0) in this solvent system. It is likely that the remaining two peaks correspond to the presence of $Ga(OH)^{+2}$ and $Ga(OH)_2^{+1}$. Since the peak at *Rf* 0.72 increases in intensity from pH 2.5 to 4.8 while the peak

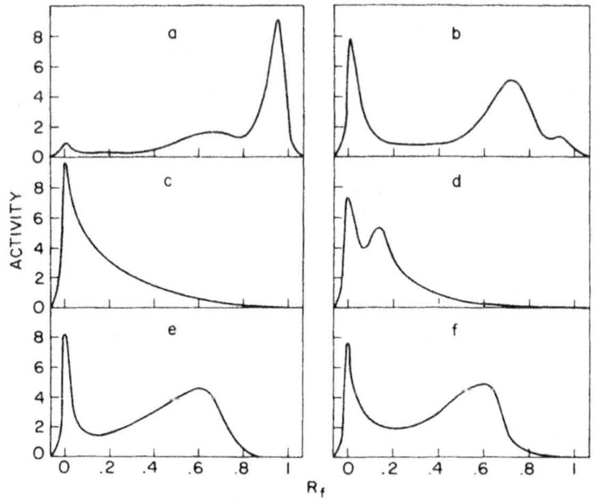

FIG. 1. Radiochromatograms of $GaCl_3$ at solvent pH 2.5 (a), 4.8 (b), 6.0 (c), 7.2 (d), 8.4 (e) and 9.6 (f).

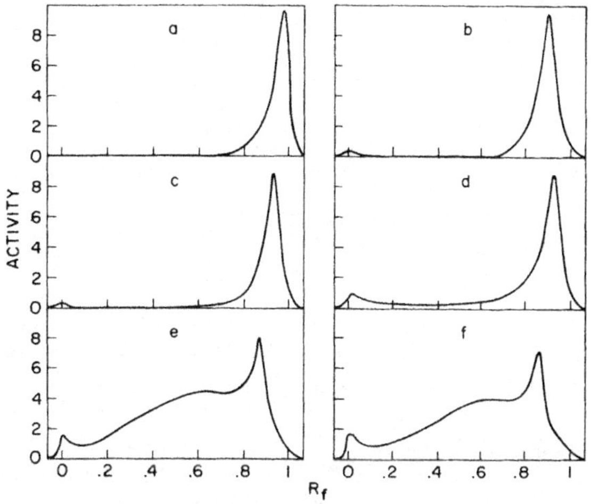

FIG. 2. Radiochromatograms of Ga-citrate at solvent pH 2.5 (a), 4.8 (b), 6.0 (c), 7.2 (d), 8.4 (e) and 9.6 (f).

at *Rf* 0.94 decreases, the former should correspond to $Ga(OH)_2^{+1}$ and the latter to $Ga(OH)^{+2}$.

As the pH of the solvent is further raised to pH 6, $Ga(OH)_3$ becomes the predominant species as shown in Fig. 1(c). However as the solvent pH is raised again to 7.2, two peaks appear as shown in Fig. 1(d). Since another species of gallium, GaO(OH), is formed during slow precipitation of $Ga(OH)_3$ or during the aging of the hydroxide,[6] it is possible that the peak with *Rf* 0.11 may correspond to the presence of this compound. At pH 8.4 and 9.6, $Ga(OH)_3$ will dissolve to form gallate ion,[4] $Ga(OH)_4^{-1}$ as shown by the appearance of the peak at *Rf* 0.65 in Figs. 1(e–f).

Figure 2 presents the radiochromatograms obtained for Ga-citrate. The behavior of this complex has been investigated elsewhere, and had been found to be stable in the pH range from about 3.0 to 7.5 at comparable citrate

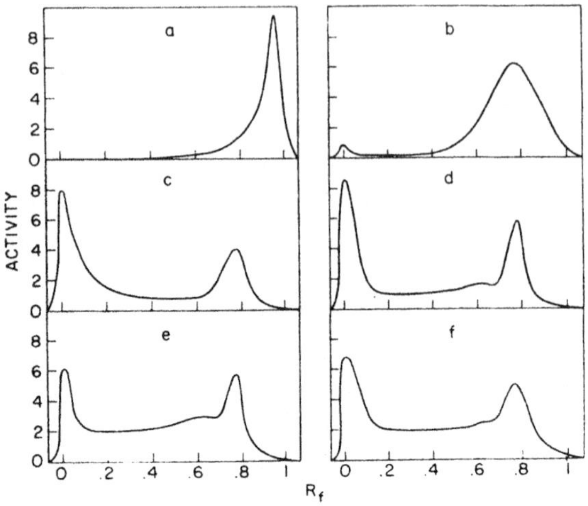

FIG. 3. Radiochromatograms of Ga-pyrophosphate at solvent pH 2.5 (a), 4.8 (b), 6.0 (c), 7.2 (d), 8.4 (e) and 9.6 (f).

concentration.[7] The same result was obtained in this study; the complex was found to be stable in the pH range of 2.5–7.2 as shown in Figs. 2(a–d). Small amounts of $Ga(OH)_3$ are apparent in these chromatograms but the majority of the activity is present in the major peak at *Rf* 0.9. At higher pH values Ga-citrate is reported to decompose to form gallate.[8] This behavior is apparent in Figs. 2e–f; at pH values of 8.4 and 9.6, most of the gallium citrate hydrolyzes to gallate (*Rf* 0.65) and $Ga(OH)_3$ (*Rf* 0.00) with only a smaller peak of Ga-citrate appearing at *Rf* 0.9.

The *Rf* value for the citrate complex varies slightly according to the pH of the solvent indicating that the chemical form of the complex is being altered by the addition of hydroxyl ions. Ga-citrate is known[7] to exist as $[C_3H_4OGa(COOH)_3]^{+2}$ and $[C_3H_4OGa(OH)_2(COOH)_3]^0$ at pH about 3; $[C_3H_4OGa(OH)_2(COO)(COOH)_2]^{-1}$ at pH about 4; and $[C_3H_4OGa(OH)_2(COO)_3]^{-3}$ at pH greater than 8.

Fig. 3 presents the radiochromatograms obtained for Ga-pyrophosphate. It is apparent that Ga-pyrophosphate is unstable in solution at pH above about 6. This is evident from the increasing presence of $Ga(OH)_3$ and $Ga(OH)_4^{-1}$. As with Ga-citrate, the *Rf* values for Ga-pyrophosphate vary, in this case, from about 0.7 to 0.9 according to the pH of the solvent.

The radiochromatograms obtained in this work for Ga-EDTA and Ga-EDTMP are not presented since they are all composed of a single peak with *Rf* 0.8–0.9. No $Ga(OH)_3$ or $Ga(OH)_4^{-1}$ was observed at any of the solvent pH values, indicating that both EDTA and EDTMP form complexes with gallium which are stable at least in the pH range from 2.5 to 9.6.

The mobile-phase solvent used in this study contains ethanol and pyridine in addition to water. Complex formations between gallium and constituents of the solvent are not expected since ethanol is inert to complex formation and gallium-pyridine complexes have not been observed in aqueous solution.[9] A gallium–pyridine complex has been prepared in non-aqueous media but the complex hydrolysis in aqueous media.[4]

CONCLUSIONS

The stability of several gallium-labelled compounds were determined in this work by using a paper chromatographic method. This technique may be used to evaluate the stability of gallium radiopharmaceuticals toward acidic, neutral and basic solution. At the concentrations studied in this work, it was found that Ga-EDTA and Ga-EDTMP are stable throughout the pH range studied, Ga-citrate is stable up to about pH 7, and Ga-pyrophosphate begins to dissociate at pH greater than about 5. This latter compound was one of a number of phosphate-containing compounds studied in our laboratory as potential bone-imaging agents when labeled with ^{68}Ga. Very poor bone localization in dogs was observed with this agent; a large fraction of the activity appeared in the liver. This result may be understood now that the instability of this complex at the pH of physiologic saline (7.3) has been established.

Acknowledgement—This study was supported in part by NIH Grant No. CA-07368.

REFERENCES

1. HNATOWICH D. J. *J. nucl. Med.* **16,** 764 (1975).
2. DEWANJEE M. K., HNATOWICH D. J. and BEH R. Submitted to *J. nucl. Med.* **17,** 1003 (1976).

3. LATIMER W. M. *Oxidation Potentials.* Prentice-Hall, New York (1948).
4. SHEKA I. A., CLAUS I. S. and MITYUREVA T. T. *The Chemistry of Gallium.* Elsevier, Amsterdam (1966).
5. BASOLO F. and PEARSON R. C. *Mechanisms of Inorganic Reactions,* pp. 141–145. Wiley, New York (1968).
6. WEISER H. B. and MILLIGANN W. O. *J. phys. Chem.* **42,** 673 (1938).
7. BLANCO R. E. and PERKINSON J. D. *J. Am. Chem. Soc.* **73,** 2696 (1951).
8. LAVROVA G. V., SHEKA I. A., TSIMMERGAKL V. A. and KONDRUK E. I. *V Ukrainskoe Respublikanskoe Soveshchanie Po Neorganicheskoi Khimii,* Kiev (1966).
9. DYMOV A. M. and SAVOSTIN A. P. *Analytical Chemistry of Gallium,* p. 8. Ann Arbor Science Publishers, Ann Arbor (1970).

Author Index

Subject Index

Printed in Dunstable, United Kingdom

91898130R00145